# Proton Exchange Membrane Fuel Cells (PEMFCs)

# Proton Exchange Membrane Fuel Cells (PEMFCs)

Editors

**Jean St-Pierre**
**Shangfeng Du**

MDPI • Basel • Beijing • Wuhan • Barcelona • Belgrade • Manchester • Tokyo • Cluj • Tianjin

*Editors*
Jean St-Pierre
Hawaii Natural Energy Institute
University of Hawaii at Manoa
Honolulu
United States

Shangfeng Du
School of Chemical Engineering
University of Birmingham
Birmingham
United Kingdom

*Editorial Office*
MDPI
St. Alban-Anlage 66
4052 Basel, Switzerland

This is a reprint of articles from the Special Issue published online in the open access journal *Molecules* (ISSN 1420-3049) (available at: www.mdpi.com/journal/molecules/special_issues/PEMFCs).

For citation purposes, cite each article independently as indicated on the article page online and as indicated below:

LastName, A.A.; LastName, B.B.; LastName, C.C. Article Title. *Journal Name* **Year**, *Volume Number*, Page Range.

**ISBN 978-3-0365-1544-1 (Hbk)**
**ISBN 978-3-0365-1543-4 (PDF)**

# Contents

# About the Editors

**Jean St-Pierre**

Jean St-Pierre, PhD, PEng is a graduate of Polytechnique, Montréal, Canada. More than 25 years of his career have been devoted to the development of proton exchange membrane fuel cells including aspects such as water management, freezing, degradation mechanisms, mathematical modeling, diagnosis and measurement methods, electrocatalysis, pure oxygen operation for space and air independent applications, and reactant stream unit operations (gas separation and fuel reforming catalysts). He previously held principal research scientist and research professor positions at, respectively, Ballard Power Systems (1995-2005) and the University of South Carolina (2006-2010). He is currently a researcher at the University of Hawaii – Manoa, and focuses on fuel cell and related technology activities. His work has led to more than 115 journal papers, book chapters, and conference proceedings, and to more than 30 granted, published, and provisional patents.

**Shangfeng Du**

Shangfeng Du (Associate Professor) leads the Low Temperature Fuel Cell and Electrolyser research group in the School of Chemical Engineering at the University of Birmingham. He received his BEng in Materials Science and Engineering from Tsinghua University, China in 1999, and his PhD in Chemical Engineering from the Institute of Process Engineering, Chinese Academy of Sciences, China in 2005. He has been a postdoctoral researcher in the Powder Metallurgy Laboratory (PML) at the Max Planck Institute for Metals Research before he joined Birmingham in 2007. His group is focused on electrochemical engineering research of catalyst electrodes that bridges the gap between novel catalyst nanostructures and high-power performance electrochemical devices. His research activities cover electrode structure design, 1D nanostructure development, and surface and interface characterization. He has been the author of one book, and published more than 60 journal papers, book chapters, and patents.

# Preface to "Proton Exchange Membrane Fuel Cells (PEMFCs)"

A proton exchange membrane fuel cell (PEMFC) spontaneously produces power from a fuel and oxygen with heat and water byproducts. Hydrogen is preferred as the fuel because it is renewable if generated by water electrolysis using electricity derived from renewable energy. PEMFCs are more efficient than an internal combustion engine because reactants are directly converted into energy through a one-step electrochemical reaction. Fuel cells coupled with hydrogen production can improve resiliency by fulfilling the energy storage needs of electric grids constrained by intermittent renewable energy generation (e.g., solar, wind). More specifically, discharge (fuel cell) and recharge (water electrolysis) durations exceeding a few days for power plant ratings below the megawatt level are not presently accessible to other energy storage technologies. Furthermore, fuel cells combined with water electrolyzers powered by renewable energy mitigate global warming concerns with reduced carbon dioxide emissions. PEMFCs are commercially available for a few applications including vehicles, buildings, and power backup systems. Improvements in cost, performance, and durability are needed to assist commercialization efforts because the technology is not yet mature.

Papers published in the *Molecules* Special Issue on PEMFCs are collected in this book. The Special Issue was initially intended to attract contributions focusing on all PEMFC scientific and technological aspects that decrease cost and increase performance and durability, including novel characterization methods, mathematical models, and accelerated stress tests to gain additional insight, as well as degradation mechanisms, innovative materials, and original designs for components, cells, stacks, and systems.

The collected papers comprise recent advancements in PEMFC technology aimed at reducing cost, improving performance, and extending durability. Almost all key materials, as well as their integration into a cell, are discussed: the bus plates that collect the electrical current, the gas diffusion medium that distributes the reactants over catalysts promoting faster reactions, and the membrane separating oxygen and hydrogen gases and closing the electrical circuit by transporting protons. Fuel cell operation below the freezing point of water and with impure reactant streams, which impacts durability, is also discussed. Papers focusing on materials and their integration appear at the beginning and follow the sequence: membrane, catalyst layer, gas diffusion layer, membrane electrode assembly, and bus plate. Subsequent papers center on operational aspects. All papers include at least one author with significant fuel cell experience. Authors originate from all organization types, universities, national laboratories, and companies, and are distributed over Asia, Europe, and North America. The varied origins of the contributors indicate a continued and widespread interest in fuel cell technology to address energy and climate issues.

**Jean St-Pierre, Shangfeng Du**
*Editors*

molecules

MDPI

*Review*

# Composite Polymers Development and Application for Polymer Electrolyte Membrane Technologies—A Review

**Gabriele G. Gagliardi [1], Ahmed Ibrahim [2], Domenico Borello [1,*] and Ahmad El-Kharouf [2,*]**

1 Department of Mechanical and Aerospace Engineering, Sapienza Università di Roma, 00184 Rome, Italy; Gabriele.gagliardi@uniroma1.it

2 School of Chemical Engineering, University of Birmingham, Birmingham B15 2TT, UK; AXI763@student.bham.ac.uk

* Correspondence: domenico.borello@uniroma1.it (D.B.); a.el-kharouf@bham.ac.uk (A.E.-K.)

Academic Editors: Jean St-Pierre and Shangfeng Du

Received: 12 February 2020; Accepted: 3 April 2020; Published: 8 April 2020

check for updates

**Abstract:** Nafion membranes are still the dominating material used in the polymer electrolyte membrane (PEM) technologies. They are widely used in several applications thanks to their excellent properties: high proton conductivity and high chemical stability in both oxidation and reduction environment. However, they have several technical challenges: reactants permeability, which results in reduced performance, dependence on water content to perform preventing the operation at higher temperatures or low humidity levels, and chemical degradation. This paper reviews novel composite membranes that have been developed for PEM applications, including direct methanol fuel cells (DMFCs), hydrogen PEM fuel cells (PEMFCs), and water electrolysers (PEMWEs), aiming at overcoming the drawbacks of the commercial Nafion membranes. It provides a broad overview of the Nafion-based membranes, with organic and inorganic fillers, and non-fluorinated membranes available in the literature for which various main properties (proton conductivity, crossover, maximum power density, and thermal stability) are reported. The studies on composite membranes demonstrate that they are suitable for PEM applications and can potentially compete with Nafion membranes in terms of performance and lifetime.

**Keywords:** composite membranes; electrolyte; PEM; fuel cells; electrolysers

## 1. Background

During the last 100 years the world average temperature has increased by almost 0.8 °C [1], becoming the most critical environmental issue of our time. Even though there are many different factors responsible, the greatest concern is greenhouse gas emissions due to human activities linked to energy production and use. In this sense, governments worldwide are acting to take measures to revise their energy mix by reducing fossil fuels usage and promoting alternative sources. The European Union, with the objectives set in the 20-20-20 pack, put forward strict targets to be reached before 2020, namely 20% reduction of greenhouse gases, 20% primary energy production from renewables, and 20% of biofuels burned in transportation. Moreover, recently a medium-long term strategy was agreed, stating that the European energy efficiency should be improved by 27% and the renewables energy input should increase by up to the 27% of the total share before 2030. Within this overall framework, it is becoming increasingly important that research and development of new technologies are intensified to allow the penetration of more efficient energy conversion systems. In this context, polymer electrolyte membrane technologies can play an important role.

### 1.1. Polymer Electrolyte Membrane Technologies

Fuel cells and elecrolysers are energy conversion systems that electrochemically convert energy from chemical (stored in a fuel) to electric and vice versa without any intermediated combustion process. This offers superior efficiency and performance compared to the incumbent combustion-based energy generation technologies [2–5]. Fuel cells are eco-friendly devices with potential zero emission at the point of use. Moreover, if the so-called energy vectors, used in fuel cells, were generated by thermochemical processes from biomass or from electrochemical processes utilising renewable electric energy sources, the resulting carbon dioxide cycle would be null. So, they are considered to be the energy conversion devices of the future. In addition, they are a silent technology, without noise or vibration, and their design flexibility allows for simple construction and a diverse range of applications including portable, stationary, and transportation.

In general, electrochemical devices, including fuel cells and electrolysers, consist of two electrodes—anode and cathode—separated by an electrolyte with the purpose to allow the passage of ions generated during the redox half reactions. At the anode side, the oxidation reaction takes place while the reduction reaction occurs at the cathode side. The electrolyte conducts the produced/required ions to complete the reactions at the electrodes and serves as a separator between the anode and cathode reactants in the fuel cell and electrolyser technologies.

This study will focus on Proton Exchange Membrane (PEM) technologies, namely, Hydrogen Polymer Electrolyte Membrane Fuel Cells ($H_2$ PEMFCs), Direct Methanol Fuel Cells (DMFCs) and Polymer Electrolyte Membrane Water Electrolysers (PEMWEs). PEM technologies fall under the low temperature fuel cells category with operating temperatures up to 90 °C. Protons (hydrogen ions) are the transported ion through the PEM structure. Below, the thermodynamic, state of art and technical challenges of the three technologies are briefly described.

#### 1.1.1. $H_2$ PEMFC

PEMFCs utilize hydrogen as a fuel and exploit the electrochemical reaction of hydrogen and oxygen to produce electrical energy. Protons pass through the membrane reaching the cathode while electrons are forced to flow through an external circuit. Protons, electrons, and oxygen react at the cathode producing water. Reactions involved in the chemical process are described below:

$$\text{At the anode}: \; H_2 \rightarrow 2H^+ + 2e^- \tag{1}$$

$$\text{At the cathode}: \; \frac{1}{2}O_2 + 2H^+ + 2e^- \rightarrow H_2O \tag{2}$$

$$\text{Overall reaction}: \; H_2 + \frac{1}{2}O_2 \rightarrow H_2O \tag{3}$$

PEMFC advantages include high efficiency, fast response to load, high power density and low operating temperature [6,7]. However, PEMFCs are expensive due to the use of expensive catalyst materials, have durability issues and are challenging for mass production [8–10]. PEMFCs applications focus on transportation, distributed/stationary and portable power generation: Toyota, Honda, and Hyundai have already introduced their fuel cell electric vehicles (FCEV) to the market.

The durability of the polymer membrane is a crucial factor affecting the lifetime of the stack. Industry requirements for automotive fuel cell stacks durability is 5000 h with a performance drop of no more than 10% [11]. Nafion is the only membrane reported to achieve this requirement. Also, at elevated temperatures (> 90 °C) the durability is further reduced due to the dehydration of the membrane and the subsequent drop in proton conductivity. However, operation at higher temperatures is desirable as it allows for quicker reaction kinetics and simpler water and heat management [12]. This means a smaller fuel cell system can be employed in vehicles to provide the same power output.

### 1.1.2. DMFC

Methanol can be used in PEM fuel cells as a replacement fuel to Hydrogen. The first physical advantage is the liquid natural state of this compound that facilitates its transport, avoiding delicate compressed gas infrastructure development or heavy metal hydrides. It has higher volumetric energy density than compressed hydrogen at 350bar, has low volatility and is almost environmentally neutral in its degradation [13] [14]. In addition, Methanol can be produced from syngas (carbon monoxide and hydrogen compound), allowing to primarily exploit renewable feedstocks as biomass or solid wastes. In recent years, this idea has attracted a lot of attention to find a possible carbon-neutral energy cycle [15]. DMFCs could find application as alternative power sources for vehicle propulsion [16] but are mainly considered for portable applications [17].

Reactions occur at the two electrodes of a DMFC upon the catalyst active area as shown Equations (4)–(6). Again, the flow of electrons through the external circuit accompanied by the flow of protons through the polymer electrolyte allows the reactions to occur and electrical power to be produced [18].

$$\text{At the anode}: CH_3OH + H_2O \rightarrow 6H^+ + 6e^- + CO_2 \tag{4}$$

$$\text{At the cathode}: \frac{3}{2}O_2 + 6H^+ + 6e^- \rightarrow 3H_2O \tag{5}$$

$$\text{Overall reaction}: CH_3OH + \frac{3}{2}O_2 \rightarrow 2H_2O + CO_2 \tag{6}$$

However, DMFCs suffer from low performance due to methanol crossover. Methanol crossover happens when methanol molecules diffuse through the membrane and are directly oxidized by oxygen on the cathode, causing a mixed potential so, consequently, a decrease in cell performance [19]. Although methanol has a high energy density (about 1.8 kWh $kg^{-1}$ or 1.7 kWh $L^{-1}$), it must be diluted in order to reduce methanol crossover. A consequence of dilution is that the cell stack dimensions must be proportionally increased, making it challenging to utilize on small and portable devices.

Moreover, it was found that methanol crossover slows down the rate of reaction at the cathode. It has been reported that the rate of reactions not only can be accelerated by adding more platinum–ruthenium catalyst that has a negative effect on the cost, but also by selecting proper membranes and oxygen tolerant cathodes [20].

Also the carbon dioxide generated during the methanol oxidation can further increase the concentration losses: $CO_2$ bubbles can obstruct GDL pores, reducing the available volume aimed to transport methanol towards the anode catalyst [21] causing a decrease in generated power by more than 40% [22], especially at small flow rate and high current density.

### 1.1.3. Electrolysers

Electrolysis is an electrochemical process where water molecules split into hydrogen and oxygen gases using the supplied direct electric current. In PEMWEs, the reverse of the PEMFCs reactions described above Equations. (1)–(3) take place, see Equations (7)–(9).

$$\text{At the anode}: H_2O \rightarrow 2H^+ + \frac{1}{2}O_2 + 2e^- \tag{7}$$

$$\text{At the cathode}: 2H^+ + 2e^- \rightarrow H_2 \tag{8}$$

$$\text{Overall reaction}: H_2O \rightarrow H_2 + \frac{1}{2}O_2 \tag{9}$$

PEMWE is one of the favorable methods for conversion of renewable energy to high purity hydrogen. The technology has great advantages such as compact design, high current density (above 2 A $cm^{-2}$), fast response, small footprint, low operating temperatures (20–80 °C) and the ability to produce ultrapure hydrogen [23]. However, high energy consumption and low hydrogen evolution rate are two important issues that hinder the large diffusion of this technology. Therefore, in order to increase the efficiency and reduce the energy consumption and costs, many researchers have dedicated their work to the development of alternative low cost materials, and efficiency increase and energy reduction [24]. Although the main challenging issue of the PEMWE technology is the enhancement of oxygen evolution rate, the improvement of membranes with reduced cost, appropriate conductivity, low gas cross-over and enhanced mechanical properties at high operating pressure is mandatory [25].

#### 1.1.4. A Short Review on PFSAs (Nafion Polymers)

Nafion is a fluoropolymer made by sulfonated polytetrafluorethylene introduced by DuPont in the mid-1960s. It is commercially available with a thickness between 25 and 250 μm. As for Teflon, its structure (Figure 1) consists of $CF_2$ (difluorocarbene) radicals, alternated to $CFOCF_2$ and with end chains of sulfonic acid $SO_2OH$.

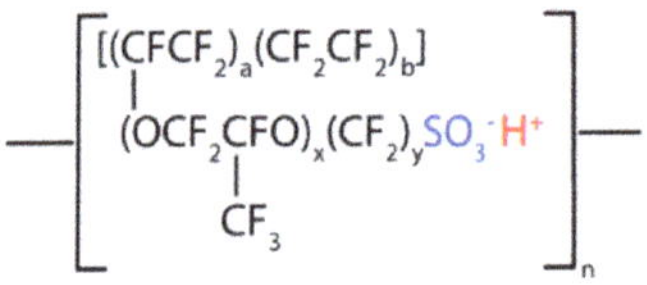

**Figure 1.** Nafion polymer structure; reproduced with permission from [26].

Main Nafion characteristics are the following:

- High proton conductivity: when the pendant sulfonyl fluoride groups ($SO_2F$) are chemically converted to sulfonic acid ($SO_3H$), the ionic functionality is introduced [27];
- Water permeability;
- High chemical resistance: only alkaline metals as sodium can damage the Nafion at STP (standard temperature and pressure); and
- High electronic resistance: forcing electrons to flow through the external circuit to generate electrical power.

The most important parameter to control, in order to keep the ion conductivity high is the relative humidity. Nafion membrane requires water to facilitate the transport of protons through either the Grotthus (hopping) mechanism or the vehicle (diffusion) mechanism [28]. This requirement has led to limiting PEMFC operating temperature to 80 °C as operation above that will lead to dehydration of the membrane and a subsequent loss in proton conductivity and fuel cell performance. However, PEMFC operating above 80 °C can benefit from increased reaction kinetics, reduced CO poisoning and simplified water and thermal management [29]. One method utilised to allow that transition into intermediate temperatures is the use of filler materials within the polymer [30]. Alternatively, there are polybenzimidazole membranes that are doped with phosphoric acids that operate at even higher temperatures [31].

Regarding the use in DMFC, Nafion has high ionic conductivity and chemical stability but high methanol permeability due to:

1. active transport of protons and water;
2. diffusion through the water-filled pores within the Nafion-structure;
3. diffusion through the Nafion itself.

Methanol crossover is an issue that leads to the much worse DMFC performance when compared with $H_2$ PEMFCs. Various approaches to minimize or eliminate methanol crossover have been carried out: increasing membrane thickness [32], increasing the cathode reactant pressure [33], decreasing both cell temperature [34] and methanol concentration [35]. Another innovative way is to modify the membrane using materials that allow obtaining the same chemical and thermal characteristics of Nafion but with lower crossover and cost.

Also, in PEMWE, the commercial Nafion membrane is commonly used as a solid electrolyte due to its chemical and thermal stability, good proton conductivity, and mechanical strength [36]. A drawback of membranes made of Nafion is that they are known to lose water, and thus ionic conductivity, at temperatures above 100 °C [37], which prohibits them from being used for higher temperature water electrolysis. However, high operating temperature offers several advantages from enhanced electrode kinetics and reduced overpotentials [38]. Moreover, higher operation pressure would also be favourable for PEM electrolysis since it would reduce the gas pressurization constraints for storage purposes [39]. Hence, it would be important to develop membranes that can sustain high performance at higher operating temperature and pressure [40].

To overcome the drawbacks of Nafion membranes, novel membranes have been developed and can be classified into three main categories, namely; (i) polymeric, (ii) ceramic, and (iii) composite membranes.

Among these three categories, composite membranes have generated great attention recently. A composite (or hybrid) material can be defined as a material that includes two or more blended compounds on the molecular scale [41]. The use of filler material mixed into the Nafion (or an alternative ionomer) can aid in providing additional properties such as mechanical reinforcement, chemical resistance and proton conductivity. For example, hydrophilic fillers would result in increased membrane water uptake, ideal for low relative humidity (RH) operation. These filler materials can also be functionalised to provide secondary functionalities (e.g., sulphonating a hydrophilic filler) or boost the functionality it already has. Another example is the use of cerium oxide as a radical scavenger to slow down membrane degradation [42]. This review paper aims to provide a summary and analysis of the published work focusing on the development of composite membranes to improve the performance of the DMFC, modify operating conditions and enhance durability for PEMFC and PEMWE.

## 2. Composite Membranes for DMFC

In this part, the range of composite membranes that have been developed to improve the performance of the DMFC at low temperature with reduced methanol crossover and low cost is reviewed.

Two categories of composite membrane materials are considered in the literature; modified Nafion; and non-perfluorinated polymers.

### 2.1. Composite Nafion-Based Membrane

Composite Nafion membranes can be loaded with organic and inorganic fillers that have been used predominantly to increase proton conductivity and to act as a barrier to methanol crossover [43,44]. The following sub-sections discuss the latest developments in the organic, inorganic and carbon nano-material filler-based membrane.

#### 2.1.1. Organic Fillers

Organic materials are commonly used as fillers in the polymeric composite membrane for fuel cells. They supply reinforcement and allow higher stability of the polymer matrix while making it more cost-effective. One of the most commonly applied organic filler is polytetrafluoroethylene (PTFE). PTFE is highly hydrophobic and although it is not suitable alone for membrane application for fuel cells [45], it can be used as a reinforcement of Nafion membrane due to its chemical stability, corrosion resistance and mechanical strength [46]. Few papers focused on testing DMFC performance using Nafion/PTFE membranes. Lin et al. [47] conducted a study on the application of this composite membrane, the authors investigated the effect of this polymer on conductivity, methanol crossover, and cell performance, and compared them with that of commercial Nafion membrane. Experimental data indicated that introducing PTFE into the Nafion polymer reduced both methanol diffusion and methanol electro-osmosis crossover in the membrane. The comparison between Nafion 117 and Nafion/PTFE was performed in a DMFC at 70 °C: Nafion/PTFE membrane was able to operate in a wider current density range achieving a maximum power output of 87.5 mW $cm^{-2}$, 1.3 times higher than Nafion 117. This positive effect of the Nafion/PTFE membrane was also obtained by Nouel et al. [48] and Yu et al. [49] who tested a fuel cell MEA made of Nafion/PTFE comparing results with Nafion 117, 115 and 112. The performance was higher than Nafion 117 and 115 but similar to 112.

In an attempt to further enhance the performance of Nafion/PTFE membrane, Chen et al. [50] included zirconium phosphate (ZrP) into the membrane structure and so the Nafion matrix was modified with both PTFE and ZrP/PTFE for comparison. The composite membranes were prepared via two processes:

1. By impregnating PTFE directly in a Nafion/$ZROCL_2$ solution and then annealing it at high temperature;
2. By impregnating the PTFE membrane in a Nafion solution, annealing at high temperature to prepare Nafion/PTFE membrane, then impregnating again in a $ZrOCL_2$ solution.

Experimental results indicated that the introduction of ZrP led to reduced methanol crossover and proton conductivity. The impact of proton conductivity is stronger than methanol crossover on DMFC performance, thus, as confirmed by tests conducted on the cell, the performance of ZrP/PTFE was lower than Nafion/PTFE.

Most research is focused on the preparation and modification of various proton conductive membranes that are inexpensive and provide better performance and properties than Nafion membranes. To this end, innovative organic materials, which have good thermal and chemical stability and can be easily modified to be used as ionic conductive membranes such as polybenzimidazole (PBI) and polyvinyl alcohols (PVA), were studied [51,52].

Shao et al. [53] and Mollà et al. [54] fabricated Nafion/PVA membranes using casting [55] and impregnation method [56,57] respectively. PVA has higher affinity for water than to methanol (i.e., 55 wt. % and 10 wt. %, respectively), so it can be potentially used for DMFC applications. Both works demonstrated that comparable DMFC performance can be obtained using these membranes. Specifically, Mollà et al. focused on the characterization of Nafion/PVA membranes with varying operating temperature (45, 70 °C), thickness of the membrane (19–47 μm) and concentration of methanol (1–2M). The performance of pristine Nafion membrane and Nafion/PVA were roughly equivalent at very low thickness while Nafion/PVA exceeded the pristine Nafion performance only at higher thickness and higher temperature. At any fixed condition; thickness, temperature and methanol concentration, the OCV of Nafion/PVA was higher than pristine Nafion indicating reduced methanol crossover.

Hobson et al. [58] presented Nafion-PBI dipped and screen-printed films to investigate the effect on membrane performance. They concluded that the modification of Nafion with PBI by both spin coating and dipping reduced the methanol permeability; however, the benefit of low methanol crossover was counterbalanced by the negative effect of the too high impedance. Since neither of the techniques produced a suitable membrane for DMFC, screen printing was investigated and here methanol permeability was reduced without an increase in impedance. The membranes were then tested in a single cell at 60 °C. Using methanol solution of 3.2 M, the cell performance was greatly improved with the current density increased by 42% combined with an increase in maximum power output by 46% as compared with the pristine Nafion membrane. Ainla et al. [59] work on Nafion-PBI membrane was in agreement with the above results. In fact, they demonstrated that the Nafion-PBI membrane has lower methanol permeability and higher conductivity than a commercial membrane. It is important to note that the utilization of these composite membranes led to lower methanol permeability and enhanced the performance only at high methanol concentration.

Conductive Polymers such as polyaniline (PANI) and polypyrrole (PPy) have recently been incorporated into Nafion membranes to reduce its methanol permeability [60,61]. Composite Nafion polypyrrole membranes were prepared by two methods: electrodeposition of polypyrrole on Nafion-coated electrodes [62] or by in situ polymerization with a chemical oxidant [63]. Zhu et al. [64] made a membrane by in situ polymerization using Fe(III) and $H_2O_2$ as oxidising agents. The electrostatic interaction between the sulphonate groups of Nafion and polypyrrole, decreased the pore volume of Nafion membrane which led to low methanol permeability. However, the electrostatic interaction between the polypyrrole chains and sulphonate groups of Nafion also decreased the proton conductivity and therefore increased membrane and cell resistances. So, the benefit of the reduced methanol crossover was neutralized particularly when DMFC worked at high current densities.

Polyaniline is a good conductive polymer that can improve both methanol oxidation and the stability of the catalyst; it can also be included in the Nafion structure through both electrochemical and chemical modification to improve its properties. Wang et al. [65] and Escudero-Cid et al. [66] assembled a composite membrane of Nafion/polyanilina and carried out DMFC performance tests including polarization curve and durability tests showing that both the ionic conductivity and methanol permeability of the Nafion membrane containing PANI decreased when compared with Nafion membrane. In particular, Wang et al. indicated that the performance of the fuel cell increased using the modified membrane especially at high methanol concentration (maximum power output at 6 M) while the power output using Nafion pristine membrane was reduced with increasing methanol concentration. Moreover, it was noted that the PANI composite membrane performed better than that with polypyrrole [67]. It is important to highlight here that the reviewed papers; investigating the use of conductive polymers, do not consider the change in the electronic conductivity of the membrane (short circuit current) in the composite membrane. This should be considered in any future work on these materials.

In recent years, two types of sulphonate fillers, sulphonate poly arylene ether ketone (SPAEK), and sulphonate poly ether ether ketone (SPEEK) have been developed and used to modify the Nafion membrane in DMFCs. Both SPAEK and SPEEK have good attributes: high proton conductivity and methanol resistance for SPAEK [68,69]; good mechanical properties, proton conductivity and good processing capacity of SPEEK polymers [70,71]. Regarding the behaviour in a methanol fuel cell [72,73], an increase by at least 30% in OCV and by 10% in highest power density were observed. The positive results suggest that these membranes could be taken into account for use in future DMFC application once durability is deeply investigated.

#### 2.1.2. Inorganic Fillers

The implementation of inorganic particles into Nafion helps enhance the thermal stability and proton conductivity of composite membranes making them more attractive and appropriate than bare Nafion membranes [74]. This section focuses on silica, metal oxides, montmorillonite and zeolite fluorinated composite membranes; materials that have received considerable attention due to their chemical and thermal properties.

Silica materials have attracted a considerable attention because they possess high surface area and high chemical stability [75]. Generally, they are prepared using different precursors such as alkoxy silanes (like tetraethyl orthosilicate (TEOS)), sodium metasilicate and fumed silica [76]. The addition of silica compounds into polymer membrane is believed to lower methanol crossover [77,78]. In this context, Ren et al. [79] prepared a composite membrane modifying Nafion polymer with tetraethyl orthosilicate (TEOS) and sulfonic TEOS using the casting method. They investigated the influence of silica into Nafion matrix and the changes in proton conductivity, methanol permeability and performance. The results showed that the proton conductivity of these composite membranes was lower than that of commercial Nafion membranes due to the hydrophobic side chain of the TEOS that reduced the water content of the membrane. However, the methanol permeability was also reduced. DMFC single cell tests were carried out at both 1 M and 5 M and at 75 °C. The polarisation curves depicted how the silica composite membrane could achieve better performance than the Nafion when using high methanol concentration because although the proton conductivity of the composite membrane decreased, the methanol permeability also reduced. Works are necessary to increase proton conductivity of those composite membranes.

Some studies have been carried out to investigate experimentally the effect of Nafion membranes with metal oxides, such as; $SiO_2$, $TiO_2$, $WO_3$, as fillers on the performance of DMFCs [80]. As a result of these experiments, the Nafion-modified composite membranes provided higher power density in comparison to the commercial Nafion 115 membrane. Regarding the application in a DMFC, Nafion/$TiO_2$ improved water uptake and reduced methanol absorbance [81] while Nafion/SiO2 showed lower methanol permeability than commercial membrane [82]. Moreover, Nafion membrane modified with both $SiO_2$ and $TiO_2$ were prepared by solvent casting method and studied by Ercelik et al. [83] that investigated the effect of these particles on proton conductivity, water uptake and performance varying temperature. The authors claimed that:

- proton conductivity of Nafion-$TiO_2$/$SiO_2$ increased with temperature. The maximum value obtained was 0.255 S $cm^{-1}$ which is 25% higher than Nafion 117 at 75 °C;
- water uptake values are similar with those of bare Nafion; and
- power densities of MEAs with composite membranes are higher than the MEAs using Nafion 115: at 80 °C and 1M of methanol concentration. The maximum power densities obtained by Nafion/$TiO_2$, Nafion/$SiO_2$ and by the MEA with bare Nafion are 641.16 W/$m^2$, 628.68 W/$m^2$ and 612.96 W/$m^2$, respectively.

The abovementioned studies demonstrated that the incorporation of inorganic particles like $SiO_2$ and $TiO_2$ provided better performance if compared with the Nafion membrane. Although, the use of metal oxides as filler has enabled many advantages, they too have some problems associated with them. The metal particles are often very difficult to disperse homogenously in the polymer membrane, it would mean that the performance of the composite will not be uniform throughout the bulk of the membrane. Moreover, metal oxides accelerate the degradation of membranes, and so durability studies are required to understand the actual benefits and drawbacks of these fillers.

Montmorillite (MMT) is a filler that has attracted much attention recently as Nafion/MMT membranes have been reported to have improved mechanical and thermal properties compared to pristine Nafion membranes [84]. But, the incorporation of this filler into the Nafion matrix does not improve the proton conductivity. Wu et al. [85] prepared the composite membrane via casting solution and reported a slight decrease (about 9%) in proton conductivity compared with pristine Nafion but the methanol crossover decreased more than 90% by loading MMT of only 1% wt [86]. As described above, the utilization of this filler does not contribute to improve the proton conductivity, thus, to minimize the loss in performance Rhee et al. [87] and Lin et al. [88] modified the montmorillonite with an organic sulfonic acid group (MMT-$SO_3H$) with varying the content of the filler. Their studies showed that the proton conductivity of the composite membranes generally declined from that of pristine Nafion membrane with the increase in the inorganic filler content, but the methanol permeability was reduced by up to 90%. The combination of these effects led to an improvement in the performance of a DMFC. In fact, the polarization curve of the MEA with Nafion 115 and composite membrane, realized at 40 °C and 2M of methanol concentration, showed that the performance of the DMFC improves initially with increasing the inorganic content, with a maximum power density at 5 wt. % loading. Curves revealed that all composite membranes achieved better performance than Nafion membrane at high current density region. However, the thermal stability of the membrane is not yet adequate and performance at higher temperature and methanol concentration deserves to be investigated.

To hinder permeation of methanol, another approach is to develop composite membranes using zeolites. Zeolites are micro porous crystalline materials containing silicon, aluminium and oxygen in their framework. They are based on an infinitely extending three-dimensional network of $AlO_4$ and $SiO_4$ tetrahedra linked by sharing oxygen atoms [89,90]. The chemical structural formula of a zeolite may be expressed by the following [91]:

$$Mx/n[(AlO_2)_x(SiO_2)_y]m\ H_2O \quad (10)$$

where M is a cation of valence n, m is the number of water molecules and the sum of x and y is the total number of tetrahedra in the unit cell.

Several authors [92,93] have claimed that zeolite membranes can be adopted for DMFC application. The approach of these studies was to take advantage of the molecular sieving property of zeolite to prevent methanol from passing through the membrane. However, a pure zeolite exhibits poor mechanical properties such as brittleness and fragility and hence is unsuitable for use as a membrane. Moreover, the performance of the zeolite composite membranes depends on the zeolite properties in terms of pore size and surface tension (hydrophobility or hydropholicity). It was reported that hydrophobic zeolites ensure low affinity to water so high permeability to methanol however hydrophilic zeolites lead to an opposite trend and therefore reduce methanol crossover [94]. When zeolites are combined with a polymer support (e.g., Nafion), the advantages of both polymer and zeolite are combined. Among the various type of zeolites, mordenite (MOR) and analcime (ANA) have attracted a lot of interest because they are hydrophilic substances which promote the adsorption of water, excluding alcohol, and provide a good proton pathway through the membrane. Prapainainara et al. [95] fabricated composite membranes with those two fillers studying and comparing their properties and performance. The authors claimed that the presence of the filler benefited the proton migration through the membrane whilst the homogeneous distribution of the filler contributed to block the flow of methanol through the membrane, leading to lower methanol permeability. The composite membrane with MOR filler gave better membrane properties, namely; higher proton conductivity and lower methanol permeability, than those using ANA filler. The best DMFC performance was achieved by MOR composite membrane with a maximum power density of 10.75 mW/cm$^2$, 1.5 times higher than ANA membrane and two times higher than a commercial Nafion 117 membrane.

To enhance the performance of MOR/Nafion, Prapainainara et al. [96] incorporated graphene oxide (GO) to the matrix. The authors used GO to modify the surface of MOR by increasing the surface hydrophilic functional groups resulting in better incorporation of MOR to Nafion and comparable chemical properties with those of pristine Nafion and MOR/Nafion. The use of GO led to better proton conductivity, 1.5 times higher than that of Nafion/MOR and Nafion 117 at 70 °C and it had the lowest methanol permeability too. The authors also tested the membrane in a single cell, obtaining a power density of 27.5 mW/cm$^2$; almost 5 times than that of Nafion 117 at the same operating condition (1 M methanol, 70 °C). However, the operation lifetime was still not good enough for commercial applications.

#### 2.1.3. Carbon Nanomaterial Fillers

Graphene oxide (GO) was used as a filler in Nafion membranes because it allows easy proton transport and good water uptake due to its high surface area. The different oxygen groups such as epoxide, hydroxide, carbonyls and carboxyls converts GO into electrically insulating and hydrophilic material [97] while retaining other properties like mechanical strength, surface area, and gas impermeability.

Graphene oxide has an excellent compatibility with Nafion so it can be adopted as a modifier to improve the selectivity (to allow the passage of specific species) and performance of such membranes. Choi, et al. [98] developed a composite GO/ Nafion membrane. The authors claimed that the compatibility between both components was guaranteed due to their strong interfacial attraction. GO enhanced thermal backbone and side chains stabilities due to the interaction between GO sheets and Nafion: the non-polar backbone of Nafion interacted with the hydrophobic structure of GO while the polar ionic clusters of Nafion with the hydrophilic groups of GO. Their study revealed that the permeability for methanol with just 0.5 wt.% of GO was reduced to 60.2% of Nafion 112 at 25 °C. However, the proton conductivity tests revealed an opposite trend showing a decrease with increasing the GO filler content and a loss of 55.3% in proton conductivity is reported with 2 wt. % GO loading. This was not completely unexpected as GO alone is not known to be an excellent proton conductor.

The authors obtained the maximum power density of 62 mW cm$^{-2}$ at 30 °C and 141 mW cm$^{-2}$ at 70 °C after optimising the GO loading in the membrane (1% wt.) as depicted in Figure 2.

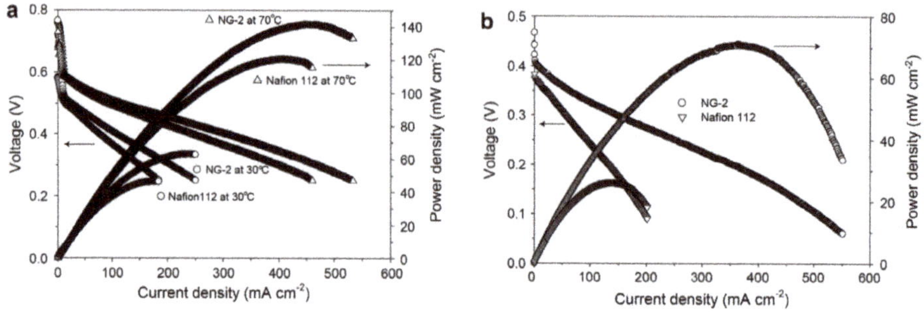


**Figure 2.** Polarization curves of DMFC obtained for Nafion 112 and GO composite membranes at (**a**) 1M methanol at 30 °C and 70 °C and (**b**) 5M methanol at 30 °C; reproduced with permission from [98].

Moreover, at high methanol concentration, where methanol crossover becomes critical, benefits provided from the incorporation of graphene oxide were more evident: the composite membrane showed much higher power density, 3 times higher than Nafion (71 mW cm$^{-2}$ vs 26 mW cm$^{-2}$).

Chien, et al. [99] prepared a composite membrane with sulphonated graphene oxide (SGO)/Nafion for DMFC to avoid the aggregation of GO in the polymer matrix. It was reported that proton conductivity increases with increasing amounts of SGO, as the SGO was distributed throughout the matrix and created more interconnected transfer channels. However, with further SGO amounts, aggregation began to predominate, thus reducing the conductivity of the composite membrane. Methanol permeability was shown to decrease in the presence of SGO as they block the methanol migration through the membrane. In DMFC test, the SGO/Nafion composite membrane exhibited higher current and power densities than commercial Nafion 115, for example;

- in 1 M methanol solution, the current density and power density for the composite membrane at 0.4 V were 102.7 mA cm$^{-2}$ and 42.9 mW cm$^{-2}$, whereas the commercial Nafion 115 revealed only 78.6 mA cm$^{-2}$ and 32.6 mW cm$^{-2}$.
- in 5 M methanol solution, the composite membrane showed values of 83.2 mA cm$^{-2}$ (at 0.4 V) and 34.6 mW cm$^{-2}$, which were better than the commercial membrane (54.1 mA cm$^{-2}$ at 0.4 V and 22.1 mW cm$^{-2}$).

Additionally, the SGO/Nafion composite membrane had a lower catalyst activation loss than Nafion 115, which indicated that the composite membrane had lower methanol crossover and faster reaction kinetics.

Yan et al. [100] proposed an innovative way to modify Nafion membrane by sandwiching a graphene oxide layer between two Nafion membranes. With the addition of a monolayer graphene film, methanol permeability decreased by 68.6% while observing only a marginal decrease in proton conductivity of 7% at 80 °C in comparison to pristine Nafion membrane. The authors tested the membrane in a DMFC varying methanol solution from 5 M to 15 M. Tests depicted that the graphene film allowed for a substantial performance improvement particularly when the passive DMFC was fed with high concentration methanol solutions enabling the passive DMFC to be operated at high concentrations.

### 2.2. *Non-Perfluorinated Polymers Composite Membranes*

Non-fluorinated membranes seem to have a promising future for DMFCs as a replacement for the expensive fluorinated membranes that have high methanol and ruthenium crossover. Aromatic polymers are considered to be one of the most promising routes to obtain high performance polymer electrolytes because of their availability, variety of chemical composition and stability in the cell environments. Specifically, poly ether ether ketone, polyvinyl alcohol, and poly arylene ether and their derivatives are currently under investigation.

Poly(ether ether ketone)s (PEEKs) [101] are semicrystalline polymers that present high thermal and chemical stability. The sulfonation of PEEK produces copolymers with sulfonic acids into the aromatic backbone; membranes made of these sulfonated polymers show useful properties for DMFCs, such as; low methanol cross-over, good ion conductivity, thermal stability and high mechanical strength [102]. The proton conductivity of SPEEK depends on the sulfonation degree [103], it generally increases with the sulfonation degree but high sulfonation results in high methanol permeability so that its application is limited [104]. The development of SPEEK composite membranes is currently being investigated by using SPEEK for the polymer matrix and modifying it in order to reduce methanol permeability at high sulfonation degree. Many researches have focused on SPEEK-based membrane with phenoxy resin [105], polyphenylsulfone (PPSU) [106], solid heteropolyacids [107], polyaniline [108], SiO2 and zirconium phosphate (ZrP) [109], zeolite [110], polypyrrole [111]. As discussed above, the utilization of montmorillonite and polybenzimidazole into Nafion matrix improved the DMFC performance; in the same way, they can be used to modify the matrix of SPEEK polymer. In fact, Gosalawit et al. [112] used a SPEEK/MMT membrane in their work and compared its performance in a DMFC single cell with pristine SPEEK membrane and Nafion bare membrane. The study confirmed that the performance was higher: current density generated from the MEAs of Nafion 117, SPEEK, SPEEK/MMT 1wt.%, SPEEK/MMT 3wt.% and SPEEK/MMT 5 wt.% membranes at the constant voltage of 0.2 V were 51, 76, 103, 96 and 94 $mAcm^{-2}$, respectively with the maximum power density of 10, 15, 21, 19 and 18 $mWcm^{-2}$. However, the thermal stability was significantly reduced. Pasupathi et al. [113] synthetized a non-perfluorinated membrane by casting SPEEK and PBI solution into a glass plate. A SPEEK/PBI membrane enhanced DMFC performance: the maximum power density obtained (45 mW $cm^2$) was two times higher than Nafion 117 at 60 °C. Moreover, SPEEK/PBI membranes were found to be extremely stable under DMFC operating conditions up to 60 °C. However, their stability dropped considerably at higher temperatures. Experiments are underway to address the stability issue of these membranes at higher temperatures.

Generally, sulfonated aromatic polymer membranes require a high sulfonation level to achieve sufficient proton conductivity resulting from the low acidity of the sulfonic groups in the aromatic rings [114]. However, such a high sulfonation level usually makes them excessively swell and even soluble in methanol/water solution which may lead to a loss in mechanical properties and become unavailable in applications [115]. Therefore, they should be modified including organic or inorganic fillers. Jiang et al. [116] investigated the performance of a SPEEK/GO membrane in which GO is sulfonated (SGO) to improve the proton conductivity of the membrane. In fact, the SPEEK/GO membranes exhibit lower ion exchange capacity (IEC) and water uptake than Nafion membrane, and the overall proton conductivity of the membranes remains low. This is due to the lack of proton conductive groups on the pristine GO. Therefore, these membranes are still not quite suitable for use as PEMs in DMFCs. However, by using higher contents of sulfonated GO, these SPEEK/SGO membranes showed even higher IEC and proton conductivity compared to Nafion 112, which makes them particularly attractive as PEMs for the DMFC applications. It is worth noting however that the swelling ratio of membranes increased with the increase of the content of the SGO. DMFCs with SPEEK/SGO showed better performance than those with the plain SPEEK or the pristine SPEEK/GO. With the optimized contents of SGO in SPEEK (3% and 5% wt.), the DMFCs exhibited 38 and 17%, respectively, higher performance than those with Nafion 112 and Nafion 115. Despite of having higher IECs and proton conductivity, the membranes with higher content of the SGO exhibited higher methanol permeability leading to a decrease of the fuel utility and the lifetime of the cathode catalysts, thus low performance. In this regard, for practical applications of the SPEEK/SGO membranes, the contents of SGO in the SPEEK matrices should be well controlled.

Sulfonated Poly (arylene ether sulfone) membranes (SPAES can be useful for methanol fuel cell through the modification of the polymer matrix by introducing inorganic/organic particles such as laponite. Laponite is made of silica tetrahedral and alumina octahedral sheets which have advantageous hygroscopic properties [117]. This inorganic compound was used by Kim et al. [118] to prepare and evaluate the behaviour of the SPAES membrane. Properties of the hybrid membranes for DMFC such as methanol permeability, and proton conductivity were investigated. Authors claimed that methanol permeability was lower than that of a SPAES pure membrane and Nafion using membranes with a small content of Laponite. This is because the presence of Laponite improved the barrier property of the membrane to methanol molecules. This was likely due to the tortuosity of layered silicate and the lower aspect ratio of the particles resulting from their exfoliation increasing methanol diffusion paths through the composite membrane. However, the proton conductivity was very low and further research should be conducted to optimise Laponite loading.

Another approach to reduce the undesired swelling property and methanol crossover of sulfonated membrane is to crosslink membranes. This method has been widely investigated by many researchers for crosslinking SPEEK membranes [119,120]. These membranes showed decreased swelling ratio and methanol crossover but decreasing proton conductivity. Following this method, Feng et al. [121] used sufonated poly (arylene ether)s as PEM materials due to their good thermal stability, high glass transistion and excellent mechanical strength. They synthesized sulfonated poly (arylene ether sulfone) copolymers with propenyl groups then crosslinked using benzoyl peroxide varying the crosslinking. Proton conductivity and methanol permeability were evaluated and compared with Nafion 117, and showed that:

- the proton conductivity of the SPAES membranes increased from 0.1 to 0.16 S/cm with the increase of temperature from 30 to 70 °C which was quite similar to that of Nafion 117: 0.11 to 0.17 S/cm from 30 to 70 °C [122];
- the methanol permeability evaluated was lower than that of Nafion117 ($2.07 \times 10^{-6}$ $cm^2/s$): the less pronounced hydrophobic/hydrophilic separation of sulfonated polyaryls compared to Nafion corresponded to narrower, less connected hydrophilic channels and larger separations of the less acidic sulfonic acid functional groups, which affected the permeability of methanol [123].

Poly vinyl alcohol (PVA) is usually synthesized from poly vinyl acetate and commonly used for adhesive, food wrapping, and desalination and pervaporation membranes [124]. Regarding the possible use of a membrane made of PVA in DMFC, it is known that PVA does not have any negative charged ions, like carboxylic and sulfonic acid groups, so the conductivity is very low as compared with Nafion membrane. Therefore, PVA membranes can be used in a fuel cell if negative ions are incorporated within their structure to increase their conductivity [125]. Moreover, several studies have shown how PVA polymer membrane leads to a reduced methanol crossover [126]. Regarding the reduction of methanol crossover, it was demonstrated that the addition of filler into PVA matrix contributes to mitigate this issue: fillers such as $SiO_2$ [127], polyrataxane [128] were reported. Yang et al. [129] used montmorillonite (MMT) as a filler and tested ion conductivity, methanol permeability, current density-potential and power density curves of the PVA/MMT composite polymer showing the following properties:

- high ionic conductivity: 0.0368 S $cm^{-1}$, performed by PVA/10wt. % MMT at 30 °C;
- methanol permeability: 3–4 × $10^{-6}$ $cm^2$ $s^{-1}$, which was lower than that of Nafion 117 membrane of $5.8 \times 10^{-6}$ $cm^2$ $s^{-1}$;
- maximum peak power density: 6.77 mW $cm^{-2}$ at ambient pressure and temperature with the PtRu anode based on Ti-mesh in a 2M $H_2SO_4$ + 2M $CH_3OH$ solution.

### *2.3. Other Composite Non-Fluorinated Membranes*

The modified polyvinylidene fluoride (PVDF) membranes using inorganic additives were prepared with a view of combining the properties of inorganic ion exchanger (high thermal stability, and excellent water holding capacity at higher temperatures) and organic support (chemically stability and high mechanical strength). Impregnation of porous polymeric film of PVDF is the method used by Pandey et al. [130,131] to synthesize PVDF/silica and PVDF/Zirconium phosphate (ZrP). Single cell DMFC tests were carried out to study the DMFC performance for the synthesized membrane. The membranes showed better thermal stability, water uptake ratio and lower methanol crossover than Nafion 117, however, performance were low because of poor proton conductivity.

PolyFuel Inc. produced polycarbon membrane for passive DMFC [132] showing a power density of 80 mW $cm^{-2}$ for thickness of 45µm, lifetime for a nearly constant runtime is 5000 h and back diffusion of water was improved by 30%, which helped mediate the dissolution of the methanol concentration in a passive DMFC.

There are also other composite membranes developed for PEMFC applications which may also have a good prospect for DMFC. These include trifluorostyrene-based membranes developed by Ballard Power System Inc [133], a butadiene/stryene rubber-based membrane developed by Hoku Scientific Inc [134] and polystyrene sufonate (PSS) membranes [135]. Table 1 summarizes the properties and the pros and cons of composite electrolyte membranes described in this review compared to those of the commercial membrane. In addition, Table 2 summarizes their DMFC best performance.

**Table 1.** Summary of DMFC composite membrane properties.

| Membrane | Preparation Method | Pros | Cons |
|---|---|---|---|
| **Nafion/PTFE** | Impregnation | Low methanol permeability | Decreased conductivity |
| **Nafion/PVA** | Casting | Low methanol permeability Easily manipulation with small thickness | Lower proton conductivity |
| **Nafion/PBI** | Screen printing | Reduced methanol permeability | High impedance |
| **Nafion/Polypyrrole** | Electrodeposition-In situ polymerization | Low methanol permeability | Decreased proton conductivity Increased resistances |
| **Nafion/Polyaniline** | In situ-polymerization | Decreased methanol permeability Increased selectivity | Decreased conductivity |
| **Nafion/SPAEK** | Casting | Low methanol permeability Higher proton conductivity | Easily breakable |
| **Nafion/SPEEK** | Casting | Decreased methanol permeability Reasonable thermal properties | Reduced proton conductivity |
| **Nafion/Metal oxides ($SiO_2$-$TiO_2$)** | Casting | Increased proton conductivity | Accelerated degradation Difficult homogeneity |
| **Nafion/Montmorillonite** | Casting | Methanol crossover decreased | Slight proton conductivity decrease |
| **Nafion/Zeolites (Analcime-Mordenite)** | Spray | Methanol crossover decreased Slight increased proton conductivity | Low tensile strength |
| **Nafion/Graphene oxide** | Casting | Methanol crossover decreased High thermal and mechanical stability | Decreased proton conductivity |
| **SPEEK** | Casting | Low methanol crossover | Poor mechanical stability |
| **SPAES/Laponite** | Casting | Low methanol crossover Enhanced tensile strength | Low proton conductivity |
| **PVA/Montmorillonite** | Casting | Low methanol crossover Cheap High proton conductivity | Filler content should be well controlled Specific operating condition and specific stack material should be used |
| **PVDF/silica-Zirconium** | Impregnation | High tensile strength Low methanol crossover | Poor proton conductivity |

**Table 2.** Summary of DMFC best performance using composite membranes.

| Membrane | Type of DMFC | Voltage (V) | Current Density ($A.cm^{-2}$) | Power Density ($mW\ cm^{-2}$) | Temperature (°C) | Methanol Concentration (M) |
|---|---|---|---|---|---|---|
| Nafion/silica [79] | Active | 0.3 | 0.2 | 60.0 | 75 | 5 |
| Nafion/TiO2 [83] | Active | 0.3 | 0.214 | 64.2 | 80 | 1 |
| Nafion/SiO2 [83] | Active | 0.3 | 0.204 | 62.9 | 80 | 1 |
| Nafion/sulfonated montmorillonite [87] | Active | 0.2 | 0.336 | 67.2 | 40 | 2 |
| Nafion/GO [98] | Active | 0.31 | 0.46 | 141.0 | 70 | 1 |
| Nafion/SGO [99] | Active | 0.40 | 0.1 | 43.0 | - | 1 |
| Sandwich Nafion/GO [100] | Passive | 0.17 | 0.15 | 25.0 | - | 5 |
| Nafion/mordenite [95] | Active | 0.18 | 0.06 | 10.8 | 70 | 4 |
| Nafion/analcime [95] | Active | 0.18 | 0.04 | 7.2 | 70 | 4 |
| Nafion/mordenite/GO [96] | Active | 0.23 | 0.12 | 27.5 | 70 | 1 |
| Nafion/polypirrole [64] | Active | 0.30 | 0.15 | 45.0 | 60 | |
| Nafion/polyaniline [65] | Active | 0.23 | 0.3 | 70.0 | 60 | 6 |
| Nafion/PVA [53] | Active | 0.26 | 0.5 | 130.0 | 70 | 1 |
| Nafion/PBI [58] | Active | 0.36 | 0.06 | 21.7 | 60 | 2 |
| Nafion/PTFE [47] | Active | 0.25 | 0.35 | 87.5 | 70 | 2 |
| Nafion/PTFE/zirconium phosphate [50] | Active | 0.20 | 0.3 | 60.0 | 80 | 2 |
| Nafion/SPAEK [72] | Active | 0.38 | 0.3 | 114.0 | 80 | 2 |
| Nafion/SPEEK [73] | Active | 0.18 | 0.15 | 27.0 | 80 | 2 |
| SPEEK [112] | Active | 0.20 | 0.076 | 15.2 | 60 | 1 |
| SPEEK/MMT [112] | Active | 0.20 | 0.1 | 20.0 | 60 | 1 |
| SPEEK/GO [116] | Active | 0.35 | 0.21 | 72.2 | 65 | 1 |
| SPEEK/PBI [113] | Active | 0.28 | 0.16 | 45.0 | 60 | 1 |
| PVA/montmorillonite [129] | Active | 0.29 | 0.023 | 6.8 | 25 | 2 |
| PVDF/zirconium phosphate [131] | Active | 0.54 | 0.060 | 32.3 | 60 | 1 |

According to the data collected into the two tables above, membranes with fillers guarantee the highest performance. This is due to the fact that the reduced alcohol permeability counterbalances the reduced proton conductivity in the composite membranes. Materials, such as; PTFE, PVA and metal oxided are not proton-conducting materials so result in a reduction in the electrolyte proton conductivity, however they increase the tortuosity of the membrane thereby leading a reduced amount of crossover. Two approaches to increase the proton conductivity were adopted to further enhance the performance, namely; by optimising the filler content or by functionalising the filler (most commonly by incorporating sulphonic groups) to increase the overall electrolyte proton conductivity.

Polarization curves, proton conductivity, water uptake and methanol crossover are tests commonly carried out for all the membranes described in this section. However, durability tests are still lacking in the literature. Therefore, only commercial Nafion provides guarantees in terms of lifetime and degradation, therefore it cannot be completely substituted up to this time. Research activities on the lifetime time and the degradation of Nafion based composite membranes should be carried out.

## 3. Composite Membranes for $H_2$ PEMFCs

### 3.1. *Inorganic Fillers*

Inorganic fillers have a long history of use as fillers in membranes for fuel cells. The general explanation of their suitability is due to their high thermal stability, mechanical strength, and water-absorbing nature. Therefore, the main aim of introducing fillers into the polymer membrane is to enhance its properties and enable its operation at elevated temperatures and/or low relative humidities. Figure 3 illustrates the change in proton conductivity and hydrogen crossover with the change in operating conditions to higher temperatures and lower relative humidities (Figure 3b) and the incorporation of fillers (Figure 3c).

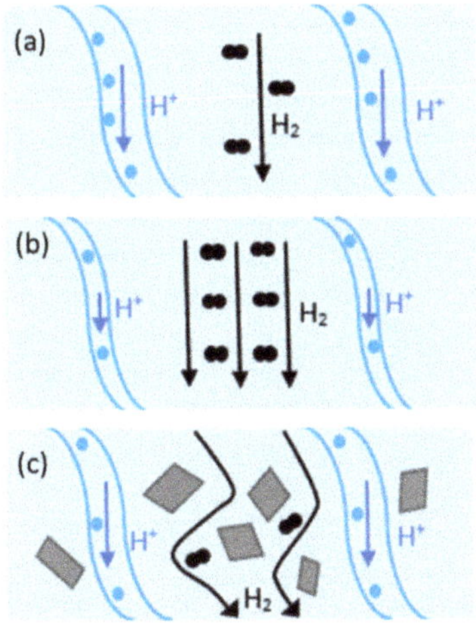


**Figure 3.** (**a**) Hydrogen and proton transport through Nafion at 80 °C (**b**) Hydrogen and proton transport through Nafion at higher temperatures (**c**) Hydrogen and proton transport through Nafion composite at higher temperatures.

At 80 °C and with high relative humidity, the membrane and its channels are fully saturated with water. The protons travel through the membrane via either Grotthus mechanism or via diffusion. There is also molecular hydrogen that passes through the membrane, known as hydrogen crossover, that then interacts with the cathode resulting in a reduced OCV.

At higher temperatures, the water within the membrane begins to evaporate. This results in a shrinkage in the channels within the membrane and less water for protons to either hop across or diffuse through the membrane, resulting in a subsequent drop in proton conductivity. Furthermore, the effect of hydrogen crossover is enhanced due to the increased operating temperature.

The addition of a filler material increases the path of the hydrogen to pass from the anode to the cathode. This increased tortuosity results in a decrease in hydrogen crossover. Also, the filler material itself can be functionalised, so for example, hydrophilic fillers can draw lots of water, hence improving proton conductivity and reducing the detrimental effects of increased temperatures.

Di Noto et al. studied the proton-conducting properties of mixed organic-inorganic membranes with Nafion mixed with various metal oxides, such as titanium, zirconium, hafnium, tantalum and tungsten oxides [136]. Thermal experimentation revealed that the composite membranes are stable below 170 °C, suggesting their possible application in PEMFCs that operate at elevated temperatures. Four different water domains were detected in the composite membrane, regarding the different interactions such as bulk water and water solvating ions interacting with the sulphonic side group. The quantity of each domain depends on the filler material. The authors proposed that the conduction mechanism within the studied membranes involves proton hopping through different fluctuating water domains [137].

Adjemian et al. [138] also investigated composite membranes with different metal oxide fillers such as titanium, silica, alumina, and zirconium for PEMFCs with varying the operating temperature from 80 to 130 °C. It was found that the membranes with titanium oxide or silica revealed better performance at higher temperature and lower humidity conditions compared to recast Nafion. Furthermore, a study into carbon monoxide tolerance revealed that; by using the composite membranes and increasing the operating temperature to 130 °C, the CO tolerance of the catalyst layer was improved to 500 ppm of carbon monoxide without failing comparted to 50 ppm at conventional operating conditions. In this study, none of the composite membrane filler materials were surface modified so potentially further performance enhancement can be achieved by functionalising the fillers.

A composite membrane with Nafion and a filler consisting of silica nanoparticles with surface modified fluoroalkyl functionalities was presented in [139]. It was noted that although the silica nanoparticles are hydrophobic, the water uptake of the membrane was not negatively affected. In addition, the composite membrane showed thermal stability up to temperatures of 240 °C. Proton conductivity tests revealed that the composite membrane with 5 wt. % silica nanoparticles with a ratio of [Nafion/ (Si80F) 0.7] had the highest conductivity at 0.083 S $cm^{-1}$ at 135 °C. Following from this, a single cell test was performed with the composite membrane at 85 °C. The composite membrane displayed a better power density compared to the recast Nafion, when the oxidant is air and oxygen (Under air: 0.38 vs 0.27, under oxygen: 0.48 vs 0.35 W $cm^{-2}$ composite and recast Nafion respectively). It would be interesting to see the behaviour of the membrane at elevated temperatures.

Following on from their work, Griffin et al. [140] fabricated and characterised a composite membrane with sulphonated zirconia dispersed in a Nafion matrix. The idea behind this is that functionalising the zirconia with sulphonic groups would boost the proton conductivity of the membrane. Proton conductivity tests at 120 °C and under anhydrous conditions revealed that the membrane had a conductivity of $3 \times 10^{-3}$ S $cm^{-1}$. This makes the membrane ideal for fuel cell applications at intermediate temperatures and under dry condition.

Saccà et al. [141] studied the influence of zirconium oxide as a filler material at different loadings of 5, 10 and 20% for Nafion composites. Recast Nafion membranes had a water uptake of 20%. The addition of zirconium oxide led to an increase in water uptake to 24, 24, and 30% for loadings of 5, 10 and 20% respectively. Fuel cell testing of the membranes in a single cell at operating temperatures of 80, and 110 °C show that at 80 °C, addition of 5% of filler makes very little difference in performance compared to recast Nafion. However, the membrane with 20% filler had a much lower potential, potentially due to excessive water uptake at 80 °C. Composite MEAs with 10% filler produced higher polarisation compared to Nafion at both temperatures with a maximum power density of 400 mW $cm^{-2}$ was achieved at 130 °C, 85%, and 0.5–0.6 V.

D'Epifanio et al. [142] took this one step further and sulphonated the zirconium oxide. Water uptake experiments at 25 °C with varying relative humidity revealed that both composite membranes outperformed recast Nafion at all relative humidities (30 to 100%). Polarisation curves at 70 °C and at three different RH (65, 83 and 100%), show that the composite membrane produced better current densities at all voltage ranges, with current densities of 1015 mA $cm^{-2}$ vs 680 mA $cm^{-2}$ at 0.6 V, respectively. The difference between the two polarisation curves was emphasized during the ohmic and mass transport region, showing that the filler reduced ohmic resistance and improved water diffusion. A final test at 30% RH showed even greater differences with current densities of 930 mA $cm^{-2}$ vs only 200 mA $cm^{-2}$.

Alberti et al. [143] attempted to improve the proton conductivity and stability of membranes at elevated temperatures and studied the effect of doping Nafion with zirconium phosphate. However, they found that the conductivity decreases with increasing filler loadings. In addition, the authors explained that the difference in proton conductivity between Nafion and their composite membrane is mostly at lower relative humidities and higher filler loadings. On the other hand, Sahu et al. embedded silica nanoparticles into Nafion via a sol-gel method [144]. Single cell tests at 100 °C and at 100% RH showed the composite membrane (doped 10 wt. % silica) produced a peak power density of 350 mW $cm^{-2}$. Moreover, composite membranes with 15 wt. % experience large mass transport losses due to flooding.

Costamagna et al. [145] then prepared zirconium phosphate Nafion composite membranes via the impregnation of Nafion 115 and recast Nafion for high temperature PEMFC use. The composite membrane from Nafion 115 produced a current density of 1000 mA $cm^{-2}$ at 0.45 V and at an operating temperature of 130 °C, which is much better compared to 250 mA $cm^{-2}$ pristine Nafion. In addition, the cell fabricated from recast Nafion reached current densities of 1500 mA $cm^{-2}$ at the same operating conditions.

Furthermore, Sahu et al. [146] presented a Nafion composite membrane with mesoporous zirconium phosphate as the filler, prepared via a co-assembly method. The single cell testing was performed at 70 °C and at varying relative humidities, 100, 50, 31 and 18%. The difference between the composite membrane and pristine Nafion membrane increases with decreasing RH (via the maximum power density peaks). In terms of filler loading, the best performing was 5 wt. %, followed by 10 and 2.5. At 18% RH, the composite membrane produced a maximum power density of 353 mW $cm^{-2}$, in comparison to pristine Nafion's 224 mW $cm^{-2}$ (both at 500 mA $cm^{-2}$).

Pineda-Delgado et al. [147] decided to study the behaviour and performance of Hafnium oxide Nafion composite membranes. The fabricated composite membranes displayed greater water uptake of 61% at 100 °C compared to 29% for recast Nafion. This improvement in water uptake led to better proton conductivity at 100 °C with 112 vs 82 mS $cm^{-1}$, for the composite and recast Nafion respectively. Following from this, the authors decided to test their membranes in a single cell set up at operating temperatures of 30, 50, 80 and 100 °C. The recast Nafion achieved a greater maximum power density at 30 and 50 °C, but at 80 and 100 °C the composite membrane performed better. At 100 °C, the composite produced a maximum power density of 0.336 W $cm^{-2}$ compared to 0.188 W $cm^{-2}$ for the recast Nafion, at a voltage of 0.46 V.

The performance of sulphonated silica Nafion composites where assessed where the filler was synthesised with a simple sol-gel calcination process [148]. Optimisation studies revealed that a 1% filler was the optimum loading, outperforming 0.5, 1.5% and recast Nafion. In-situ fuel cell testing under reduced humidity also confirmed the initial ex-situ results. The authors attributed the enhanced performance to efficient proton transport due to the well-defined phases in the membrane structure which was seen with TEM.

One method to improve the dispersion of filler material within the polymer matrix is to swell the polymer membrane in a solution of the filler [149,150]. Xu et al. employed this technique by swelling the Nafion membrane with silica to achieve a composite membrane, in comparison to the traditional solution casting technique. They highlighted that this method maintains the ordered nanophase-separation structure of Nafion. This was shown in water uptake tests, where the swelled composite showed a higher water uptake but lower swelling, in comparison to the recast membrane. Fuel cell testing at 110 °C and 20% RH showed that the swelled composite produced a maximum power density of 113 mA $cm^{-2}$, in comparison to 80 mW $cm^{-2}$ for recast Nafion with no filler. The performance was explained due to the lower internal resistance of the composite membrane.Saccà et al. [151] introduced titanium oxide of different loadings (5, 10 and 15 wt. %) into Nafion for the purpose of operating fuel cells at a reduced humidity. SEM images revealed that the dispersion of filler throughout the cross-section of the membrane show that the lower filler loadings are better dispersed. The higher loading membranes showed the presence of filler agglomerates. Water uptake testing at different temperatures showed that there is a small drop initially when the filler material is introduced. In addition, the higher loading membranes are less influenced by the increasing temperatures. A similar trend was also observed for swelling, with the composite membranes having lower swelling percentages. However, excessive introduction of filler material can result in the membrane becoming stiffer and more fragile. Fuel cell testing revealed that the 10 wt. % composite membrane was the best performing, with it being closest in polarisation behaviour to recast Nafion.

Saccà et al. [152] continued their work by studying the characteristics of Nafion-Titanium oxide membrane for PEMFCs operating at medium temperatures. Introduction of 3 wt. % of titania powder increased the water uptake from 20% for recast Nafion to 29%. The composite membrane outclassed commercial Nafion at all fuel cell operating temperatures (80, 90, 110 and 130 °C). At 110 °C (and 0.56 V), maximum power densities were 0.514 W $cm^{-2}$ and 0.354 W $cm^{-2}$ for the composite and commercial membrane respectively. As well as the better polarisation performance, the cell resistance of the composite membrane decreased with temperature up until 110 °C, where it starts to increase (0.106 Ω $cm^{-2}$). This is in comparison to the commercial membrane whose resistance begins to increase after 100 °C (0.088 Ω $cm^{-2}$). Interestingly, experiments with steam reforming fuel (with

10 ppm CO, 20% $CO_2$, 75% $H_2$ and 1% $CH_4$) at 110 °C showed similar OCV values compared to pure hydrogen. Morever, the maximum current density decreased from 1300 mA $cm^{-2}$ for pure hydrogen to 800 mA $cm^{-2}$ for the synthetic fuel.

Amjadi et al. [153] also studied the influence of titanium oxide as a filler in Nafion composites, with two types of composites prepared via different methods, a solution casted and an in-situ sol-gel synthesis. EDX mapping across the composite membrane revealed that the sol-gel composite had a better dispersion of particles, which ultimately led to improved properties. One example is water uptake, where both composites had improved uptake capabilities compares to recast Nafion. However, due to the agglomeration and reduced uniformity in the casted composite, a decrease in surface area of the filler reduced the achieved water uptake. The introduction of filler led to a drop in proton conductivity, which was explained by the disruption of proton pathways in the membrane. Fuel cell testing at 110 °C showed that the composite membrane was able to reach a maximum current density of nearly 600 mA $cm^{-2}$, compared to just over 200 mA $cm^{-2}$ for Nafion 117.

Furthermore, Matos et al. [154] studied the influence of particle shape (spherical nanoparticles, high surface area mesoporous particles, and nanotubes) of titania for the application of PEMFCs operating at temperatures up to 130 °C. Water uptake tests revealed that any addition of spherical or high surface area (HSA) titania led to a decrease in water uptake compared the recast Nafion, with greater decrease at higher filler loading. However, the water uptake for the titania nanotube composite membrane increased, reaching a maximum of nearly 60% at 15% loading, compared to 42% for recast Nafion. The authors state that this is because of the "nanotubular" structure in which water molecules being able to exist inside the nanotube. Single cell tests (Figure 4) were performed at 80 and 130 °C. Nafion outperformed the composite membranes at 80 °C, however, the membrane degraded significantly once the temperature increased. All three composite membranes displayed a smaller amount of decrease in polarisation at 130 °C in comparison to recast Nafion. However, increasing filler loading in all three prospective filler materials (nanoparticle, mesoporous particles and nanotubes) led to a decrease in polarisation, particularly in the ohmic region, which the authors explain due to the decreasing proton conductivity with greater filler loading.

Zhengbang et al. [155] synthesised titanium oxide nanowires as a filler material in Nafion for PEMFCs operating at a higher temperature in addition to reinforcing the mechanical properties of the membrane. Addition of the nanowires led to a subsequent drop in water uptake and swelling, with increasing loading leading to increased reductions. The reduced swelling would help maintain mechanical integrity at higher operating temperatures. Fuel cell testing at 90 °C showed that the composite membrane experienced a smaller drop in polarisation when the humidity was reduced, in comparison to Nafion where the change in polarisation was much greater. Humidity stress tests revealed that the composite membrane had less stress (which becomes smaller with increased loading) than the recast Nafion, which experienced a high level of humidity related stress indicating lower lifetime.

Ketpang et al. [156] further developed their idea of tubular inorganic fillers by studying the effect of titanium oxide nanotubes as a filler. The composite membranes had a higher water uptake compared to recast Nafion, with recast Nafion achieving 21.8%, Nafion-TiNT-10 33.7%, Nafion-TiNT-20 31.3% and the Nafion composite with 50% titanium oxide nanotubes achieving a water uptake of 29.6%. In addition, FT-IR analysis after drying the membranes at 110 °C revealed that the composite membranes still had water (from electrostatic interaction) through peaks that corresponded to -OH stretching (3455 $cm^{-1}$) and –HOH- (1625 $cm^{-1}$) bending vibration. Proton conductivity measurements at 80 °C and 100% RH confirmed that the filler improved the proton conductivity compared to recast Nafion (97 mS $cm^{-1}$). The highest proton conductivity measurement was achieved by the composite with 10% filler (155 mS $cm^{-1}$), with the 20% (142 mS $cm^{-1}$) and 50% (121 mS $cm^{-1}$) having slightly decreased conductivity. The composite membranes also outperformed the pristine Nafion at variable RH. Fuel cell experiments at 80 °C and 100% RH (Figure 5) show that the composite membranes perform much better than the recast membrane, with current densities at 0.6 V of 1777, 1609, 1498 and

1357 mA $cm^{-2}$ for the composite membranes with filler of 10, 20, and 50% titanium oxide nanotube (TNT) content and recast Nafion, respectively.

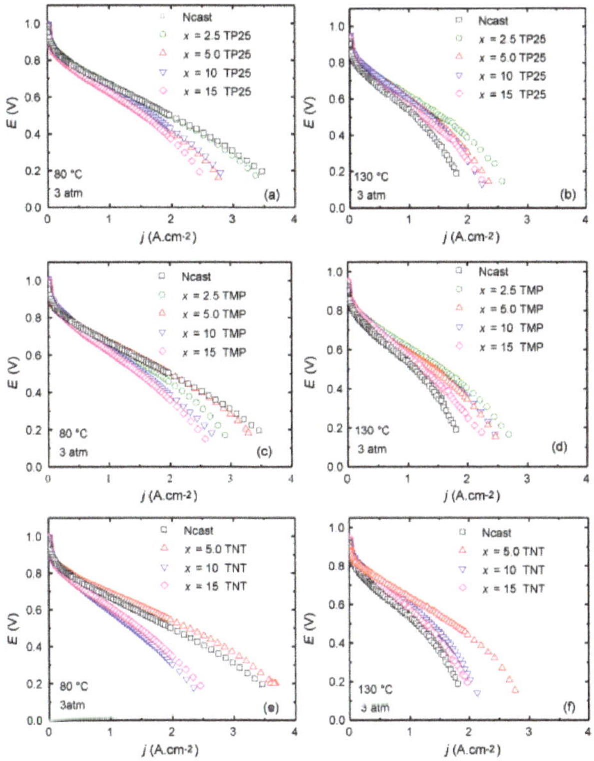


**Figure 4.** Polarization (*I–V*) curves for the (**a**,**b**) TP25, (**c**,**d**) TMP, and (**e**,**f**) titanium oxide nanotube (TNT) composites measured at 80 and 130 °C; reproduced with permission from [154].

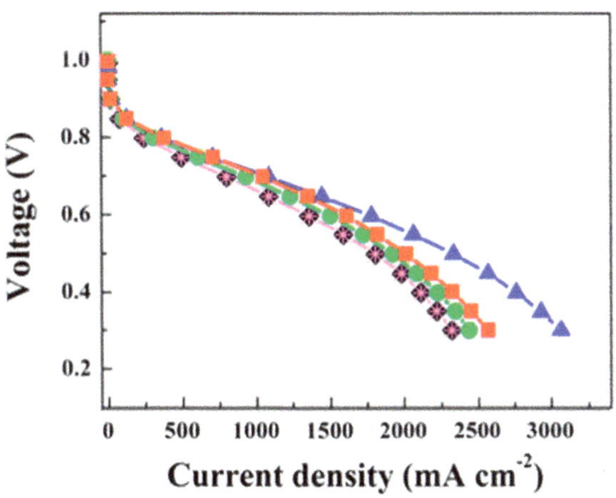


**Figure 5.** Polarization plots of Nafion 212 (black), recast Nafion (pink), Nafion TNT 10% (blue), Nafion TNT 20% (red), Nafion TNT 50% (green) at 80 °C and 100% RH; reproduced with permission from [156].

The OCV also ranged from 0.97 to 1.03 V, indicating low crossover. Similar to the zirconium oxide nanotube, the titanium oxide nanotube composite displayed greater current densities at lower voltages. A 100 h stability test at 0.5 V, 80 °C and 18% RH showed that the composite membrane's maximum power density decreased from 470 to 442 mW $cm^{-2}$, whereas Nafion 212 only managed to produce a maximum power density of 55 mW $cm^{-2}$ which degraded to 22 mW $cm^{-2}$ after 100 h, an impressive difference in performance.

Jun et al. [157] then fabricated a Nafion composite with functionalised titanium oxide nanotubes and 3-mercaptopropyl-tri-methoxysilane (MPTMS) was used to functionalise the inorganic filler, to further improve proton conductivity. Nanotubes are a promising filler due to their high surface area and internal space, in addition to providing mechanical strength. In addition, the water uptake was greater, with 27.2 to 23.7%, for functionalised nanotubes to functionalised nanoparticles, respectively. Proton conductivity measurements at 120 °C and varying relative humidities show that the functionalised titania nanotubes exhibited greater conductivities than recast Nafion, at all humidities, with the deviation being greater at lower RH.

A Nafion composite comprising of porous zirconium oxide nanotubes were fabricated by Ketpang et al. for the purpose of high temperature PEMFCs [158]. The tubular structure of the filler was used to improve water transport, which should result in improved water uptake and proton conductivity. The performance of these composite membranes was tested at 80 °C at varying relative humidities of 100, 50 and 18%. It was found that the addition of the filler resulted in improved power densities at 0.6 V, implying that the filler lowers the ohmic resistance. In addition, the composite membrane revealed greater current densities at low voltages (0.3 V), this was explained due to the more efficient back diffusion from the cathode to the anode, mitigating flooding. A further 200 h durability test of the membrane (with 1.5 wt. % of filler) at the same operating conditions (80 °C and 18% RH) displayed a small decrease in OCV, from 0.99 to 0.92 V after 200 h. The authors have shown that use of a porous nanotube morphology can improve water transport and have potential advantages in low relative humidity application.

Research into composite membranes extended beyond of the use of Nafion to other proton-conducting polymers. Marani et al. decided to combine sulphonated poly(ether ether ketone) (SPEEK) with titania nanosheets (an alternate material structure) for the application in PEMFCs operating at temperatures of 140 °C [159]. The authors studied the effect of treating the composite membranes with either water or with acid prior to use in addition to the effect of inorganic filler loading. It was found that acid treated membranes (with the lower filler loading of 1.67%) had the greatest proton conductivity in comparison to pristine SPEEK, with values of $4.14 \times 10^{-2}$ $Scm^{-1}$ at 140 °C and at 100% relative humidity to $1.76 \times 10^{-2}$ $Scm^{-1}$, respectively. This is because the acid washing displaced the tetrabutylammonium (TBA+), which was used to create the stable suspension of Titania nanosheets. However, acid treated membranes with higher loading displayed a porous structure and extreme swelling indicating chemical instability and high degradation rate.

Devrim et al. [160] fabricated a composite membrane with titanium oxide and sulphonated polysulfone as the polymer matrix. The degree of sulphonation of the polymer was varied and higher levels of sulphonation led to a higher water uptake, with a sulphonation degree of 15% providing a water uptake of 7%, compared to 33% for a sulphonation degree of 40%. Adding titanium oxide to the sulphonated polymer (40% sulphonation degree) resulted in a drop in water uptake, to 29%. The authors explain that this is because the introduction of the filler reduces the membranes' free volume and ability to swell sufficiently. Proton conductivity values increased with increasing levels of sulphonation and temperature, with the composite of 40% sulphonated polysulfone/titanium oxide producing a conductivity of 0.098 S $cm^{-1}$. Single cell tests at varying operating temperatures from 60 to 85 °C reveal that the pristine sulphonated polysulfone undergoes excessive swelling above 70 °C, leading to lower power output. The composite membrane outperforms the pristine reference membrane as the filler provides mechanical reinforcement to the membrane, preventing excessive swelling and deformation. The sulphonated polysulfone membrane produced a maximum power density of 0.16 W $cm^{-2}$ at 85 °C, compared to 0.24 W $cm^{-2}$ for the composite membrane.

Sambandan et al. synthesised silica and functionalised sulphonated silica composite membranes with SPEEK as polymer of choice [161]. Water uptake results show that the composite membrane had lower water uptakes compared to SPEEK. Fuel cell testing at 80 °C and 75%, in addition to 50% RH show that the composite membranes, particularly those with functionalised filler have polarisation curves similar to that of recast Nafion. Proton conductivity results for the functionalised composite membrane were 0.05 S $cm^{-1}$ and 0.02 $S^{-1}$ cm with the same operational parameters to the fuel cell, respectively. Therese at al. [162] prepared a SPEEK/PAI (poly amide imide) membrane with sulphonated silica filler. The PAI was added to the SPEEK to improve the mechanical strength and chemical resistance at higher operating temperatures. The composite membrane produced a proton conductivity of $8.12 \times 10^{-2}$ S $cm^{-1}$ at 90 °C. The idea of combining more than one polymer for composite membranes in an interesting one as instead of trying to choose one optimum polymer to work with, several can be blended.

Sahin et al. [163] produced a SPEEK cerium phosphate composite membrane to improve fuel cell performance and to increase oxidative stability. Fenton testing revealed that the composite membrane lost 10% in weight over 80 h, whereas the SPEEK membrane was completely destroyed. Proton conductivity also increased with filler content until 10% loading, where it begins to decrease.

Carbone et al. [164] fabricated a SPEEK composite membrane with amino-functionalised silica filler for elevated temperature operation in PEMFCs. Two types of SPEEK were synthesised, with 35 and 52% degree of sulphonation. The addition of the functionalised filler did not change the water uptake or swelling (at 100 °C) of the 35% sulphonated SPEEK. However, the 52% SPEEK water uptake and swelling dropped significantly with the addition of filler (from 400% to 120% water uptake and from 4 to 1.5 degree of swelling ratio). The authors explained that this is due to the strong sulphonic-aminic groups. Fuel cell testing at 120 °C showed that the composite membrane with 52% degree of sulphonation and with 20 wt. % of filler produced a peak power density of 246 mW $cm^{-2}$ (around 400 mA $cm^{-2}$) compared to 179 mW $cm^{-2}$ (around 320 mA cm-2) for 52% SPEEK without filler.

The same authors decided to continue this line of work and studied the effects of a zeolite filler (H-BETA) inside a SPEEK matrix for medium temperature fuel cells [165]. The introduction of zeolite reduced IEC of the SPEEK membrane (around 50% degree of sulphonation) from 1.55 to 1.47 (5% filler), 1.4 (10% filler) and 1.31 meq $g^{-1}$ (15% filler). At 80 °C, the pristine SPEEK outperforms the three composite membranes but at 120 °C all three composite membranes outperform the SPEEK reference membrane. The composite membrane also had a higher OCV than the reference SPEEK. The authors explain this as the zeolite providing necessary mechanical reinforcement as well as retaining water in the membrane that would otherwise be removed at elevated temperatures.

Moreover, Ozdemir et al. investigated the addition of different inorganic fillers (silicon dioxide, titanium dioxide and zirconium phosphate) to PBI for high temperature PEMFCs [166]. The properties that the authors were looking for included improved acid uptake and greater acid retention (lower levels of leaching). All three prospective fillers led to decreased acid leaching, from pristine PBI lost 85.2% of its doped acid compared to SiO2/PBI at 81.5%, TiO2/PBI at 77.4% and ZrP/PBI at 75.9%. Also, $SiO_2$/PBI and ZrP/PBI displayed improved proton conductivity values compared to pristine PBI. Both membranes produced their highest conductivity at 180 °C, with 0.113 and 0.200 S $cm^{-1}$, respectively. However, $TiO_2$/PBI displayed proton conductivities lower than pristine PBI. This was explained due to the non-uniform dispersion of filler (agglomeration) within the Nafion, which was observed on the SEM images. All four membranes conductivities increased with increasing temperature (140, 165 and 180 °C).

Lee et al. fabricated a PBI composite with sulfophenylated titanium oxide nanoparticles for fuel cells operating at elevated temperatures [167]. As expected, the introduction of the filler material improved acid retention and proton conductivity. The composite membrane produced a peak power output of 621 mW $cm^{-2}$, whereas pristine PBI produced 471 mW $cm^{-2}$, at 150 °C. One thing to note was that the membranes were very thin for composites, with film thicknesses of around 15 μm before acid doping and 22 μm after and therefore hydrogen crossover tests would be interesting to perform to understand the difference in crossover between pristine and composite membranes.

Ooi et al. investigated improving the acid retention and oxidative stability of PBI membranes operating at increased temperatures [168]. This was achieved by preparing a composite membrane which composed of partially fluorinated PBI and a filler of cesium hydrogen sulfate-silicotungstic acid ($CsHSO_4$–$H_4SiW_{12}O_{40}$, CHS-WSiA). The synthesised composite exhibited greater acid retention rates, which was attributed to the fluorinated PBI and the filler material. This retention was examined in a fuel cell 24 h stability test, where a voltage of 0.614 V at a constant current of 0.2 A $cm^{-2}$ was produced with no drop. A longer test would be interesting to validate the durability of the membrane.

Devrim et al. [169] prepared a silica polybenzimidazole (PBI) composite membrane for high temperature PEMFCs. The silica nanoparticles improved the acid retention and the proton conductivity. Proton conductivity results measured at 140, 165 and 180 °C revealed that the composite membrane had greater conductivity than pristine PBI, 0.0675 to 0.0600, 0.0866 to 0.0765, 0.1027 to 0.0944 $Scm^{-1}$ for PBI/$SiO_2$ respectively at the three temperatures. The addition of silica also reduced the degree of acid leaching from 41.5 (for pristine PBI) to 36.3% due to increased covalent bonding between the inorganic filler and acid. Single cell testing was also performed at the three temperatures previously stated, under hydrogen and air at 1 atmosphere. At 140 °C the pristine PBI outperformed the composite but at the two higher temperatures the composite membrane produced a greater maximum power density. The best performance was from the composite membrane at 165 °C, producing a maximum power density of 0.24 to 0.2 $Wcm^{-2}$ for the pristine PBI, at 0.6 V. The authors have shown the novelty of using inorganic filler to retain acid in PBI for high temperature PEMFC applications.

Plackett et al. [170] tested laponite clay as a filler in PBI for high temperature fuel cells. Two sets of fillers were prepared, by functionalising the clay with an imidazole group and another with quaternary ammonium group. Water uptake results showed that no difference was made when the organic filler was introduced in the PBI matrix, but the composite membranes did experience less acid swelling. The composite membranes achieved an OCV of 1.02 V (at room temperature, 0.96 V at 125 °C and 0.91 V at 200 °C), which implies low or almost non-existent hydrogen crossover, which was confirmed in permeability tests.

Aili et al. [171] doped silica with phosphotungstic acid for use as a filler in phosphoric acid doped polybenzimidazole for high temperature PEMFCs. This composite had a lower swelling rate due to its lower uptake of phosphoric acid. Durability testing at 200 °C revealed that the composite membrane had a decay rate of 27 μV $h^{-1}$, whereas the membrane without the filler decayed at a rate of 129 μV $h^{-1}$.

Other inorganic materials are used as they have the potential to improve the durability of the membrane. Rodgers et al. [172] used platinum nanoparticles as a filler to remove radicals formed during fuel cell operation and therefore reduce degradation. Membranes with 0, 10, 30, and 50 mol % of platinum were prepared, and their performance was evaluated in a 100 h fuel cell test at 90 °C and 100% RH. The highest degradation (through fluoride emission) was observed for the 10 mol % platinum composite. The authors explain that this is because of the low distribution and density of the platinum particles throughout the membrane.

Pearman et al. [173] studied the influence of cerium oxide as a radical scavenger in PEMFCs. Two forms of cerium oxide were used as fillers within a PFSA polymer structure, a synthesised version with 2–5 nm sizing, and a commercial version with 20–150 nm. The addition of cerium oxide resulted in a 50% reduction in OCV decay rate (from a 94 h test), from 0.9 mV $hr^{-1}$ for pristine Nafion to an average of around 0.4 mV $hr^{-1}$. However, the weight percentage of cerium oxide seemed to make no difference in the decay rate. Electron microscopy images show that less platinum particles were present in the composite membrane in comparison to the recast. A 500 h OCV hold test with pre and post-test polarisation curves, depicted in Figure 6, demonstrated that the composite membranes had a much smaller deviation in polarisation compared to the baseline Nafion membrane. The authors followed this work up by studying the proton conductivity of the composite membranes [174]. Unfortunately, the composite membranes did not perform as well as the baseline Nafion, with long term conductivity testing resulting in a continuous decrease in conductivity (80 °C, 70% RH for 4 days). The authors discovered that this is due to the excess acidity of the PFSA, humidification and gas flow reducing the cerium oxide into (III) which then binds to the sulphonic groups, inhibiting proton conduction. This was confirmed by reprotonation via sulphuric acid and the proton conductivity went back to its original value.

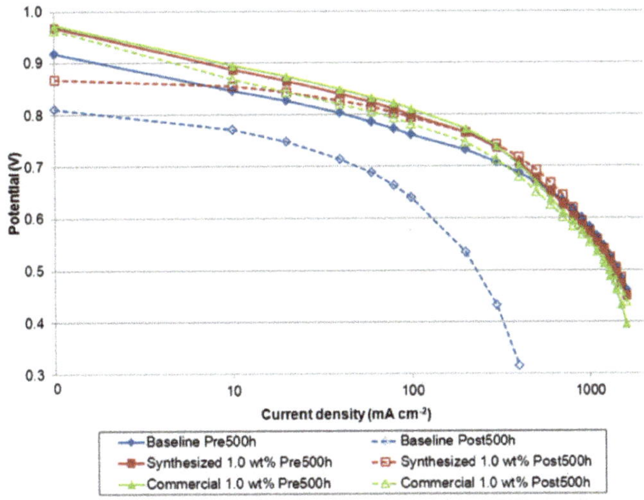


**Figure 6.** Pre- and post 500 h test performance curves; reproduced with permission from [173].

Lee et al. [175] prepared cerium oxide impregnated sulfonated poly(arylene ether sulfone) (SPES, 50% degree of sulphonation) membranes aiming for improved fuel cell durability. The addition of cerium oxide led to a drop in water uptake, IEC and proton conductivity. With increasing loading resulting in lower water uptake, IEC and proton conductivity. However, the ex-situ Fenton reagent tests was performed to study the oxidative stability of the composite membranes. Introduction of the cerium oxide led to a decrease in degradation of the membrane. In addition, single cell accelerated OCV hold testing (90 °C, 30% RH, 0.5 bar) showed that the composite membrane with 2% cerium oxide was stable for up to 2200 h, compared to 670 h for the pristine SPES.

Elakkiya et al. decided to enhance the proton conductivity of composite membranes by using sulphonated $TiO_2$ coated in polyaniline within a SPES polymer matrix [176]. Water uptake and proton conductivity improved with the addition of the filler however, no in-situ testing was performed. It would be interesting to see what effect the polyaniline has on fuel cell performance, and if the sulphonated filler improves performance at elevated temperatures/reduced humilities.

Lee et al. [177] synthesized sulphonated silicon dioxide within SPAEK. As expected, the addition of the filler improved fuel cell performance (at 60 °C and at both 100 and 70% RH) but the functionalised filler also outperformed the composite with non-functionalised silicon dioxide. This is due to the sulphonic groups retaining more water and allowing sufficient proton transport.

### 3.2. Carbon Nanomaterial Fillers

In recent year carbon nanomaterials have become the go-to filler, particularly graphene oxide due to its abundance of oxygen containing functional groups [178].

These oxygen-containing functional groups attract water molecules and are able to retain higher levels of water in comparison to pristine Nafion. In addition, graphene oxide's flat structure means that these functional groups are easily accessible. The inclusion of filler materials can also improve the mechanical strength of the composite membrane.

Kumar et al. [179] prepared a GO/Nafion membrane for PEMFCs operation. Addition of GO in 2, 4, and 6% loading to recast Nafion led to a subsequent increase in water up from 21.1 to 27.9, 37.2 and 36.1% respectively. Additionally, IEC changed from 0.891 to 1.21, 1.38 and 1.26 meq $g^{-1}$ respectively. The authors argued that there is an optimum quantity of filler and any addition would result in increased membrane stiffness and subsequently reduced water uptake. Fuel cell tests at 100 and 25% RH show that the 4% GO composite membrane outperformed the reference recast Nafion by nearly 4 times (212 mW $cm^{-2}$ to 56 mW $cm^{-2}$).

Sahu et al. [180] instead functionalised graphene with sulfonic acid groups inside a Nafion matrix for low relative humidity operation. This is interesting as the use of graphene oxide as a filler is due to its abundant oxygen containing functional groups, which make it more hydrophilic. This is in comparison to graphene, which is hydrophobic and hard to disperse in water, however the sulfonation procedure would have reduced the hydrophobicity of the graphene filler. This is shown in the water uptake and IEC tests. Recast Nafion has a water uptake of 20.1%, the addition of graphene slightly increases it to 21.4%. However, the introduction of sulphonated graphene, in 0.5, 1 and 1.5% loading results in improved water uptakes of 24.5, 27.3 and 29.2% respectively. The IEC values are: 0.88, 0.89, 0.92, 0.96 and 0.95 meq $g^{-1}$ respectively. A similar trend was also observed with the proton conductivity, with the 1% sulphonated graphene having the best performance, which is also hinted at by it having the highest IEC. Fuel cell testing at 70 °C and 20% RH revealed that the composite membrane with sulphonated graphene (1%) produced a maximum power density of 300 mW $cm^{-2}$, whereas recast Nafion and Nafion-graphene (1%) produced peak power densities of 220 mW and 246 mW $cm^{-2}$ respectively.

Lee et al. [181] prepared Nafion/GO and a novel Pt on graphene/Nafion composite membranes for low humidity PEMFCs. The idea behind using platinum on graphene as a filler is to use platinum as a reaction site to produce water and "self-humidify" the membrane. Water uptake experiments showed that the GO composite membrane outperformed the pristine Nafion sample. In comparison, the Pt/Graphene filler led to a drop in water uptake. The authors explained that this is because of the less hydrophilic nature of platinum as well as the GO being reduced to graphene in the synthesis step. However, the Pt/Graphene membrane had a greater proton conductivity compared to the other two membranes, which was explained via the electronic tunnel effect. The GO composite had a lower proton conductivity due to the filler impeding the ionic pathways, but this issue was resolved when the loading was greater than 3%, resulting in an increase in proton conductivity. The GO/Nafion membrane was tested at 80 °C and under a range of RH. At 40% RH, the peak power densities of the membranes with different GO loadings were all around 0.5-0.6 W $cm^{-2}$. On the other hand, the Pt/Graphene membrane gave disappointing current output under anhydrous conditions, with peak current densities of 0.27, 0.36 and 0.14 A $cm^{-2}$ for 0.5, 3 and 4% loading respectively.

The authors followed up this work with designing a composite membrane with platinum on graphene in addition to silicon dioxide to improve the "self-humidifying" capabilities of the membrane by using the silica to retain the water produced by the platinum-graphene [182]. The water uptake and proton conductivity of these novel membranes increased with increasing silica content. Maximum water uptake of 30% was achieved with 3% Pt-G and 3% silica content. Fuel cell experiments showed that the addition of silica improved the polarisation curve. However, performance dropped with too much silica at low RH, which the authors explain is possibly due to the filler blocking the ionic pathways. Filler optimisation was concluded by the authors, as increases in Pt-G loading also resulted in a drop in performance.

Yang et al. [183] fabricated a composite membrane with platinum deposited on titania, which is then incorporated with graphene oxide into a Nafion polymer matrix. The composite membranes displayed a better IEC than recast Nafion, with increased until 20% GO is reached, where the IEC began to decrease beyond that. The proton conductivity followed a similar trend to the IEC, which also decrease past 20% GO loading. Fuel cell testing with varying levels of RH showed that the Pt-$TiO_2$ improved the fuel cell performance, however the authors noted that this was still not sufficient at zero RH. Adding the GO led to an even greater improvement in cell polarisation. Nafion/0.8Pt–$TiO_2$/0.2GO generated a current density of 0.54 A $cm^{-2}$ at 0.6 V at 0% RH, which compared to Nafion/Pt–$TiO_2$ that produced a current density of 0.01 A $cm^{-2}$ at the same RH. Furthermore, the introduction of GO not only improved the current density generation, but also helped alleviate significant OCV loss when the humidity was lowered.

Kim et al. [184] fabricated a GO/Nafion composite where the GO is modified with phosphotungstic acid ($H_3[PW_{12}O_{40}]{\cdot}29H_2O$) to aid water uptake and proton conduction at low relative humidity PEMFC operation. Fuel cell testing at 80 °C and 20% RH showed that the composite membrane with modified GO produced a maximum power density of 841 mW $cm^{-2}$, which is a great improvement in comparison to non-acid doped Nafion/GO which generated 488 mW $cm^{-2}$, and 208 mW $cm^{-2}$ for recast Nafion. Polarization curves are shown in Figure 7.

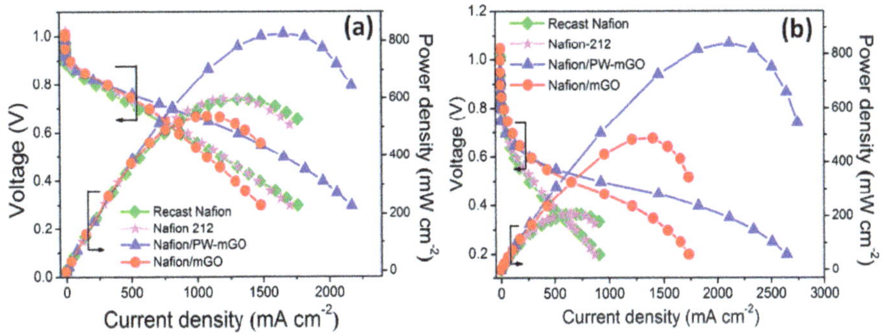


**Figure 7.** Polarization and power density plots of Nafion-212, recast Nafion, Nafion/mGO and Nafion/PW-mGO operating (**a**) under 100% RH at 80 °C and (**b**) under 20% RH at 80 °C. One weight percent filler content was used in composite membranes, and catalyst loading in the anode and cathode was kept 0.5 mg $cm^{-2}$; reproduced with permission from [184].

Maiti et al. [185] synthesised a composite Nafion membrane comprising of graphene oxide and an ionic liquid (dihydrogen phosphate functionalised imidazolium) for high temperature PEMFCs. TGA shows that the ionic liquid is stable up to 230 °C, which is more than enough for high temperature operation. The composite membranes displayed greater proton conductivity compared to Nafion 117 throughout the entire temperature range tested (70–110 °C). This improved performance was carried through to the single cell test where the composite membrane generated higher current densities in comparison to the Nafion 117 MEA (at 110 °C and dry conditions).

Branco et al. [186] investigated the performance of multilayer membranes for IT-PEFC applications. Multilayer membranes with two external Nafion outer layers and an inner layer of graphene oxide and another with sulphonated polyindene were fabricated with solution casting and hot-pressing methods. The solution casting protocol involves heating the first Nafion layer at 100 °C for two hours to remove the solvents in the Nafion dispersion, followed by the addition of the graphene oxide solution/sulphonated polyindene (in deionized water) and another two hours at the same temperature. Lastly, the final Nafion layer was added and the multilayer membrane was heat treated for one hour at 120 °C. The increased temperature is to anneal the polymer. Multilayer membranes that were casted displayed better performance and proton conductivity than the hot-pressed multilayer membranes. This was explained by the casted membranes having better interface interaction compared to the hot-pressed membranes, which suffered delamination. Multilayer membranes with sulphonated polyindene showed higher performance than Nafion at 120 °C.

Ibrahim et al. [187] studied the behaviour of GO composite membranes fabricated via solution casting with different thicknesses at intermediate operating temperatures. The composite membranes had improved mechanical strength and a higher water uptake in comparison to pristine Nafion. In-situ fuel cell testing of the membranes as MEAs revealed that the 30 μm composite membrane at 100 and 120 °C outperformed the 50 μm Nafion membrane at 80 °C. This is most likely due to the reduction in thickness and the GO filler retaining more water, hence reducing the drop in proton conductivity.

Kumar et al. [188] sulphonated GO and incorporated it into the polymer matrix of SPEEK. The SGO improved the water of SPEEK from 57.58% to 60%. In addition, the composite membrane outperformed SPEEK at temperatures from 30 °C to 80 °C (at 100% RH) and at 80 °C (with varying RH from 30 to 50%). Fuel cell testing at 80 °C, 30% RH humidified hydrogen and dry oxygen showed that the composite membrane produced a maximum power density of 378 mW $cm^{-2}$, a large increase in comparison to SPEEK which produced 250 mW $cm^{-2}$.

Sulphonated carbon nanotubes were used as a filler within a SPEEK matrix to offset the effect of high levels of sulphonation compromising the durability of the membrane [189]. The composite membrane had better proton conductivity and fuel cell performance compared to its pristine counterpart. A point of consideration is that the filler was functionalised to prevent the disruption of the proton transport channels, which is something that should be considered when incorporating a filler.

Uregen et al. [190] fabricated a graphene oxide/polybenzimidazole membrane for the operation at high temperatures. The introduction of graphene oxide improved the proton conduction in comparison to pristine PBI as well as reducing the quantity of acid leaching (from 85% for pristine PBI, to 70% for the composite membrane). Fuel cell testing at 165 °C and with dry hydrogen and air revealed that the pristine PBI and GO composite membrane had maximum power densities of 0.31 and 0.38 W $cm^{-2}$ respectively. However, the authors noted that there could potentially be degradation of the GO functional groups at operating temperatures above 165 °C. A 500 h durability test showed that the performance loss of the composite membrane was lower, at 3.8% in comparison to 8.3% for the PBI membrane. This could be due to the reduced hydrogen crossover and acid leaching.

Xue et al. [191] decided to functionalise their graphite oxide, once again in a PBI polymer matrix for high temperature PEMFCs. Isocyanate functional groups were modified onto graphite oxide to improve the dispersion in water and organic media. This resulted in greater proton conductivity and less swelling. A similar study but with the GO sulphonated was studied by Xu et al. [192]. The proton conductivity of the membranes was increased from 0.023 S $cm^{-1}$ for pristine PBI to 0.027 S $cm^{-1}$ for GO/PBI and 0.052 S $cm^{-1}$ for SGO/PBI. The respective activation energies for proton conduction fell from 16.1 kJ $mol^{-1}$ to 11.4 kJ $mol^{-1}$ to 9.3 kJ $mol^{-1}$ respectively. Fuel cell testing at 175 °C and under anhydrous conditions with hydrogen and oxygen showed that the addition of GO or SGO result in an increase in maximum power density, from 0.22 for PBI, to 0.38 for GO/PBI and 0.6 W $cm^{-2}$ for SGO/PBI. The same trend was observed under air. This work was followed by the same authors studying the same filler and polymer but this time functionalised the GO with an ionic liquid (1-(3-Aminopropyl)-3-methylimidazolium groups) [193]. The composite membrane had a higher proton conductivity in comparison to the reference PBI membrane. Fuel cell tests at 175 °C with dry inlet fuel showed that the addition of the ionic liquid improved peak power densities from 0.26 W $cm^{-2}$ for PBI to 0.32 W $cm^{-2}$ for the composite. The authors stated that this is due to the improved proton conduction within the composite membrane.Abouzari-Lotf et al. [194] designed a composite membrane for high temperature fuel cells by combining PBI that has been functionalized with 2,6-Pyridine with phosphonated grapene oxide. The use of the filler was in order to reduce the extent of acid leaching and to increase long term stability as increasing acid content can mechanically compromise the polymer. The addition of 1.5% phosphonated graphene oxide significantly increased the proton conductivity from $19.6 \times 10^{-3}$ S $cm^{-1}$ for pyridine PBI to $76.4 \times 10^{-3}$ S $cm^{-1}$ at 140 °C.

Kannan et al. [195] presented a composite PBI membrane consisting of phosphonic acid functionalised multi-walled carbon nanotubes as the filler material. Proton conductivity tests revealed that the composite membrane achieved 0.11 S $cm^{-1}$, whereas the pristine PBI produced a conductivity of 0.07 S $cm^{-1}$. Figure 8 show fuel cell testing at 140 °C with dry inlet feeds showed that the composite membrane (1% filler loading) outperformed both the pristine PBI membrane and an additional composite that contained non-functionalised nanotubes (peak power densities of 780, 600 and 590 mW $cm^{-2}$ respectively). In addition, the mechanical stability was also improved due to the architecture of the carbon nanotubes, achieving a higher yield strength.

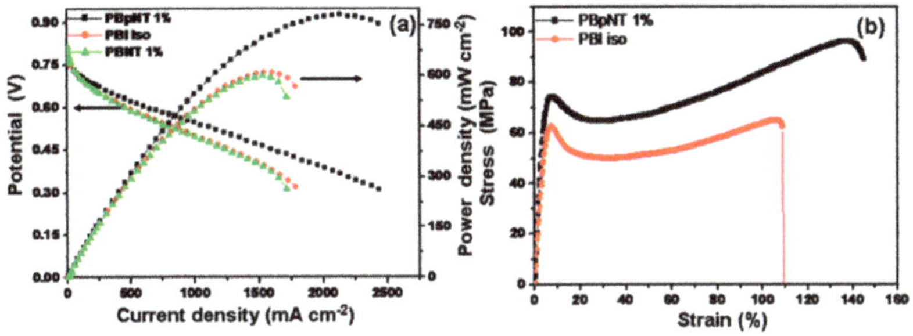


**Figure 8.** (**a**) Polarization plots of PBI iso, PBpNT, and PBNT composite membranes measured at 140 °C by passing dry $H_2$ and $O_2$ at a flow rate of 0.2 slpm. The cells were conditioned at 0.6 V for 30 min. (**b**) Stress–strain curve for the pristine PBI and PBpNT composite membrane; reproduced with permission from [195].

A further more detailed study involved various characterisation techniques and an investigation on variable loading [196]. Thermal analysis revealed that the membranes are stable up to 250 °C, and from 250 °C to 400 °C only lose 10% mass. The composite membranes all produced a proton conduction greater than pristine PBI throughout the experimental temperature range. The phosphoric acid uptake was similar for both the pristine membrane and the composite of different loadings. However, the activation energy dropped from 40.9 kJ $mol^{-1}$ for PBI to 25.1 kJ $mol^{-1}$ for the composite membrane with 1% functionalised CNTs. The composite membranes experience a smaller drop in activation compared to the reference membrane and the authors explained this as the catalysts (platinum) having a higher exchange current density on CNT than carbon. Additionally, the composite membrane produced higher current densities without a sudden drop due to concentration limitations, being able to reach nearly 3000 mA $cm^{-2}$, in comparison to nearly 2000 mA $cm^{-2}$ for pristine PBI.

Yang et al. [197] used GO as a filler functionalised with triazole groups in order to aid dispersion and to improve proton conduction within PBI for high temperature fuel cells. SEM imaging revealed that the modified GO composite had a much better dispersion in comparison to the non-modified GO composite which had the presence of agglomerates. This improved homogeneity led to an increase in proton conductivity as well as an improvement in its mechanical properties. Fuel cell testing at 160 and 180 °C showed that the composite membrane outperformed the PBI reference membrane at the same acid doping level, with maximum power densities of 537 to 506 mW $cm^{-2}$ respectively.

Cao et al. [198] fabricated a graphene oxide poly (ethylene oxide) (PEO) composite membrane for the purpose of PEMFCs operation. The conductivity of the composite membrane increased from 0.086 S cm to 0.134 S cm with increasing temperature (from 25 °C to 60 °C). However, the authors explained that increasing the temperature above that results in the membrane softening. The composite membrane produced a maximum power density of 53 mW $cm^{-2}$ at 60 °C with full humidity.

Lee et al. [199] prepared a SPAES composite with GO grafted onto sulfonated poly(arylene thioether sulfone) as the filler. This was done due to the inherent lower proton conductivity of hydrocarbon-based polymers in comparison to PFSA. This membrane exhibited improved mechanical strength and oxidative resistance, as well as better proton conductivity in comparison to pristine SPAES. Fuel cell testing to understand the performance of this composite as an MEA would be very interesting, and whether grafting the GO makes a difference in performance.

Dai et al. [200] developed novel composite membranes consisting of carbon dots of different sizes and with varying levels of hydrophilicity within a matrix of polyvinylpyrrolidone (PVP) and polyethersulfone (PES). AFM and TEM characterization showed that carbon dots with a size of 2–5 nm showed no aggregation and good uniformity. Single cell tests at 150 °C and under anhydrous conditions revealed that the composite membrane had a higher peak power density in comparison to pristine PES-PVP, 166 to 113 mW $cm^{-2}$, respectively. The idea of altering the hydrophilicity of the filler material is an interesting technique to improving the performance of the composite membrane.Ahmed et al. prepared a chitosan membrane with sulphonated multiwall carbon nanotube filler [201]. As chitosan has a lower proton conductivity than Nafion there is a greater need for using fillers to improve its proton conductivity. The mechanical strength and proton conductivity increased but water uptake decreased, and the authors explain that this is due to the decrease in -$NH_2$ functional groups. It would be interesting to see how these membranes perform in a fuel cell in comparison to pristine chitosan and Nafion.

### 3.3. *Acids and Ionic Liquids Fillers*

Ionic liquids have been extensively used in fuel cells operating at higher temperatures due to their high thermal degradation temperature. For example, Choi et al. fabricated two types of composite membranes doped with phosphotungstic acid, one with 1100EW Nafion and the other with 750EW [202]. At 120 °C, both the composite membranes performed better than the reference Nafion MEA, achieving a voltage of 0.51 and 0.55 V (1100EW and 750EW) at 400 mA $cm^{-2}$, compared to 0.47 V for the reference Nafion. In addition, the ohmic resistance was smaller than that of the reference Nafion, at 0.32, 0.21 and 0.13 Ω $cm^{-2}$ for Nafion, 1100EW composite and 750EW composite respectively.

Lee et al. designed a composite membrane of a sulphonated polymer doped with fluorohydrogenate ionic liquid [203]. The ionic liquid was used due to their high thermal stability as the application was geared towards intermediate temperature operation with dry conditions. Single cell testing at 130 °C revealed an OCV of 1 V for 5 h. In addition, the ionic conductivity of the prepared composite membrane increased with temperature, from 11.3 mS $cm^{-2}$ at 25 °C to 34.7 mS $cm^{-2}$ at 130 °C.

Ramani et al. [204] introduced heteropolyacids into Nafion for PEMFCs operating at higher temperatures and reduced relative humidity. Additives studied included phosphotungstic (PTA), silicotungstic (STA), phosphomolybdic acid (PMA) and silicomolybdic acid (SMA). Water uptake results revealed that there is no significant different between recast Nafion and the Nafion/PTA composite membrane at a range of relative humidities. Single cell tests at 120 °C and 35% RH of Nafion/PTA, Nafion/STA and Nafion/SMA showed relative performance, with Nafion/PTA and Nafion/STA reaching a maximum current density of around 800 mA $cm^{-2}$. The same authors prepared a composite membrane of Nafion with a heteropolyacid (HPA), phosphotungstic acid (PTA) [205]. The MEA was "stabilized" via high temperature heat treatment (200 °C at 30 atm). In order to allow the membrane to not disintegrate and to prevent the HPA from dissolving, the MEA was ion exchanged in caesium carbonate, swapping the protons for much larger caesium ions. TGA experiments showed that this stabilized membrane degraded at higher temperatures in comparison to its proton exchanged counterpart. Fuel cell testing at 120 °C and at 35% RH showed that both the stabilized and reference membrane have similar polarisation behaviour. However, the specific area resistance was lower for the stabilized membrane and the authors explained that this is because of the lower contact resistance from the high temperature heat treatment. The work was followed by looking into the effect of extent of ion exchange. Composite membranes with 2, 1, and 0 protons left after substitution were prepared [206]. Weight loss measurements to assess the stability of PTA in Nafion were performed, with increasing proton substitution leading to less weight loss after protonation. Pristine PTA had a weight loss of around 27%, which decreased to less than 5% for the PTA modified to have its protons removed. Water uptake experiments interestingly showed that there is no difference between pristine PTA composite and the substituted protons. In addition, membranes were also ion exchanged using different cations, but this led to no change. The authors proposed that they think that any improvement in proton conductivity in the fabricated membranes will be exclusively because of the Grotthus mechanism, with negligible contribution via the vehicular mechanism.

Another avenue to reduce leaching of the HOA was to use metal dioxides as a support, in a similar fashion to carbon for the electrocatalysts [207]. TGA analysis showed that the addition of the PTA and metal dioxide (silica in this case), increased the membrane decomposition temperature from 270 to 305 °C, and this is because of the silica partially immobilising the side chains of the Nafion. FT-IR before and after protonation treatment revealed that the PTA did not wash out of the composite membranes that were prepared via sol-gel technique. In-situ resistance measurements at 120 °C and 35 °C show that the composite membranes (with PTA supported on silica) have an area specific resistance of 0.16 Ω $cm^{-2}$ in comparison to 0.19 Ω $cm^{-2}$ for Nafion.

Lee et al. [208] prepared a membrane with a protic ionic liquid diethylmethylammonium trifluoromethanesulfonate ([dema] [TfO]), within a sulphonated polyimides (SPI) structure for anhydrous PEMFCs. 300 °C was estimated as the composite membrane's thermal decomposition temperature, which is much higher than its intended PEMFC operating temperature. Fuel cell operation at 80 °C and under dry conditions, illustrated in Figure 9, revealed that the composite membrane produces a maximum power density of 100 mW $cm^{-2}$ at a current density of 240 mA $cm^{-2}$. What is interesting is that at 30 °C, a maximum power density of 68 mA $cm^{-2}$ at 300 mA $cm^{-2}$, showing some promise of a room temperature, anhydrous PEFC.

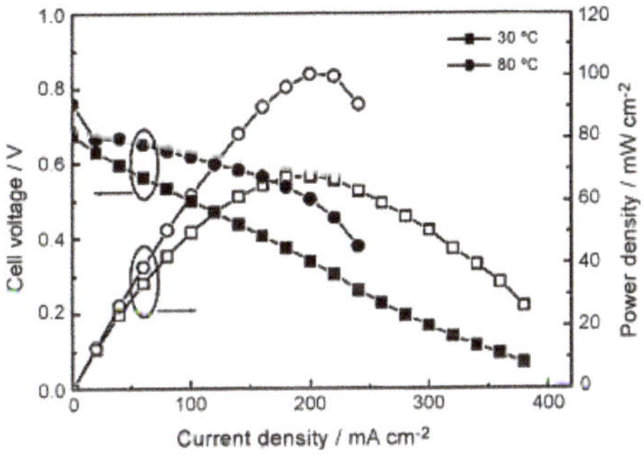


**Figure 9.** Polarization curves of a $H_2/O_2$ fuel cell using a SPI-1.51(50) composite membrane without humidification at gas utilization ratios of 30% for $H_2$ and 15% for $O_2$ (■,□) operation at 30 °C, (●,○) operation at 80 °C; reproduced with permission from [208].

Yi et al. [209] fabricated a SPEEK ionic liquid composite membrane, based on an imidazolium ionic liquid, for increased temperature and anhydrous fuel cell operation. Two composite membranes were prepared, one with 1-butyl-3-methylimidazolium tetrafluoroborate (BuMeImBF4) and the other with 1-decyl-3-methylimidazolium tetrafluoroborate (DeMeImBF4). In addition, the degree of sulphonation of the SPEEK was chosen to be 67%. Proton conductivity measurements of SPEEK/BuMeImBF4 showed that the conductivity increased with increasing temperature, reaching a maximum of $8.4 \times 10^{-3}$ S $cm^{-1}$ at 170 °. The authors explain that the increase is due to the reduction in viscosity of the ionic liquid enhances its mobility. Thermal analysis via TGA showed that the composite membrane's sulphonic groups degrades at 340 °C, which is much greater than that of 250 °C of pristine SPEEK. Leaching tests were performed to understand how much of the ionic liquid would be retained in the composite membrane. Proton conductivity tests of the composite membranes before and after immersion in water showed that the proton conductivity of SPEEK/BuMeImBF4 was undetectable after 1 h of immersion. SPEEK/DeMeImBF4 fared better, being detectable after 2 h of immersion but eventually the conductivity also became undetectable.

Yasuda et al. [210] synthesised sulfonate polyimide (SPI)/ionic liquid composite membranes with altered polymer structures, to study the effect of the positions of the sulfonic groups, ultimately for anhydrous application. Composite membranes with random and block polymers were made and characterised to understand their behaviour. The degradation onset temperature for all the membranes were 250 °C and above, this means that they are suitable for operating in higher temperature fuel cells. The authors stated that the distribution of the ionic groups and the flexibility of the sulfonic groups are important determinants in ion conduction. Fuel cell experiments at 120 °C and with no humidity showed that the random chain SPI and homopolymer SPI produced maximum power densities of 100 and 70 mW cm$^{-2}$ respectively.

Malik et al. [211] prepared a SPEEK/ethylene glycol/ionic liquid composite membrane for high temperature application. The ethylene glycol was added to use as a crosslinker to help alleviate the quantity of leaching. The composite membranes had a high thermal stability, with the sulphonic groups beginning to degrade at 240 °C. The composite membranes had a lower leaching weight loss in comparison to the non-crosslinked membranes, however the proton conductivity of the crosslinked membranes was lower. The authors explained that some of the sulphonic groups where used in the crosslinking resulting in lower conductivity.

From the above analysis of the literature, it can be seen that membrane fillers are very versatile, in terms of chemical structure, size, dimensions, etc. Numerous different characteristics to adapt to specific application whether that is operating at high temperature, low relative humidity, increasing mechanical strength, preventing acid leaching, increasing proton conductivity, or producing self-humidifying membranes. On top of that they can also be functionalised to either boost these characteristics or provide a secondary functionality. The abovementioned studies indicate promising performance for composite membranes; however, highlight the need for further research to improve the lifetime and durability of these membranes. Table 3. summarises hydrogen PEM performance of composite polymer electrolyte membrane described in this review.

**Table 3.** Summary of hydrogen PEMFC best performance using composite membranes.

| Membrane | Power Density (mW cm$^{-2}$) | Temperature (°C) | RH% |
|---|---|---|---|
| Nafion/Silica [144] | 350 | 100 | 100 |
| Nafion/Silica particles [139] | 380 | 85 | 100 |
| Nafion/Hafnium oxide [147] | 336 | 100 | - |
| Nafion/Titanium oxide [152] | 514 | 110 | - |
| Nafion/Titanium oxide nanotubes [156] | 1020 | 80 | - |
| Nafion/Zirconium oxide [141] | 400 | 130 | 85 |
| Nafion/Sulphonated zirconium oxide [142] | 609 | 70 | 83 |
| Nafion/mesoporous zirconium pshosphate [146] | 353 | 70 | 18 |
| Nafion/ zirconium pshosphate [145] | 450 | 130 | - |
| Nafion/GO [179] | 212 | 100 | 25 |
| Nafion/SGO [180] | 300 | 70 | 20 |
| Nafion/GO/$TiO_2$ [183] | 324 | | 0 |
| Nafion/GO/Phosphotungstic acid [184] | 841 | 80 | 20 |
| Nafion/Phosphotungstic acid [202] | 220 | 120 | - |
| PBI [169] PBI/$SiO_2$ [169] | 200240 | 165165 | - |
| PBI/GO [190] | 388 | 165 | 0 |
| PBI/SGO [191] | 600 | 175 | 0 |
| SulfonatedPolysulfone [160] | 160 | 85 | - |
| SPolysulfone/titanium oxide [160] | 240 | 85 | - |
| SPEEK [164] | 179 | 120 | - |
| SPEEK/GO [188] | 378 | 80 | 30 |
| SPEEK/silica [164] | 246 | 120 | - |
| SPI/ionic liquid [210] | 100 | 120 | 0 |
| SPI/demaTfO [208] | 100 | 80 | 0 |

Membranes with fillers that were functionalised (most commonly with sulphonic groups) displayed a better performance in terms of proton conductivity and cell polarisation at elevated temperatures. This is attributed to the water retaining capabilities of these functional groups. Different fillers demonstrated different performances, and this is because of their chemical structure in addition to their physical structure (nanoparticle, flat, nanotubes). Therefore, when selecting a filler material not only should the material itself and possible functionalising be considered, but also the shape of the filler itself. Another point of consideration is the polymer that the filler will be embedded in. As we have shown, composite membranes were made using different polymers such as Nafion, SPEEK and SPAES. However, only Nafion meets industry standards regarding lifetime and durability. This implies that composite membranes should use a Nafion matrix in addition to Nafion ionomer in the GDE. Overall, membrane performance has to be looked at from a variety of experiments, such as; cell polarisation and power, long term durability and ex-situ tests to name a few. A membrane that performs well in-situ might degrade quickly during thermal/humidity cycling and be unsuitable.

Another point of consideration is the interaction between the composite membrane and the catalyst layer. Conventional Nafion membranes use gas diffusion electrodes that consist of Nafion ionomer binder. However the addition of a filler material could potentially affect this interaction between the membrane and catalyst layer, for example the filler is added to only improve the membrane performance but if some of the filler is dispersed closer to the edges of the membrane then this could interfere with the anode and cathode functions (hydrophilic fillers situated close to the cathode could cause flooding more easily). Also, the ionomer and membrane might not consist of the same material, further complicating this interaction, which would be exacerbated during manufacturing of MEAs with composite membranes and binders of different materials. To the authors knowledge, the use of composite ionomers in the catalyst layer is not studied and requires further research.

Composite membranes are tested for their performance in-situ (fuel cell testing) and ex-situ (proton conductivity, water uptake etc.), however their durability during fuel cell testing is an area of research not fully explored. In order to be competitive with conventional Nafion membranes, the composite membrane must not only be able to perform better but also perform adequately over long periods. Composite membranes are developed aiming for harsher operating conditions (higher temperature and lower humidity) and therefore their durability must be investigated and demonstrated in-situ. It is also important to study the change in the membrane degradation mechanisms due to the presence of fillers, for instance; how does the filler affect the membrane mechanical properties due to the humidity cycling and what is the impact of the filler on the catalyst stability or dissolution into the membrane.

## 4. Composite Membranes for Electrolysers

Composite membranes with metal oxides as fillers ($SiO_2$, $TiO_2$, or $WO_2$) showed promising properties for high temperature operation of PEM water electrolysers allowing achieving high performance with respect to a commercial membrane. Baglio et al. [212] and Antonucci et al. [213] focused their work on Nafion-$TiO_2$ and Nafion-$SiO_2$ respectively, to allow efficient operation at high temperature, above 100 °C. Both works claimed that the high temperature operating conditions were allowed by the better water retention and more uniform distribution of water across the composite membrane due to the presence of inorganic hygroscopic fillers inside the polymeric matrix. This resulted in reduced ohmic resistance and therefore better electrolyser performance [214]. The performance of composite membranes was better than that of Nafion membrane under high temperature and high pressure so the application of this technology is very promising especially when high electrical efficiency is required. As evidence of this result, Figure 10 illustrates characteristics curve of cell equipped with commercial Nafion and composite Nafion-$SiO_2$ membranes at high temperature and pressure.

These alternative composite membranes also showed a decrease of the cross-over of the gases through the membrane. However, a slight decay of performance was observed during the experiment; thus, a further amelioration of membrane is necessary to improve the stability and lifetime.

Another way to produce electrolyte membranes with high conductivity and durability for water electrolysers is using perfluorosulfonic acid with shorter and non-branched pendant side-chain with higher crystallinity than longer side-chain perfluorosulfonic acid. Aricò et al. [215] used the Aquivion short side chain perfluorosolfonic membrane using Nafion 111 for comparison. Authors claimed that although those membranes showed high conductivity, mechanical stability and dimensional properties, they are not appropriate for water electrolysis application. To reinforce those membranes, organic fillers can be included in the Aquivion matrix. Boaretti et al. [216] included SPEEK reinforcement which led to an improvement in the mechanical strength but resulting in low proton conductivity. Another approach to reinforce proton exchange membrane is to physically separate the properties of mechanical strength and proton transport embedding a porous nanofibre web into the matrix. Aquivion membranes reinforced with electrospun polysulfone (PSU) fibre webs were prepared by Giancola et al. [217]. The fibrous reinforcement strongly enhanced the mechanical strength and also reduced hydrogen crossover. However, the addition of the reinforcing fibre in membranes had little effect on the cell electrochemical performance: the cell voltage at 2 A $cm^{-2}$ was 1.760 V which is slightly higher than the performance obtained with a non-reinforced membrane (1.758 V). Therefore, increased mechanical and dimensional stability and reduced hydrogen crossover of the composite membrane are promising properties for electrolysis application but with little effect on performance.

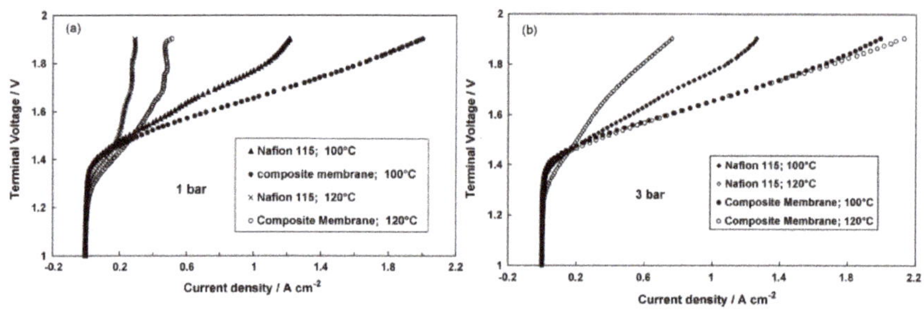


**Figure 10.** Comparison of voltage and current density of Nafion 115 and composite Nafion-$SiO_2$ membrane at 100 and 120 °C and at 1 bar abs (**a**) and 3 bar abs (**b**); reproduced with permission from [213].

Ion-Ebrasu et al. [218] produced a composite membrane by spray coating graphene on commercial PEM material. They exploited the properties of this material to enhance the efficiency of PEM electrolysers and reducing costs by achieving high surface area to volume ratio, good mechanical and thermal properties. The composite membrane showed an improved behaviour in term of thermal and electrochemical characterization when compared to pristine commercial membrane: the interaction of graphene with fluorinated membrane led to an increased conductivity and a better water adsorption. In spite of all this benefits, further experimental work has to be carried out to investigate the behaviour of these graphene-modified membranes under current voltage measurements.

Linkous et al. [219] evaluated different types of engineering polymers and identified a few options that could withstand the conditions found in PEMWEs. Among them, polybenzimidazoles (PBI), poly(ether ether ketones) (PEEK), poly(ether sulfones) (PES) and sulfonated polyphenyl quinoxaline (SPPQ), were selected to be used for PEM electrolysis. In particular, SPEEK polymer is considered to have high strength and it is an easy membrane forming material. High degree of sulfonation enables high proton conductivity. In fact, Linkous et al. observed that high degree of sulfonation (65%) led to a higher proton conductivity that exceeded Nafion by 29%. However, these alternative membranes showed low durability and low current densities compared to standard Nafion membranes and tent to swell excessively or even dissolve at elevated temperature. An alternative would be to reinforce the SPEEK membrane with other polymer structures and/or fillers [220]. Song. et al. [221] prepared a composite membrane including tungstophosphoric acid(TPA) to increase proton conductivity and $CeO_2$ (Cs) to improve the durability of the membrane into the SPEEK matrix. The composite membrane showed better mechanical and electrochemical properties than Nafion 117 membrane: proton conductivity, tensile strength, and elongation were enhanced. However, the cell voltages of the MEA using Nafion 117 and SPEEK-Cs/TPA membrane were 1.91V and 1.82 V at 1 A $cm^{-2}$

operating at 80 °C under atmospheric pressure; thus, they may replace Nafion 117 due to their mechanical characteristic, electrochemical properties once performance becomes comparable with that of Nafion117.

It can be noted that despite the promising characteristics of composite membranes, little research has been conducted into using them in PEMWEs. It is important to note that in addition to Nafion, and similar to what is reported for fuel cells, other polymers have been explored for PEMWE application. The polymers include: SPEEK, SPSU and PBI which have been employed with varying levels of performance and lifetime achieved. Again, these polymers can also be modified and made into composite membranes for PEMWEs for achieving both higher temperature and pressure operation. Table 4 summarizes the properties and the pros and the cons of composite electrolyte membranes described in this section compared to those of the commercial membrane. In addition, Table 5 summarizes their best power output obtained:

**Table 4.** Summary of electrolyser composite membrane properties.

| Membrane | Preparation Method | Pros | Cons |
|---|---|---|---|
| **Nafion/Metal oxide** | Casting | Better water retention | Improvement registered at temperature greater than 100 °C and high pressure- Poor stability |
| **Aquivion** | Not present | Better water retention | Acceptable performance at certain temperature range, low humidity and high pressure |
| **Aquivion reinforced with polysulfone** | Casting | Low hydrogen crossover- Good mechanical stability | Performance not so much higher than non-reinforced membrane |
| **SPEEK** | Casting | Higher proton conductivity | Low durability and low performance at elevated temperature |
| **SPEEK-TPA** | Casting | Better chemical and mechanical stability | Performance slightly lower |

**Table 5.** Summary of PEM electrolyser best performance using composite membranes.

| Membrane | Voltage (V) | Current Density ($A\ cm^{-2}$) | Temperature (°C) | Pressure |
|---|---|---|---|---|
| Nafion/$SiO_2$ [213] | 1.9 | 2.1 | 120 | 3 bar |
| Nafion/$TiO_2$ [212] | 2 | 1.46 | 120 | 3 bar |
| SPEEK [219] | 2.5 | 0.21 | 60 | atmospheric |
| SPEEK/TPA/Ce [221] | 1.82 | 1 | 80 | atmospheric |
| Aquivion/PSU [217] | 1.76 | 2 | 80 | 0.1 Mpa absolute |
| SPSf [222] | 1.8 | 1.08 | 80 | atmospheric |

As illustrated in the tables above, despite having better water retention, composite membranes are not yet a suitable alternative to the commercial one in terms of performance and durability. Several efforts should be made to achieve performance industrially reasonable.

## 5. Conclusions

This review analysed several composite membranes developed in recent years for the use in PEM technologies to overcome the drawbacks of the commercial perfluorosulfonated membranes. Composite fluorinated, with organic and inorganic fillers, and non-fluorinated membranes have been scrutinized for DMFC, hydrogen PEMFC, and PEMWE.

All materials reported in this paper show promising characteristics and results, so it is not possible to indicate which one is the best. It can be noted that papers reporting high performance are dealing with the incorporation of fillers into the Nafion matrix, suggesting that Nafion cannot be completely replaced yet. Beyond the use of organic fillers like PBI and PANI, whose effects are evident only at high methanol concentration, carbon and inorganic fillers are the most promising materials. Low weight percentage of graphene oxide contributes to sensibly lower methanol crossover leading to better performance also a low methanol concentration. Moreover, GO composite membrane extends the operating temperature range for hydrogen PEMFC due to the fact that GO retains more water, so it decreases the loss in proton conductivity. Among all the composite membranes described in this review paper, inorganic fillers are the most versatile materials: their good thermal stability, improved water uptake and reduced methanol absorbance, provided high power density for DMFC and PEMFC but also allow high temperature and pressure operating conditions for electrolysis. Ionic liquids can be potentially used at intermediate temperatures once performance increases. Despite these positive results, durability tests are necessary to understand the real capacity of those fillers. Other materials like SPEEK and PVA are used to completely substitute Nafion. They seem to be a promising alternative to obtain high performance membranes. Research activities on their potentialities are still ongoing.

**Funding:** This research received no external funding.

**Acknowledgments:** Ahmed Ibrahim would like to acknowledge the EPSRC for the funding of the CDT in Fuel Cells and their Fuels, EP/L015749/1, which contributed to the research conducted in this paper.

**Conflicts of Interest:** The authors declare no conflict of interest.

## References

1. Hansen, J.; Ruedy, R.; Sato, M.; Lo, K. Global surface temperature change. *Rev. Geophys.* **2010**, *48*, 4. [CrossRef]
2. Barbir, F.; Gomez, T. Efficiency and economics of proton exchange membrane (PEM) fuel cells. *Int. J. Hydrog. Energy* **1996**, *21*, 891–901. [CrossRef]
3. EG&G Technical Services Inc. *Fuel Cell Handbook*, 7th ed.; U.S. Department of Energy/Office of Fossil Energy/National Energy Technology Laboratory: Morgantown, WV, USA, 2004.
4. Larminie, J.; Dicks, A. *Fuel System Explained*; John Wiley & Sons Ltd.: Chichester, UK, 2009.
5. Thomas, S.; Zalbowitz, M. *The Polymer Electrolyte Membrane Fuel Cell*; Fuel Cells Green Power/Los Alamos National Laboratory U.S.: Los Alamos, NM, USA, 2000.
6. Rikukawa, M.; Sanui, K. Proton-conducting polymer electrolyte membranes based on hydrocarbon polymers. *Prog. Polym. Sci.* **2000**, *25*, 1463–1502. [CrossRef]
7. Asensio, J.A.; Sanchez, E.M.; Gomez-Romero, P. Proton-conducting membranes based on Benzimidazole polymers for high-temperature PEM fuel cells. *Chem. Soc. Rev.* **2010**, *39*, 3210–3239. [CrossRef]
8. Li, Q.; He, R.; Jensen, J.O.; Bjerrum, N.J. Approaches and recent development of polymer electrolyte membranes for fuel cells operating above 100 °C. *Chem. Mater.* **2003**, *15*, 4896–4915. [CrossRef]
9. Villers, D.; Jacques-Bedard, X.; Dodelet, J.P. Fe-based catalysts for oxygen reduction in PEM fuel cells pretreatment of the carbon support. *J. Electrochem. Soc.* **2004**, *151*, 1507–1515. [CrossRef]
10. Emery, M.; Frey, M.; Guerra, M.; Haugen, G.; Hintzer, K.; Lochhaas, K.H.; Pham, P.; Pierpont, D.; Schaberg, M.; Thaler, A.; et al. The development of new membranes for proton exchange membrane fuel cell. *ESC Trans.* **2007**, *11*, 3–14.
11. USDRIVE. *Fuel Cell Technical Team Roadmap*; USDRIVE: Washington, DC, USA, 2013.
12. Chandan, A.; Hattenberger, M.; El-kharouf, A.; Du, S.; Dhir, A.; Self, V.; Pollet, B.G.; Ingram, A.; Bujalski, W. High temperature (HT) polymer electrolyte membrane fuel cells (PEMFC)—A review. *J. Power Sources* **2013**, *231*, 264–278. [CrossRef]
13. Dicks, A.; Larminie, J. *Fuel Cells Systems Explained*, 2nd ed.; John Wiley & Sons Ltd.: Hoboken, NJ, USA, 2003.
14. Element 1 Powering Innovation. *E1 Methanol Handbook. Element 1 Hydrogen Generators*; Element 1 Powering Innovation: Bend, OR, USA, 2013.
15. Ancona, V.; Barra Caracciolo, A.; Campanale, C.; De Caprariis, B.; Grenni, P.; Uricchio, V.F.; Borello, D. Gasification treatment of poplar biomass produced in a contaminated area restored using plant assisted bioremediation. *J. Environ. Manag.* **2019**, *239*, 137–141. [CrossRef]
16. Cameron, D.S.; Hards, G.A.; Harrison, B.; Potter, R.J. Direct methanol fuel cells: Recent developments in the search for improved performance. *Platin. Met. Rev.* **1987**, *31*, 173–181.
17. Kamarudin, S.K.; Achmad, F.; Daud, W.R.W. Overview on the application of direct methanol fuel cell (DMFC) for portable electronic devices. *Int. J. Hydrog. Energy* **2009**, *34*, 6902–6916. [CrossRef]

18. Goor, M.; Menkin, S.; Peled, E. High power direct methanol fuel cell for mobility and portable applications. *Int. J. Hydrog. Energy* **2019**, *44*, 3138–3143. [CrossRef]
19. Hosseinpour, M.; Sahoo, M.; Perez-Page, M.; Baylis, S.R.; Patel, F.; Holmes, S.M. Improving the performance of direct methanol fuel cells by implementing multilayer membranes blended with cellulose nanocrystals. *Int. J. Hydrog. Energy* **2019**, *44*, 30409–30419. [CrossRef]
20. Wang, Z.B.; Wang, X.P.; Zuo, P.J.; Yang, B.Q.; Yin, G.P.; Feng, X.P. Investigation of the performance decay of anodic PtRu catalyst with working time of direct methanol fuel cells. *J. Power Sources* **2008**, *181*, 93–100. [CrossRef]
21. Kulikovsky, A. Bubbles in the anode channel and performance of a DMFC: Asymptotic solutions. *Electrochim. Acta* **2006**, *51*, 2003–2011. [CrossRef]
22. Calabriso, A.; Borello, D.; Romano, G.P.; Cedola, L.; Del Zotto, L.; Santori, S.G. Bubbly flow mapping in the anode channel of a direct methanol fuel cell via PIV investigation. *Appl. Energy* **2017**, *185*, 1245–1255. [CrossRef]
23. Rashid, M.M.; Al Mesfer, M.K.; Naseem, H.; Danish, M. Hydrogen Production by Water Electrolysis: A Review of Alkaline Water Electrolysis, PEM Water Electrolysis and High Temperature Water Electrolysis. *Int. J. Eng. Adv. Technol.* **2015**, *4*, 2249–8958.
24. Barbir, F. PEM electrolysis for production of hydrogen from renewable energy sources. *Sol. Energy* **2005**, *78*, 661–669. [CrossRef]
25. Aricò, A.S.; Siracusano, S.; Briguglio, N.; Baglio, V.; Di Blasi, A.; Antonucci, V. Polymer electrolyte membrane water electrolysis: Status of technologies and potential applications in combination with renewable power sources. *J. Appl. Electrochem.* **2013**, *43*, 107–118. [CrossRef]
26. Ogungbemi, E.; Ijaodola, O.; Khatib, F.N.; Wilberforce, T.; El Hassan, Z.; Thompson, J.; Ramadan, M.; Olabi, A.G. Fuel cell membranes—Pros and cons. *Energy* **2019**, *172*, 155–172. [CrossRef]
27. Curtin, D.E.; Lousenberg, R.D.; Henry, T.J.; Tangeman, P.C.; Tisack, M.E. Advanced materials for improved PEMFC performance and life. *J. Power Sources* **2004**, *131*, 41–48. [CrossRef]
28. Agmon, N. The Grotthuss mechanism. *Chem. Phys. Lett.* **1995**, *244*, 456–462. [CrossRef]
29. Sun, X.; Simonsen, S.C.; Norby, T.; Chatzitakis, A. Composite membranes for high temperature PEM fuel cells and electrolysers: A critical review. *Membranes* **2019**, *9*, 83. [CrossRef] [PubMed]
30. Bakangura, E.; Wu, L.; Ge, L.; Yang, Z.; Xu, T. Mixed matrix proton exchange membranes for fuel cells: State of the art and perspectives. *Prog. Polym. Sci.* **2015**, *57*, 103–152. [CrossRef]
31. Li, Q.; He, R.; Jensen, J.O.; Bjerrum, N.J. PBI-based polymer membranes for high temperature fuel cells—Preparation, characterization and fuel cell demonstration. *Fuel Cells* **2004**, *4*, 147–159. [CrossRef]
32. Liu, J.G.; Zhao, T.S.; Liang, Z.X.; Chen, R. Effect of membrane thickness on the performance and efficiency of passive direct methanol fuel cells. *J. Power Sources* **2006**, *153*, 61–67. [CrossRef]
33. Heinzel, A.; Barragàn, V.M. A review of the state-of-the-art of the methanol crossover in direct methanol fuel cells. *J. Power Sources* **1999**, *84*, 70–74. [CrossRef]
34. Han, J.; Liu, H. Real time measurements of methanol crossover in a DMFC. *J. Power Sources* **2007**, *164*, 166–173. [CrossRef]
35. Ahmed, M.; Dincer, I. A review on methanol crossover in direct methanol fuel cells: Challenges and achievements. *Int. J. Energy Res.* **2011**, *35*, 1213–1228. [CrossRef]
36. Napoli, L.; Franco, J.; Fasoli, H.; Sanguinetti, A. Conductivity of Nafion® 117 membrane used in polymer electrolyte fuel cells. *Int. J. Hydrog. Energy* **2014**, *39*, 8656–8660. [CrossRef]
37. Ivanchev, S.S. Fluorinated proton-conduction Nafion-type membranes, the past and the future. *Russ. J. Appl. Chem.* **2008**, *81*, 569–584. [CrossRef]
38. Ursua, A.; Gandia, L.M.; Sanchis, P. Hydrogen Production from Water Electrolysis: Current Status and Future Trends. *Proc. IEEE* **2012**, *100*, 410–426. [CrossRef]
39. Ayers, K.E.; Anderson, E.B.; Capuano, C.; Carter, B.; Dalton, L.; Hanlon, G.; Manco, J.; Niedzwiecki, M. Research Advances Towards Low Cost, High Efficiency PEM Electrolysis. *ECS Trans.* **2010**, *33*, 3–15.
40. Schalenbach, M.; Sfolten, D. High-pressure water electrolysis: Electrochemical mitigation of product gas crossover. *Electrochim. Acta* **2015**, *156*, 321–327. [CrossRef]
41. Kickelbick, G. *Hybrid. Materials: Synthesis, Characterization, and Applications*; Wiley—VCH: Weinheim, Germany, 2007; ISBN 978-3-527-31299-3.

42. Trogadas, P.; Parrondo, J.; Ramani, V. Degradation Mitigation in Polymer Electrolyte Membranes Using Cerium Oxide as a Regenerative Free-Radical Scavenger. *Electrochem. Solid State Lett.* **2008**, *11*, 113–116. [CrossRef]
43. Bracco, C.M.; Sharma, S.; De Camargo Forte, M.M.; Steinberger-Wilckens, R. New approaches towards novel composite and multilayer membranes for intermediate temperature-polymer electrolyte fuel cells and direct methanol fuel cells. *J. Power Sources* **2016**, *316*, 139–159.
44. Dhanapal, D.; Xiao, M.; Wang, S.; Meng, Y. A Review on Sulfonated Polymer Composite/Organic-Inorganic Hybrid Membranes to Address Methanol Barrier Issue for Methanol Fuel Cells. *Nanomaterials* **2019**, *9*, 668. [CrossRef]
45. Karimi, M.B.; Mohammadi, F.; Hooshyari, K. Recent approaches to improve Nafion performance for fuel cell applications: A review. *Int. J. Hydrog. Energy* **2019**, *44*, 28919–28938. [CrossRef]
46. Liu, X.; Fang, S.; Ma, Z.; Zhang, Y. Structure Design and Implementation of the Passive μ-DMFC. *Micromachines* **2015**, *6*, 230–238. [CrossRef]
47. Lin, H.L.; Yu, T.L.; Huangb, L.N.; Chena, L.C.; Shen, K.S.; Jung, G.B. Nafion/PTFE composite membranes for direct methanol fuel cell applications. *J. Power Sources* **2005**, *150*, 11–19. [CrossRef]
48. Nouel, K.M.; Fedkiw, P.S. Nafion®-based composite polymer electrolyte membranes. *Electrochim. Acta* **1998**, *43*, 2381–2387. [CrossRef]
49. Yu, T.L.; Lin, H.L.; Shen, K.S.; Chang, Y.C.; Jung, G.B. Nafion/PTFE Composite Membranes for Fuel Cell Applications. *J. Polym. Res.* **2004**, *11*, 217–223. [CrossRef]
50. Chen, L.C.; Yu, T.L.; Lin, H.L.; Yeh, S.H. Nafion/PTFE and zirconium phosphate modified Nafion/PTFE composite membranes for direct methanol fuel cells. *J. Membr. Sci.* **2008**, *307*, 10–20. [CrossRef]
51. Vijayakumar, V.; Kim, K.; Nam, S.Y. Recent Advances in Polybenzimidazole (PBI)-based Polymer Electrolyte Membranes for High Temperature Fuel Cell Applications. *Appl. Chem. Eng.* **2019**, *30*, 643–651.
52. Wong, C.Y.; Wong, W.Y.; Loh, K.S.; Daud, W.R.W.; Lim, K.L.; Khalid, M.; Walvekar, R. Development of Poly(Vinyl Alcohol)-Based Polymers as Proton Exchange Membranes and Challenges in Fuel Cell Application: A Review. *Polym. Rev.* **2020**, *60*, 171–202. [CrossRef]
53. Shao, Z.G.; Wang, X.; Hsing, I.M. Composite Nafion/polyvinyl alcohol membranes for the direct methanol fuel cell. *J. Membr. Sci.* **2002**, *210*, 147–153. [CrossRef]
54. Mollà, S.; Compan, V. Performance of composite Nafion/PVA membranes for direct methanol fuel cells. *J. Power Sources* **2011**, *196*, 2699–2708. [CrossRef]
55. Shao, Z.G.; Hsing, I.M. Nafion Membrane Coated with Sulfonated Poly(vinyl alcohol)-Nafion Film for Direct Methanol Fuel Cells. *Electrochem. Solid State Lett.* **2002**, *5*, 185. [CrossRef]
56. Lin, H.L.; Wang, S.H.; Chiu, C.K.; Yu, T.L.; Chen, L.C.; Huang, C.C.; Cheng, T.H.; Lin, J.M. Preparation of Nafion/poly(vinyl alcohol) electro-spun fiber composite membranes for direct methanol fuel cells. *J. Membr. Sci.* **2010**, *365*, 114–122. [CrossRef]
57. Mollà, S.; Compan, V. Polyvinyl alcohol nanofiber reinforced Nafion membranes for fuel cell applications. *J. Membr. Sci.* **2011**, *372*, 191–200. [CrossRef]
58. Hobson, L.J.; Nakano, Y.; Ozu, H.; Hayase, S. Targeting improved DMFC performance. *J. Power Sources* **2002**, *104*, 79–84. [CrossRef]
59. Ainla, A.; Brandell, D. Nafion®–polybenzimidazole (PBI) composite membranes for DMFC applications. *Solid State Ion.* **2007**, *178*, 581–585. [CrossRef]
60. Ausejo, J.G.; Cabedo, L.; Gamez-Perez, J.; Molla', S.; Giménez, E.; Compañ, V. Modification of Nafion Membranes with Polyaniline to Reduce Methanol Permeability. *J. Electrochem. Soc.* **2015**, *162*, 325–333. [CrossRef]
61. Ben Jadi, S.; El Guerraf, A.; Bazzaoui, E.A.; Wang, R.; Martins, J.I.; Bazzaoui, M. Synthesis, characterization, and transport properties of Nafion-polypyrrole membrane for direct methanol fuel cell (DMFC) application. *J. Solid State Electrochem.* **2019**, *23*, 2423–2433. [CrossRef]
62. Penner, R.M.; Martin, C.R. Electronically Conductive Composite Polymer Membranes. *J. Electrochem. Soc.* **1986**, *133*, 310. [CrossRef]
63. Sata, T.; Funakoshi, T.; Akai, K. Preparation and Transport Properties of Composite Membranes Composed of Cation Exchange Membranes and Polypyrrole. *Macromolecules* **1996**, *29*, 4029–4035. [CrossRef]
64. Zhu, J.; Sattler, R.R.; Garsuch, A.; Yepez, O.; Pickup, P.G. Optimisation of polypyrrole/Nafion composite membranes for direct methanol fuel cells. *Electrochim. Acta* **2006**, *51*, 4052–4060. [CrossRef]

65. Wang, C.H.; Chenc, C.C.; Hsud, H.C.; Dud, H.Y.; Chend, C.P.; Hwangd, J.Y.; Chend, L.C.; Shihb, H.C.; Stejskal, J.; Chena, K.H. Low methanol-permeable polyaniline/Nafion composite membrane for direct methanol fuel cells. *J. Power Sources* **2009**, *190*, 279–284. [CrossRef]
66. Escudero-Cid, R.; Montiel, M.; Sotomayor, L.; Loureiro, B.; Fatás, E.; Ocón, P. Evaluation of polyaniline-Nafion® composite membranes for direct methanol fuel cells durability tests. *Int. J. Hydrog. Energy* **2015**, *40*, 8182–8192. [CrossRef]
67. Huanga, Q.M.; Zhanga, Q.L.; Huanga, H.L.; Li, W.S.; Huanga, Y.J.; Luoc, J.L. Methanol permeability and proton conductivity of Nafion membranes modified electrochemically with polyaniline. *J. Power Sources* **2008**, *184*, 338–343. [CrossRef]
68. Wang, B.; Hong, L.; Li, Y.; Zhao, L.; Wei, Y.; Zhao, C.; Na, H. Considerations of the Effects of Naphthalene Moieties on the Design of Proton-Conductive Poly(arylene ether ketone) Membranes for Direct Methanol Fuel Cells. *ACS Appl. Mater. Interfaces* **2016**, *8*, 24079–24088. [CrossRef] [PubMed]
69. Kim, T.; Choi, J.; Kim, S. Blend membranes of Nafion/sulfonated poly(aryl ether ketone) for direct methanol fuel cell. *J. Membr. Sci.* **2007**, *300*, 28–35. [CrossRef]
70. Bauer, B.; Jones, D.J.; Rozière, J.; Tchicaya, L.; Alberti, G.; Casciola, M.; Massinelli, L.; Peraio, A.; Besse, S.; Ramunni, E. Electrochemical characterisation of sulfonated polyetherketone membranes. *J. New Mat. Elect. Syst.* **2000**, *3*, 93–98.
71. Fu, T.Z.; Wang, J.; Ni, J.; Cui, Z.M.; Zhong, S.L.; Zhao, C.J.; Na, H.; Xing, W. Sulfonated poly(ether ether ketone)/aminopropyltriethoxysilane/phosphotungstic acid hybrid membranes with non-covalent bond: Characterization, thermal stability, and proton conductivity. *Solid State Ion.* **2008**, *179*, 2265–2273. [CrossRef]
72. Ru, C.; Gu, Y.; Duan, Y.; Zhao, C.; Na, H. Enhancement in proton conductivity and methanol resistance of Nafion membrane induced by blending sulfonated poly(arylene ether ketones) for direct methanol fuel cells. *J. Membr. Sci.* **2019**, *573*, 439–447. [CrossRef]
73. Tsai, J.; Cheng, H.; Kuo, J.; Huang, Y.; Chen, C. Blended Nafion®/SPEEK direct methanol fuel cell membranes for reduced methanol permeability. *J. Power Sources* **2009**, *189*, 958–965. [CrossRef]
74. Ye, G.; Hayden, C.A.; Goward, G.R. Proton Dynamics of Nafion and Nafion/SiO2 Composites by Solid State NMR and Pulse Field Gradient NMR. *Macromolecules* **2007**, *40*, 1529–1537. [CrossRef]
75. Peighambardoust, S.J.; Rowshanzamir, S.; Amjadi, M. Review of the proton exchange membranes for fuel cell applications. *Int. J. Hydrog. Energy* **2010**, *35*, 9349–9384. [CrossRef]
76. Mishra, A.K.; Bose, S.; Kuila, T.; Kim, N.H.; Lee, J.H. Silicate-based polymer-nanocomposite membranes for polymer electrolyte membrane fuel cells. *Prog. Polym. Sci.* **2012**, *37*, 842–869. [CrossRef]
77. Junoh, H.; Jaafar, J.; Nordin, N.A.H.M.; Ismail, A.F.; Othman, M.H.D.; Rahman, M.A.; Aziz, F.; Yusof, N.; Salleh, W.N.W. Porous Proton Exchange Membrane Based Zeolitic Imidazolate Framework-8 (ZIF-8). *J. Membr. Sci. Res.* **2019**, *5*, 65–75.
78. Lufrano, F.; Baglio, V.; Di Blasi, O.; Staiti, P.; Antonucci, V.; Arico, A.S. Design of efficient methanol impermeable membranes for fuel cell applications. *Phys. Chem. Chem. Phys. E PCCP* **2012**, *14*, 2718. [CrossRef] [PubMed]
79. Ren, S.; Sun, G.; Li, C.; Liang, Z.; Wu, Z.; Jin, W.; Qin, X.; Yang, X. Organic silica/Nafion® composite membrane for direct methanol fuel cells. *J. Membr. Sci.* **2006**, *282*, 450–455. [CrossRef]
80. Shao, Z.G.; Xu, H.; Li, M.; Hsing, I.M. Hybrid Nafion–inorganic oxides membrane doped with heteropolyacids for high temperature operation of proton exchange membrane fuel cell. *Solid State Ion.* **2006**, *177*, 779–785. [CrossRef]
81. Hammami, R.; Ahamed, Z.; Charradi, K.; Beji, Z.; Assaker, I.B.; Naceur, J.B.; Auvity, B.; Squadrito, G.; Chtourou, R. Elaboration and characterization of hybrid polymer electrolytes Nafion–TiO2 for PEMFCs. *Int. J. Hydrog. Energy* **2003**, *38*, 11583–11590. [CrossRef]
82. Kim, J.H.; Kim, S.K.; Namb, K.; Kim, D.W. Composite proton conducting membranes based on Nafion and sulfonated SiO2 nanoparticles. *J. Membr. Sci.* **2012**, *415*, 696–701. [CrossRef]
83. Ercelik, M.; Ozden, A.; Devrim, Y.; Colpan, C.O. Investigation of Nafion based composite membranes on the performance of DMFCs. *Int. J. Hydrog. Energy* **2017**, *42*, 2658–2668. [CrossRef]
84. Paul, D.R.; Robeson, L.M. Polymer nanotechnology: Nanocomposites. *Polym. Nanotechnol.* **2008**, *49*, 3187–3204. [CrossRef]
85. Wu, X.; Wu, N.; Shi, C.; Zheng, Z.; Qi, H.; Wang, Y. Proton conductive montmorillonite-Nafion composite membranes for direct ethanol fuel cells. *Appl. Surf. Sci.* **2016**, *388*, 239–244. [CrossRef]

86. Song, M.; Park, S.; Kim, Y.; Kim, K.; Min, S.; Rhee, H. Characterization of polymer-layered silicate nanocomposite membranes for direct methanol fuel cells. *Electrochim. Acta* **2004**, *50*, 639–643. [CrossRef]
87. Rhee, C.; Kim, H.; Chang, H.; Lee, J. Nafion/Sulfonated Montmorillonite Composite: A New Concept Electrolyte Membrane for Direct Methanol Fuel Cells. *Chem. Mater.* **2005**, *17*, 1691–1697. [CrossRef]
88. Lin, Y.; Yen, C.; Hung, C.; Hsiao, Y.; Ma, C. A novel composite membranes based on sulfonated montmorillonite modified Nafion® for DMFCs. *J. Power Sources* **2007**, *168*, 162–166. [CrossRef]
89. Liebau, F. Zeolites and clathrasils—Two distinct classes of framework silicates. *Zeolites* **1983**, *3*, 191–193. [CrossRef]
90. Smith, J. Definition of a zeolite. *Zeolites* **1984**, *4*, 309–310. [CrossRef]
91. Kešelj, D.; Lazić, D.; Škudrić, B.; Penavin-Škudrić, J.; Perušić, M. The possibility of hydrothermal synthesis of nay zeolite using different mineral acids. *Int. J. Latest Res. Sci. Technol.* **2015**, *4*, 37–41.
92. Sun, X.; Yang, C.; Xia, Z.; Qi, F.; Sun, H.; Sun, Q. Molecular sieve as an effective barrier for methanol crossover in direct methanol fuel cells. *Int. J. Hydrog. Energy* **2020**, *45*, 8994–9003. [CrossRef]
93. Tricoli, V.; Nannetti, F. Zeolite–Nafion composites as ion conducting membrane materials. *Electrochim. Acta* **2003**, *48*, 2625–2633. [CrossRef]
94. Makertihartha, I.G.B.N.; Zunita, M.; Rizki, Z.; Dharmawijaya, P.T. Recent advances on zeolite modification for direct alcohol fuel cells (DAFCs). *AIP Conf. Proc.* **2017**, *1818*. [CrossRef]
95. Prapainainara, P.; Dua, Z.; Kongkachuichaya, P.; Holmes, S.M.; Prapainainar, C. Mordenite/Nafion and analcime/Nafion composite membranes prepared by spray method for improved direct methanol fuel cell performance. *Appl. Surf. Sci.* **2017**, *421*, 24–41. [CrossRef]
96. Prapainainara, P.; Pattanapisutkun, N.; Prapainainar, C.; Kongkachuich, P. Incorporating graphene oxide to improve the performance of Nafion-mordenite composite membranes for a direct methanol fuel cell. *Int. J. Hydrog. Energy* **2019**, *44*, 362–378. [CrossRef]
97. Pumera, M. Electrochemistry of graphene, graphene oxide and other graphenoids: Review. *Electrochem. Commun.* **2013**, *36*, 14–18. [CrossRef]
98. Choi, B.; Huh, Y.; Park, Y.; Jung, D.; Hong, W.; Park, H. Enhanced transport properties in polymer electrolyte composite membranes with graphene oxide sheets. *Carbon* **2012**, *50*, 5395–5402. [CrossRef]
99. Chien, H.; Tsai, L.; Huang, C.; Kang, C.; Lin, J.; Chang, F. Sulfonated graphene oxide/Nafion composite membranes for high-performance direct methanol fuel cells. *Int. J. Hydrog. Energy* **2013**, *38*, 13792–13801. [CrossRef]
100. Yan, X.H.; Wu, R.; Xu, J.B.; Luo, Z.; Zhao, T.S. A monolayer graphene—Nafion sandwich membrane for direct methanol fuel cells. *J. Power Sources* **2016**, *311*, 188–194. [CrossRef]
101. Fathima, N.N.; Aravindhan, R.; Lawrence, D.; Yugandhar, U.; Moorthy, T.S.R.; Nair, N.U. SPEEK polymeric membranes for fuel cell application and their characterization: A review. *J. Sci. Ind. Res.* **2007**, *66*, 209–219.
102. Li, L.; Zhang, J.; Wang, Y. Sulfonated polyether ether ketone membranes cured with different methods for direct methanol fuel cells. *J. Mater. Sci. Lett.* **2003**, *22*, 1595–1597. [CrossRef]
103. Li, L.; Zhang, J.; Wang, Y. Sulfonated poly( ether ether ketone) membranes for direct methanol fuel cell. *J. Membr. Sci.* **2003**, *226*, 159–167. [CrossRef]
104. Xue, S.; Yin, G. Methanol permeability in sulfonated poly(etheretherketone) membranes: A comparison with Nafion membranes. *Eur. Polym. J.* **2006**, *42*, 776–785. [CrossRef]
105. Cai, H.; Shao, K.; Zhong, S.; Zhao, C.; Zhang, G.; Li, X.; Na, H. Properties of composite membranes based on sulfonated poly(ether ether ketone)s (SPEEK)/phenoxy resin (PHR) for direct methanol fuel cells usages. *J. Membr. Sci.* **2007**, *297*, 162–173. [CrossRef]
106. Di Vona, M.L.; D'Epifanio, A.; Marani, D.; Trombetta, M.; Traversa, E.; Licoccia, S. SPEEK/PPSU-based organic–inorganic membranes: Proton conducting electrolytes in anhydrous and wet environments. *J. Membr. Sci.* **2006**, *279*, 186–191. [CrossRef]
107. Zaidi, S.M.J.; Mikhailenko, S.D.; Robertson, G.P.; Guiver, M.D.; Kaliaguine, S. Proton conducting composite membranes from polyether ether ketone and heteropolyacids for fuel cell applications. *J. Membr. Sci.* **2000**, *173*, 17–34. [CrossRef]
108. Nagarale, R.K.; Gohil, G.S.; Shahi, V.K. Sulfonated poly(ether ether ketone)/polyaniline composite proton-exchange membrane. *J. Membr. Sci.* **2006**, *280*, 389–396. [CrossRef]
109. Zhang, H.; Fan, X.; Zhang, J.; Zhou, Z. Modification research of sulfonated PEEK membranes used in DMFC. *Solid State Ion.* **2008**, *179*, 1409–1412. [CrossRef]

110. Sengul, E.; Erdener, H.; Akay, R.G.; Yucel, H.; Bac, H.N.; Eroglu, I. Effects of sulfonated polyether-etherketone (SPEEK) and composite membranes on the proton exchange membrane fuel cell (PEMFC) performance. *Int. J. Hydrog. Energy* **2009**, *34*, 4645–4652. [CrossRef]
111. Li, X.; Liu, C.; Xu, D.; Zhao, C.; Wang, Z.; Zhang, G.; Na, H.; Xing, W. Preparation and properties of sulfonated poly(ether ether ketone)s (SPEEK)/polypyrrole composite membranes for direct methanol fuel cells. *J. Power Sources* **2006**, *162*, 1–8. [CrossRef]
112. Gosalawit, R.; Chirachanchai, S.; Shishatskiy, S.; Nunes, S.P. Sulfonated montmorillonite/sulfonated poly(ether ether ketone) (SMMT/SPEEK) nanocomposite membrane for direct methanol fuel cells (DMFCs). *J. Membr. Sci.* **2008**, *323*, 337–346. [CrossRef]
113. Pasupathi, S.; Ji, S.; Bladergroen, B.J.; Linkov, V. High DMFC performance output using modified acid–base polymer blend. *Int. J. Hydrog. Energy* **2008**, *33*, 3132–3136. [CrossRef]
114. Li, W.; Manthiram, A. Sulfonated poly(arylene ether sulfone) as a methanol-barrier layer in multilayer membranes for direct methanol fuel cells. *J. Power Sources* **2012**, *195*, 962–968. [CrossRef]
115. Lee, C.H.; Park, H.B.; Chung, Y.S.; Lee, Y.M.; Freeman, B.D. Water Sorption, Proton Conduction, and Methanol Permeation Properties of Sulfonated Polyimide Membranes Cross-Linked with N,N-Bis(2-hydroxyethyl)-2-aminoethanesulfonic Acid (BES). *Macromolecules* **2006**, *39*, 755–764. [CrossRef]
116. Jiang, Z.; Zhao, X.; Manthiram, A. Sulfonated poly(ether ether ketone) membranes with sulfonated graphene oxide fillers for direct methanol fuel cells. *Int. J. Hydrog. Energy* **2013**, *38*, 5875–5884. [CrossRef]
117. Greaves, C.R.; Bond, S.P.; McWhinne, W.R. Conductivity studies on modified laponites. *Polyhedron* **1995**, *14*, 3635–3639. [CrossRef]
118. Kim, D.; Hwang, H.; Jung, S.; Nam, S. Sulfonated poly(arylene ether sulfone)/Laponite-SO3H composite membrane for direct methanol fuel cell. *J. Ind. Eng. Chem.* **2012**, *18*, 556–562. [CrossRef]
119. Mikhailenko, S.D.; Robertson, G.P.; Guiver, M.D.; Kaliaguine, S. Properties of PEMs based on cross-linked sulfonated poly(ether ether ketone). *J. Membr. Sci.* **2006**, *285*, 306–316. [CrossRef]
120. Zhong, S.; Cui, X.; Cai, H.; Fu, T.; Zhao, C.; Na, H. Crosslinked sulfonated poly(ether ether ketone) proton exchange membranes for direct methanol fuel cell applications. *J. Power Sources* **2007**, *164*, 65–72. [CrossRef]
121. Feng, S.; Shang, Y.; Xie, X.; Wang, Y.; Xu, J. Synthesis and characterization of crosslinked sulfonated poly(arylene ether sulfone) membranes for DMFC applications. *J. Membr. Sci.* **2009**, *335*, 13–20. [CrossRef]
122. Prapainainar, C.; Holmes, S.M. Proton conductivity of Nafion® membrane in actual direct methanol fuel cell Operation. In *Sustainability in Energy and Buildings: Research Advances, Volume 2: Special Edition Mediterranean Green Energy Forum 2013*; Future Technology Press: Shoreham-by-Sea, UK, 2013; Volume 2, pp. 31–35.
123. Kreuer, K.D. On the development of proton conducting polymer membranes for hydrogen and methanol fuel cells. *J. Membr. Sci.* **2001**, *185*, 29–39. [CrossRef]
124. Moritani, T.; Kajitani, K. Functional modification of poly(vinyl alcohol) by copolymerization: 1. Modification with car□ylic monomers. *Polymer* **1997**, *38*, 2933–2945. [CrossRef]
125. Rhim, J.; Park, H.; Lee, C.; Jun, J.; Kimb, D.; Lee, Y. Crosslinked poly(vinyl alcohol) membranes containing sulfonic acid group: Proton and methanol transport through membranes. *J. Membr. Sci.* **2004**, *238*, 143–151. [CrossRef]
126. Wang, Z.; Zheng, H.; Chen, Q.; Zhang, S.; Yang, F.; Kang, J.; Chen, J.; Cao, Y.; Xiang, M. Preparation and characterization of PVA proton exchange membranes containing phosphonic acid groups for direct methanol fuel cell applications. *J. Polym. Res.* **2019**, *26*, 200.
127. Panero, S.; Fiorenza, P.; Navarra, M.A.; Romanowska, J.; Scrosati, B. Silica-Added, Composite Poly(vinyl alcohol) Membranes for Fuel Cell Application. *J. Electrochem. Soc.* **2005**, *152*, 2400. [CrossRef]
128. Son, J.H.; Kang, Y.S.; Won, J. Poly(vinyl alcohol)-based polymer electrolyte membranes containing polyrotaxane. *J. Membr. Sci.* **2006**, *281*, 345–350. [CrossRef]
129. Yang, C.; Lee, Y.; Yang, J. Direct methanol fuel cell(DMFC) based on PVA/MMT composite polymer membranes. *J. Power Sources* **2009**, *188*, 30–37. [CrossRef]
130. Pandey, J.; Shukla, A. PVDF supported silica immobilized phosphotungstic acid membrane for DMFC application. *Solid State Ion.* **2014**, *262*, 811–814. [CrossRef]
131. Pandey, J.; Seepan, M.; Shukla, A. Zirconium phosphate based proton conducting membrane for DMFC application. *Int. J. Hydrog. Energy* **2015**, *40*, 9410–9421. [CrossRef]
132. Neburchilov, V.; Martin, J.; Wang, H.; Zhang, J. A review of polymer electrolyte membranes for direct methanol fuel cells. *J. Power Sources* **2007**, *169*, 221–238. [CrossRef]

133. Stone, C.; Steck, A.E.; Choudhury, B. Graft polymeric membranes and ion-exchange membranes formed therefrom. U.S. Patent 6,723,758, 26 September 2004.
134. Taft, K.M.; Kurano, M.R. Composite electrolyte for fuel cells. U.S. Patent 6,630,265, 7 October 2003.
135. Chen, S.L.; Krishnan, L.; Srinivasan, S.; Benziger, J.; Bocarsly, A.B. Ion exchange resin/polystyrene sulfonate composite membranes for PEM fuel cells. *J. Membr. Sci.* **2004**, *243*, 327–333. [CrossRef]
136. Di Noto, V.; Gliubizzi, R.; Negro, E.; Vittadello, M.; Pace, G. Hybrid inorganic–organic proton conducting membranes based on Nafion and 5 wt.% of MxOy (M = Ti, Zr, Hf, Ta and W.): Part, I. Synthesis, properties and vibrational studies. *Electrochim. Acta* **2007**, *53*, 1618–1627. [CrossRef]
137. Di Noto, V.; Lavina, S.; Negro, E.; Vittadello, M.; Conti, F.; Piga, M.; Pace, G. Hybrid inorganic–organic proton conducting membranes based on Nafion and 5 wt% of MxOy (M = Ti, Zr, Hf, Ta and W). Part II: Relaxation phenomena and conductivity mechanism. *J. Power Sources* **2009**, *187*, 57–66. [CrossRef]
138. Adjemian, K.T.; Dominey, R.; Krishnan, L.; Ota, H.; Majsztrik, P.; Zhang, T.; Mann, J.; Kirby, B.; Gatto, L.; Velo-Simpson, M.; et al. Function and Characterization of Metal Oxide−Nafion Composite Membranes for Elevated-Temperature H2/O2 PEM Fuel Cells. *Chem. Mater.* **2006**, *18*, 2238–2248. [CrossRef]
139. Di Noto, V.; Boaretto, N.; Negro, E.; Pace, G. New inorganic–organic proton conducting membranes based on Nafion and hydrophobic fluoroalkylated silica nanoparticles. *J. Power Sources* **2010**, *195*, 7734–7742. [CrossRef]
140. Giffin, G.A.; Piga, M.; Lavina, S.; Navarra, M.A.; D'Epifanio, A.; Scrosati, B.; Di Noto, V. Characterization of sulfated-zirconia/Nafion® composite membranes for proton exchange membrane fuel cells. *J. Power Sources* **2012**, *198*, 66–75. [CrossRef]
141. Saccà, A.; Gatto, I.; Carbone, A.; Pedicini, R.; Passalacqua, E. ZrO2–Nafion composite membranes for polymer electrolyte fuel cells (PEFCs) at intermediate temperature. *J. Power Sources* **2006**, *163*, 47–51. [CrossRef]
142. D'Epifanio, A.; Navarra, M.A.; Weise, F.C.; Mecheri, B.; Farrington, J.; Licoccia, S.; Greenbaum, S. Composite Nafion/Sulfated Zirconia Membranes: Effect of the Filler Surface Properties on Proton Transport Characteristics. *Chem. Mater.* **2009**, *22*, 813–821. [CrossRef] [PubMed]
143. Alberti, G.; Casciola, M.; Capitani, D.; Donnadio, A.; Narducci, R.; Pica, M.; Sganappa, M. Novel Nafion–zirconium phosphate nanocomposite membranes with enhanced stability of proton conductivity at medium temperature and high relative humidity. *Electrochim. Acta* **2007**, *52*, 8125–8132. [CrossRef]
144. Sahu, A.K.; Selvarani, G.; Pitchumani, S.; Sridhar, P.; Shukla, A.K. A Sol-Gel Modified Alternative Nafion-Silica Composite Membrane for Polymer Electrolyte Fuel Cells. *J. Electrochem. Soc.* **2007**, *154*, 123–132. [CrossRef]
145. Costamagna, P.; Yang, C.; Bocarsly, A.B.; Srinivasan, S. Nafion® 115/zirconium phosphate composite membranes for operation of PEMFCs above 100 °C. *Electrochim. Acta* **2002**, *47*, 1023–1033. [CrossRef]
146. Sahu, A.K.; Pitchumani, S.; Sridhar, P.; Shukla, A.K. Co-assembly of a Nafion–Mesoporous Zirconium Phosphate Composite Membrane for PEM Fuel Cells. *Fuel Cells* **2009**, *9*, 139–147. [CrossRef]
147. Pineda-Delgado, J.L.; Gutierrez, C.K.; Rivas, S.; Arjona, N.; Arriaga, L.G.; Chávez-Ramirez, A. Synthesis and evaluation of HfO2 as a prospective filler in inorganic–organic hybrid membranes based on Nafion for PEM fuel cells. *Nanotechnology* **2019**, *30*, 105707. [CrossRef]
148. Oh, K.; Kwon, O.; Son, B.; Lee, D.H.; Shanmugam, S. Nafion-sulfonated silica composite membrane for proton exchange membrane fuel cells under operating low humidity condition. *J. Membr. Sci.* **2019**, *583*, 103–109. [CrossRef]
149. Xu, G.; Wu, Z.; Wei, Z.; Zhang, W.; Wu, J.; Li, Y.; Li, J.; Qu, K.; Cai, W. Non-destructive fabrication of Nafion/silica composite membrane via swelling-filling modification strategy for high temperature and low humidity PEM fuel cell. *Renew. Energy* **2020**, *153*, 935–939. [CrossRef]
150. Xu, G.; Wei, Z.; Li, S.; Li, J.; Yang, Z.; Grigoriev, S.A. In-situ sulfonation of targeted silica-filled Nafion for high-temperature PEM fuel cell application. *Int. J. Hydrog. Energy* **2019**, *44*, 29711–29716. [CrossRef]
151. Saccà, A.; Carbone, A.; Gatto, I.; Pedicini, R.; Freni, A.; Patti, A.; Passalacqua, E. Composites Nafion-titania membranes for Polymer Electrolyte Fuel Cell (PEFC) applications at low relative humidity levels: Chemical physical properties and electrochemical performance. *Polym. Test.* **2016**, *56*, 10–18. [CrossRef]
152. Saccà, A.; Carbone, A.; Passalacqua, E.; D'Epifanio, A.; Licoccia, S.; Traversa, E.; Sala, E.; Traini, F.; Ornelas, R. Nafion–TiO2 hybrid membranes for medium temperature polymer electrolyte fuel cells (PEMFCs). *J. Power Sources* **2005**, *152*, 16–21. [CrossRef]

153. Amjadi, M.; Rowshanzamir, S.; Peighambardoust, S.J.; Hosseini, M.G. Investigation of physical properties and cell performance of Nafion/TiO2 nanocomposite membranes for high temperature PEM fuel cells. *Int. J. Hydrog. Energy* **2010**, *35*, 9252–9260. [CrossRef]
154. Matos, B.R.; Santiago, E.I.; Rey, J.F.Q.; Ferlauto, A.S.; Traversa, E.; Linardi, M.; Fonseca, F.C. Nafion-based composite electrolytes for proton exchange membrane fuel cells operating above 120 °C with titania nanoparticles and nanotubes as fillers. *J. Power Sources* **2011**, *196*, 1061–1068. [CrossRef]
155. Zhengbang, W.; Tang, H.; Mu, P. Self-assembly of durable Nafion/TiO2 nanowire electrolyte membranes for elevated-temperature PEM fuel cells. *J. Membr. Sci.* **2011**, *369*, 250–257. [CrossRef]
156. Ketpang, K.; Shanmugam, S.; Suwanboon, C.; Chanunpanich, N.; Lee, D. Efficient water management of composite membranes operated in polymer electrolyte membrane fuel cells under low relative humidity. *J. Membr. Sci.* **2015**, *493*, 285–298. [CrossRef]
157. Jun, Y.; Zarrin, H.; Fowler, M.; Chen, Z. Functionalized titania nanotube composite membranes for high temperature proton exchange membrane fuel cells. *Int. J. Hydrog. Energy* **2011**, *36*, 6073–6081. [CrossRef]
158. Ketpang, K.; Son, B.; Lee, D.; Shanmugam, S. Porous zirconium oxide nanotube modified Nafion composite membrane for polymer electrolyte membrane fuel cells operated under dry conditions. *J. Membr. Sci.* **2015**, *488*, 154–165. [CrossRef]
159. Marani, D.; D'Epifanio, A.; Traversa, E.; Miyayama, M.; Licoccia, S. Titania Nanosheets (TNS)/Sulfonated Poly Ether Ether Ketone (SPEEK) Nanocomposite Proton Exchange Membranes for Fuel Cells. *Chem. Mater.* **2009**, *22*, 1126–1133. [CrossRef]
160. Devrim, Y.; Erkan, S.; Baç, N.; Eroğlu, I. Preparation and characterization of sulfonated polysulfone/titanium dioxide composite membranes for proton exchange membrane fuel cells. *Int. J. Hydrog. Energy* **2009**, *34*, 3467–3475. [CrossRef]
161. Sambandam, S.; Ramani, V. SPEEK/functionalized silica composite membranes for polymer electrolyte fuel cells. *J. Power Sources* **2007**, *170*, 259–267. [CrossRef]
162. Therese, J.B.A.J.H.; Gayathri, R.; Selvakumar, K.; Prabhu, M.R.; Sivakumar, P. Incorporation of sulfonated silica nano particles into polymer blend membrane for PEM fuel cell applications. *Mater. Res. Express* **2019**, *6*, 11.
163. Sahin, A.; Tasdemir, H.M.; Ar, I. Improved performance and durability of sulfonated polyether ether ketone/cerium phosphate composite membrane for proton exchange membrane fuel cells. *Ionics* **2019**, *25*, 5163–5175. [CrossRef]
164. Carbone, A.; Pedicini, R.; Saccà, A.; Gatto, I.; Passalacqua, E. Composite S-PEEK membranes for medium temperature polymer electrolyte fuel cells. *J. Power Sources* **2008**, *178*, 661–666. [CrossRef]
165. Carbone, A.; Saccà, A.; Gatto, I.; Pedicini, R.; Passalacqua, E. Investigation on composite S-PEEK/H-BETA MEAs for medium temperature PEFC. *Int. J. Hydrog. Energy* **2008**, *33*, 3153–3158. [CrossRef]
166. Özdemir, Y.; Üregen, N.; Devrim, Y. Polybenzimidazole based nanocomposite membranes with enhanced proton conductivity for high temperature PEM fuel cells. *Int. J. Hydrog. Energy* **2017**, *42*, 2648–2657. [CrossRef]
167. Lee, S.; Seo, K.; Ghorpade, R.V.; Nam, K.H.; Han, H. High temperature anhydrous proton exchange membranes based on chemically-functionalized titanium/polybenzimidazole composites for fuel cells. *Mater. Lett.* **2020**, *263*, 127167. [CrossRef]
168. Ooi, Y.X.; Ya, K.Z.; Maegawa, K.; Tan, W.K.; Kawamura, G.; Muto, H.; Matsuda, A. CHS-WSiA doped hexafluoropropylidene-containing polybenzimidazole composite membranes for medium temperature dry fuel cells. *Int. J. Hydrog. Energy* **2019**, *44*, 32201–32209. [CrossRef]
169. Devrim, Y.; Devrim, H.; Eroglu, I. Polybenzimidazole/SiO2 hybrid membranes for high temperature proton exchange membrane fuel cells. *Int. J. Hydrog. Energy* **2016**, *41*, 10044–10052. [CrossRef]
170. Plackett, D.; Siu, A.; Li, Q.; Pan, C.; Jensen, J.O.; Nielsen, S.F.; Permyakova, A.A.; Bjerrum, N.J. High-temperature proton exchange membranes based on polybenzimidazole and clay composites for fuel cells. *J. Membr. Sci.* **2011**, *383*, 78–87. [CrossRef]
171. Aili, D.; Zhang, J.; Sondergaard, T.; Zhu, H.; Yang, T.; Liu, J.; Forsyth, M.; Pan, C.; Jensen, J.O.; Cleemann, L.N.; et al. Exceptional durability enhancement of PA/PBI based polymer electrolyte membrane fuel cells for high temperature operation at 200 °C. *J. Mater. Chem. A* **2016**, *4*, 4019–4024. [CrossRef]

172. Rodgers, M.P.; Pearman, B.P.; Bonville, L.J.; Cullen, D.A.; Mohajeri, N.; Slattery, D.K. Evaluation of the Effect of Impregnated Platinum on PFSA Degradation for PEM Fuel Cells. *J. Electrochem. Soc.* **2013**, *160*, 1123–1128. [CrossRef]
173. Pearman, B.P.; Mohajeri, N.; Brooker, R.P.; Rodgers, M.P.; Slattery, D.K.; Hampton, M.D.; Cullen, D.A.; Seal, S. The degradation mitigation effect of cerium oxide in polymer electrolyte membranes in extended fuel cell durability tests. *J. Power Sources* **2013**, *225*, 75–83. [CrossRef]
174. Pearman, B.P.; Mohajeri, N.; Slattery, D.K.; Hampton, M.D.; Seal, S.; Cullen, D.A. The chemical behavior and degradation mitigation effect of cerium oxide nanoparticles in perfluorosulfonic acid polymer electrolyte membranes. *Polym. Degrad. Stab.* **2013**, *98*, 1766–1772. [CrossRef]
175. Lee, H.; Han, M.; Choi, Y.W.; Bae, B. Hydrocarbon-based polymer electrolyte cerium composite membranes for improved proton exchange membrane fuel cell durability. *J. Power Sources* **2015**, *295*, 221–227. [CrossRef]
176. Elakkiya, S.; Arthanareeswaran, G.; Ismail, A.F.; Das, D.B.; Suganya, R. Polyaniline coated sulfonated TiO2 nanoparticles for effective application in proton conductive polymer membrane fuel cell. *Eur. Polym. J.* **2019**, *112*, 696–703. [CrossRef]
177. Lee, K.H.; Chu, J.Y.; Kim, A.R.; Yoo, D.J. Effect of functionalized SiO2 toward proton conductivity of composite membranes for PEMFC application. *Energy Res.* **2019**, *43*, 5333–5345. [CrossRef]
178. He, D.; Tang, H.; Kou, Z.; Pan, M.; Sun, X.; Zhang, J.; Mu, S. Engineered Graphene Materials: Synthesis and Applications for Polymer Electrolyte Membrane Fuel Cells. *Adv. Mater.* **2017**, *29*, 1–8. [CrossRef]
179. Kumar, R.; Xu, C.; Scott, K. Graphite oxide/Nafion composite membranes for polymer electrolyte fuel cells. *RSC Adv.* **2012**, *2*, 8777–8782. [CrossRef]
180. Sahu, A.K.; Ketpang, K.; Shanmugam, S.; Kwon, O.; Lee, S.; Kim, H. Sulfonated Graphene-Nafion Composite Membranes for Polymer Electrolyte Fuel Cells Operating under Reduced Relative Humidity. *J. Phys. Chem. C* **2016**, *120*, 15855–15866. [CrossRef]
181. Lee, D.C.; Yang, H.N.; Park, S.H.; Kim, W.J. Nafion/graphene oxide composite membranes for low humidifying polymer electrolyte membrane fuel cell. *J. Membr. Sci.* **2014**, *452*, 20–28. [CrossRef]
182. Lee, D.C.; Yang, H.N.; Park, S.H.; Park, K.W.; Kim, W.J. Self-humidifying Pt–graphene/SiO2 composite membrane for polymer electrolyte membrane fuel cell. *J. Membr. Sci.* **2015**, *474*, 254–262. [CrossRef]
183. Yang, H.N.; Lee, W.H.; Choi, B.S.; Kim, W.J. Preparation of Nafion/Pt-containing TiO2/graphene oxide composite membranes for self-humidifying proton exchange membrane fuel cell. *J. Membr. Sci.* **2016**, *504*, 20–28. [CrossRef]
184. Kim, Y.; Ketpang, K.; Jaritphun, S.; Park, J.S.; Shanmugam, S. A polyoxometalate coupled graphene oxide-Nafion composite membrane for fuel cells operating at low relative humidity. *J. Mater. Chem. A* **2015**, *3*, 8148–8155. [CrossRef]
185. Maiti, J.; Kakati, N.; Woo, S.P.; Yoon, Y.S. Nafion® based hybrid composite membrane containing GO and dihydrogen phosphate functionalized ionic liquid for high temperature polymer electrolyte membrane fuel cell. *Compos. Sci. Technol.* **2018**, *155*, 189–196. [CrossRef]
186. Branco, C.M. Multilayer Membranes for Intermediate Temperature. Ph.D. Thesis, University of Birmingham, Birmingham, UK, 2017.
187. Ibrahim, A.; Hossain, O.; Chaggar, J.; Steinberger-Wilckens, R.; El-kharouf, A. GO-Nafion composite membrane development for enabling intermediate temperature operation of polymer electrolyte fuel cell. *Int. J. Hydrog. Energy* **2019**, *45*, 5526–5534. [CrossRef]
188. Kumar, R.; Mamlouk, M.; Scott, K. Sulfonated polyether ether ketone-sulfonated graphene oxide composite membranes for polymer electrolyte fuel cells. *RSC Adv.* **2014**, *4*, 617–623. [CrossRef]
189. Kim, A.R.; Gabunada, J.C.; Yoo, D.J. Amelioration in physicochemical properties and single cell performance of sulfonated poly(ether ether ketone) block copolymer composite membrane using sulfonated carbon nanotubes for intermediate humidity fuel cells. *Int. J. Energy Res.* **2019**, *43*, 2974–2989. [CrossRef]
190. Üregen, N.; Pehlivanoğlu, K.; Özdemir, Y.; Devrim, Y. Development of polybenzimidazole/graphene oxide composite membranes for high temperature PEM fuel cells. *Int. J. Hydrog. Energy* **2017**, *42*, 2636–2647. [CrossRef]
191. Xue, C.; Zou, J.; Sun, Z.; Wang, F.; Han, K.; Zhu, H. Graphite oxide/functionalized graphene oxide and polybenzimidazole composite membranes for high temperature proton exchange membrane fuel cells. *Int. J. Hydrog. Energy* **2014**, *39*, 7931–7939. [CrossRef]

192. Xu, C.; Cao, Y.; Kumar, R.; Wu, X.; Wang, X.; Scott, K. A polybenzimidazole/sulfonated graphite oxide composite membrane for high temperature polymer electrolyte membrane fuel cells. *J. Mater. Chem.* **2011**, *21*, 11359–11364. [CrossRef]
193. Xu, C.; Liu, X.; Cheng, J.; Scott, K. A polybenzimidazole/ionic-liquid-graphite-oxide composite membrane for high temperature polymer electrolyte membrane fuel cells. *J. Power Sources* **2015**, *274*, 922–927. [CrossRef]
194. Abouzari Lotf, E.; Zakeri, M.; Nasef, M.M.; Miyake, M.; Mozarmnia, P.; Bazilah, N.A.; Emelin, N.F.; Ahmad, A. Highly durable polybenzimidazole composite membranes with phosphonated graphene oxide for high temperature polymer electrolyte membrane fuel cells. *J. Power Sources* **2019**, *412*, 238–245. [CrossRef]
195. Kannan, R.; Aher, P.P.; Palaniselvam, T.; Kurungot, S.; Kharul, U.K.; Pillai, V.K. Artificially designed membranes using phosphonated multiwall carbon nanotube-polybenzimidazole composites for polymer electrolyte fuel cells. *J. Phys. Chem. Lett.* **2010**, *1*, 2109–2113. [CrossRef]
196. Kannan, R.; Kagalwala, H.N.; Chaudhari, H.D.; Kharul, U.K.; Kurungot, S.; Pillai, V.K. Improved performance of phosphonated carbon nanotube–polybenzimidazole composite membranes in proton exchange membrane fuel cells. *J. Mater. Chem.* **2011**, *21*, 7223–7231. [CrossRef]
197. Yang, J.; Liu, C.; Gao, L.; Wang, J.; Xu, Y.; He, R. Novel composite membranes of triazole modified graphene oxide and polybenzimidazole for high temperature polymer electrolyte membrane fuel cell applications. *RSC Adv.* **2015**, *5*, 101049–101054. [CrossRef]
198. Cao, Y.C.; Xu, C.; Wu, X.; Wang, X.; Xing, L.; Scott, K. A poly (ethylene oxide)/graphene oxide electrolyte membrane for low temperature polymer fuel cells. *J. Power Sources* **2011**, *196*, 8377–8382. [CrossRef]
199. Lee, H.; Han, J.; Kim, K.; Kim, J.; Kim, E.; Shin, H.; Lee, J.C. Highly sulfonated polymer-grafted graphene oxide composite membranes for proton exchange membrane fuel cells. *J. Ind. Eng. Chem.* **2019**, *74*, 223–232. [CrossRef]
200. Dai, Y.; Wang, J.; Tao, P.; He, R. Various hydrophilic carbon dots doped high temperature proton exchange composite membranes based on polyvinylpyrrolidone and polyethersulfone. *J. Colloid Interface Sci.* **2019**, *553*, 503–511. [CrossRef]
201. Ahmed, S.; Ali, M.; Cai, Y.; Lu, Y.; Ahmad, Z.; Khannal, S.; Xu, S. Novel sulfonated multi-walled carbon nanotubes filled chitosan composite membrane for fuel-cell applications. *J. Appl. Polym. Sci.* **2019**, *136*, 47603. [CrossRef]
202. Choi, P.; Li, B.; Argawal, L.; Pearman, B.P.; Mohajeri, N.; Rodgers, M.P.; Slattery, D.; Bonville, L.; Kunz, H.R.; Fenton, J.M. Effect of Equivalent Weight of Phosphotungstic Acid-Incorporated Composite Membranes on the High Temperature Operation of PEM Fuel Cells. *ECS Trans.* **2008**, *16*, 2157–2164.
203. Lee, J.S.; Nohira, T.; Hagiwara, R. Novel composite electrolyte membranes consisting of fluorohydrogenate ionic liquid and polymers for the unhumidified intermediate temperature fuel cell. *J. Power Sources* **2007**, *171*, 535–539. [CrossRef]
204. Ramani, V.; Kunz, H.; Fenton, J. Investigation of Nafion®/HPA composite membranes for high temperature/low relative humidity PEMFC operation. *J. Membr. Sci.* **2004**, *232*, 31–44. [CrossRef]
205. Ramani, V.; Kunz, H.R.; Fenton, J.M. Stabilized composite membranes and membrane electrode assemblies for elevated temperature/low relative humidity PEFC operation. *J. Power Sources* **2005**, *152*, 182–188. [CrossRef]
206. Ramani, V.; Kunz, H.R.; Fenton, J.M. Stabilized heteropolyacid/Nafion® composite membranes for elevated temperature/low relative humidity PEFC operation. *Electrochim. Acta* **2005**, *50*, 1181–1187. [CrossRef]
207. Ramani, V.; Kunz, H.R.; Fenton, J.M. Metal dioxide supported heteropolyacid/Nafion® composite membranes for elevated temperature/low relative humidity PEFC operation. *J. Membr. Sci.* **2006**, *279*, 506–512. [CrossRef]
208. Lee, S.Y.; Yasuda, T. Fabrication of protic ionic liquid/sulfonated polyimide composite membranes for non-humidified fuel cells. *J. Power Sources* **2010**, *195*, 5909–5914. [CrossRef]
209. Yi, S.; Zhang, F.; Li, W.; Huang, C.; Zhang, H.; Pan, M. Anhydrous elevated-temperature polymer electrolyte membranes based on ionic liquids. *J. Membr. Sci.* **2011**, *366*, 349–355. [CrossRef]
210. Yasuda, T.; Nakamura, S.I.; Honda, Y.; Kinugawa, K.; Lee, S.Y.; Watanabe, M. Effects of polymer structure on properties of sulfonated polyimide/protic ionic liquid composite membranes for nonhumidified fuel cell applications. *ACS Appl. Mater. Interfaces* **2012**, *4*, 1783–1790. [CrossRef]
211. Malik, R.S.; Verma, P.; Choudhary, V. A study of new anhydrous, conducting membranes based on composites of aprotic ionic liquid and cross-linked SPEEK for fuel cell application. *Electrochim. Acta* **2015**, *152*, 352–359. [CrossRef]

212. Baglio, V.; Ornelas, R.; Matteucci, F.; Martina, F.; Ciccarella, G.; Zama, I.; Arriaga, L.G.; Antonucci, V.; Aricò, A.S. Solid Polymer Electrolyte Water Electrolyser Based on Nafion-TiO2 Composite Membrane for High Temperature Operation. *Fuel Cells* **2009**, *9*, 247–252. [CrossRef]
213. Antonucci, V.; Di Blasi, A.; Baglio, V.; Ornelas, R.; Matteucci, F.; Ledesma-Garcia, J.; Arriaga, L.G.; Aricò, A.S. High temperature operation of a composite membrane-based solid polymer electrolyte water electrolyser. *Electrochim. Acta* **2008**, *53*, 7350–7356. [CrossRef]
214. Mazzapioda, L.; Navarra, M.A.; Trequattrini, F.; Paolone, A.; Elamin, K.; Martinelli, A.; Palumbo, O. Composite Nafion Membranes with CaTiO3−δ Additive for Possible Applications in Electrochemical Devices. *Membranes* **2019**, *9*, 143. [CrossRef] [PubMed]
215. Aricò, A.S.; Di Blasi, A.; Brunaccini, G.; Sergi, F.; Dispenza, G.; Andaloro, L.; Ferraro, M.; Antonucci, V.; Asher, P.; Buche, S.; et al. High Temperature Operation of a Solid Polymer Electrolyte Fuel Cell Stack Based on a New Ionomer Membrane. *Fuel Cells* **2010**, *10*, 1013–1023. [CrossRef]
216. Boaretti, C.; Paquini, L.; Sood, R.; Giancola, S.; Donnadio, A.; Roso, M.; Modesti, M.; Cavaliere, S. Mechanically stable nanofibrous sPEEK/Aquivion® composite membranes for fuel cell applications. *J. Membr. Sci.* **2018**, *545*, 66–74. [CrossRef]
217. Giancola, S.; Zaton, M.; Reyes-Carmona, A.; Dupont, M.; Donnadio, A.; Cavaliere, S.; Roziere, J.; Jones, D.J. Composite short side chain PFSA membranes for PEM water electrolysis. *J. Membr. Sci.* **2019**, *570*, 69–76. [CrossRef]
218. Ion-Ebrasu, D.; Pollet, B.G.; Spinu-Zaulet, A.; Soare, A.; Carcadea, E.; Varlam, M.; Caprarescu, S. Graphene modified fluorinated cation-exchange membranes for proton exchange membrane water electrolysis. *Int. J. Hydrog. Energy* **2019**, *44*, 10190–10196. [CrossRef]
219. Linkous, C.A.; Anderson, H.R.; Kopitzke, R.W.; Nelson, G.L. Development of new proton exchange membrane electrolytes for water electrolysis at higher temperatures. *Int. J. Hydrog. Energy* **1998**, *23*, 525–529. [CrossRef]
220. Cao, N.; Zhou, C.; Wang, Y.; Ju, H.; Tan, D.; Li, J. Synthesis and Characterization of Sulfonated Graphene Oxide Reinforced Sulfonated Poly (Ether Ether Ketone) (SPEEK) Composites for Proton Exchange Membrane Materials. *Membranes* **2018**, *11*, 516. [CrossRef]
221. Song, M.; Ha, S.; Park, D.; Ryu, C.; Kang, A.; Moon, S.; Chung, J. Development and characterization of covalently cross-linked SPEEK/Cs-TPA/$CeO_2$ composite membrane and membrane electrode assembly for water electrolysis. *Int. J. Hydrog. Energy* **2013**, *38*, 10502–10510. [CrossRef]
222. Siracusano, S.; Baglio, V.; Lufrano, F.; Staiti, P.; Aricò, A.S. Electrochemical characterization of a PEM water electrolyzer based on a sulfonated polysulfone membrane. *J. Membr. Sci.* **2013**, *448*, 209–214. [CrossRef]

molecules

*Article*

# Full Parametric Study of the Influence of Ionomer Content, Catalyst Loading and Catalyst Type on Oxygen and Ion Transport in PEM Fuel Cell Catalyst Layers

**Robert Alink, Rajveer Singh, Patrick Schneider, Kläre Christmann, Johannes Schall, Roman Keding and Nada Zamel ***

Fraunhofer Institute for Solar Energy Systems, ISE, Heidenhofstrasse 2, 79110 Freiburg, Germany; racingrobi@yahoo.de (R.A.); Rajveer.Singh@ise.fraunhofer.de (R.S.); patrick.david.schneider@ise.fraunhofer.de (P.S.); klaere.christmann@ise.fraunhofer.de (K.C.); johannes.schall@ise.fraunhofer.de (J.S.); roman.keding@ise.fraunhofer.de (R.K.)

* Correspondence: nada.zamel@ise.fraunhofer.de

Academic Editors: Jean St-Pierre, Shangfeng Du and César Augusto Correia de Sequeira
Received: 20 December 2019; Accepted: 25 March 2020; Published: 27 March 2020

**Abstract:** To advance the technology of polymer electrolyte membrane fuel cells, material development is at the forefront of research. This is especially true for membrane electrode assembly, where the structuring of its various layers has proven to be directly linked to performance increase. In this study, we investigate the influence of the various ingredients in the cathode catalyst layer, such as ionomer content, catalyst loading and catalyst type, on the oxygen and ion transport using a full parametric analysis. Using two types of catalysts, 40 wt.% Pt/C and 60 wt.% Pt/C with high surface area carbon, the ionomer/carbon content was varied between 0.29–1.67, while varying the Pt loading in the range of 0.05–0.8 mg cm$^{-2}$. The optimum ionomer content was found to be dependent on the operating point and condition, as well as catalyst loading and type. The data set provided in this work gives a starting point to further understanding of structured catalyst layers.

**Keywords:** cathode catalyst layer; I/C ratio; diffusion limitation; conductivity limitation

## 1. Introduction

Further advancement of polymer electrolyte membrane (PEM) fuel cells, particularly for use in the automotive industry, must be achieved as a balance between cost and functionality. The catalyst layer as the heart of the cell controls the half-cell reactions and their products. Its structure governs the various transport phenomena simultaneously taking place and affects its overall activity, stability and life time. Throughout the years, the optimization of the structure of the catalyst layer, with special attention given to the cathode, has been achieved via systematic optimization of its components [1]. The importance of this optimization stems from the heterogeneous and complex structure of the layer, which must fulfill its main four requirements: (1) existence of a three-phase interface on which the reaction occurs, (2) continuous path for efficient transport of protons, (3) a continuous pore network for the transport of reactants/products and for efficient water removal, and (4) continuous passage for the conduction of electrons between the catalyst layer and the current collector. Efforts to increase the stability and durability of the catalyst layer are hence ongoing with one such effort focusing on the differences in structure between conventional and non-conventional catalyst layers [1]. Examples of non-conventional catalyst layers are dispersed catalyst layers, either on the membrane or the gas diffusion layer, ultrathin catalyst layers and nano-structured thin film (NSTF) catalyst layers [2–6]. Although much work is found on non-conventional catalyst layers, conventional layers are still at the forefront of research due to other challenges facing their non-conventional counterparts [1] and hence, the discussion within this manuscript is focused on conventional catalyst layers.

The structure of conventional catalyst layers specifically is shaped by the ink, which is composed of a catalyst deposited on a support (e.g., Pt/C), ionomer and a dispersing solvent. The dispersion medium governs the ink properties, which ultimately govern the physical and mass transport properties of the catalyst layer. In [7], a thorough review, focused on the analysis of the colloidal ink, was put forward, where the authors highlighted the importance of understanding the ink properties from a nanoscale in order to understand the macroscale effects. Hence, understanding the interaction between the layer's ingredients, its structure and performance is important to the advancement of PEM fuel cells. Another way to optimize the layer performance is to introduce gradients to adjust to inhomogeneous conditions that occur during fuel cell performance. These gradients can occur in all three dimensions, including the in-plane (x-y) and through-plane (z) dimensions of the catalyst layer.

The effect of catalyst layer structuring on the performance of PEM fuel cells was investigated by various groups in literature. This was mainly carried out by the effect of graded layers taking into consideration specific ingredients. Chen et al. [8] investigated the performance of a cell manufactured with cathode catalyst layers having two layers of different Pt/C ratio and Nafion content. They showed that the region where the reaction occurs can significantly affect the performance of the cell. Allocating more Nafion and Pt/C in the sublayer closer to the membrane was shown to significantly improve the performance. This finding, however, does not agree with the experimental study by Yoon et al. [9]. In [9], the authors prepared multi-layer structured cathodes by spray-drying, where they varied the ionomer content in the thickness of the catalyst layer. In their study, they found that a structured catalyst layer does not affect the overall performance. To further understand the interaction between Pt and ionomer, a comprehensive numerical model of a single cell to investigate the spatial distribution of Pt loading and ionomer content in the through-plane direction was developed by Xing et al. [10]. They found that an optimal distribution is influenced by the voltage. They suggested that understanding these dependencies can ultimately help in the reduction of the Pt loading. Herden et al. [11], [12] investigated the effect of varying the ionomer equivalent weight in the in-plane direction on the performance of an automotive cell. The measurements were carried out on a segmented automotive cell, where the current density and temperature distributions were recorded. To do so, a membrane electrode assembly (MEA) with 772 and 825 g/mol ionomer equivalent weight was assembled in a segmented automotive cell, where the temperature and current density distributions were recorded. The performance of this MEA was then compared to two MEAs with homogeneous ionomer equivalent weights (one with 772 g $mol^{-1}$ and another with 825 g $mol^{-1}$). They showed that the structuring of the cathode catalyst layer with varying ionomer equivalent weight is important for water management within the cathode. Similarly, the through-plane variation of the ionomer equivalent weight was investigated by Shahgaldi et al. [13], [14] using in-house produced catalyst coated membranes (CCMs). In their work, they showed that the catalyst layer performance can be enhanced by the systematic design of the layer. By choosing the proper ionomer/Pt-gradient, the morphological and microstructural characteristics of the catalyst layer can lead to a reduced ionic resistance with improved mass transport capability, catalyst activity and Pt utilization. In another publication, the same authors [15] discussed the impact of the manufacturing process of the CCM, namely using the decal method, on the performance of the cell. The effect of the manufacturing process on catalyst coated membranes produced in the lab was also investigated in [16]. Sassin et al. [16] examined various production parameters using direct deposition of the catalyst layer on the membrane via ultrasonic spray deposition. They used this parametric analysis to investigate how production of the catalyst layer can ultimately affect the performance of the cell. In the work by Yu et al. [17], the use of reactive spray deposition would be beneficial in the production of low Pt loaded catalyst layers. They investigated the effect of I/C ratio in the catalyst layer on the performance of layers directly deposited on the membrane. Ex situ analyses, such as mercury porosimetry and Nitrogen adsorption, were used to investigate the pore distribution of the layers investigated. They found that regardless of Pt loading, the Brunauer–Emmett–Teller (BET) surface area and pore volume of the layers decreased with the increase in I/C ratio. They suggested that based on the results of their work, an enhanced performance at a low I/C ratio, the use of reactive spray deposition would be beneficial in the production of low Pt loaded catalyst layers.

Other works have focused on understanding the effect of a single ingredient on the performance of the cell. The use of short side chain (SSC) ionomer in the cathode catalyst layer was also shown to improve the performance of the cell in [18], especially at dry conditions. The benefit of using a SSC ionomer under dry conditions was shown to extend to operation at below zero temperatures [19]. In [19], the authors investigated the effect of side chain length on the durability of the catalyst layer subjected to freeze–thaw cycles between 30 and −40 °C They compared the degradation mechanism that occurs due to such temperature cycling and found that degradation in the presence of a long side chain ionomer is initiated by ionomer swelling and pore expansion, and then proceeds mainly through pore expansion. Whereas the degradation of catalyst layers with SSC ionomer was initiated by ionomer swelling and pore expansion, and proceeded through the detachment of large-scale CL flakes, and morphology and microstructure changes thereafter. Further, Shukla et al. [20] investigated cell performance, Tafel slope, reaction order and local oxygen transport resistance in order to obtain a relationship between Pt loading and performance. Pt loading is an important factor in determining the catalyst layer activity and hence, overall performance of the cell.

Although much work has been carried out to investigate the performance of the cell dependent on the Pt loading or ionomer content/type, most of the work focuses on the variation of either parameter rather than the relationship between the various ingredients. This is also often carried out under a very limited range of variation. Hence, in this work, we produce a full parametric study of the three parameters: ionomer content (ultimately the ionomer to carbon ratio), platinum loading and Pt/C ratio (40 wt% and 60 wt% platinum on carbon). Using the decal transfer method, cathode catalyst coated membranes are produced in-house with Pt loadings from 0.05–0.8 mg cm$^{-2}$ and an I/C ratio of 0.29–1.67. The data collected allow for full understanding of the loss mechanisms for different I/C ratios, dependent on the platinum loadings and Pt/C ratio, and can ultimately be used for further analysis of gradient catalyst layers and as a data base for catalyst layer modeling and optimization.

## 2. Experimental

### 2.1. Production of Catalyst Coated Membranes

A homogeneous suspension composed of platinum on carbon (40 wt.% Pt/C and 60 wt% Pt/C), Aquivion® (D79-25BS, liquid dispersion, 25% in water, PFSA eq. wt. 790 g.mol$^{-1}$ $SO_3H$, stabilized $CF_3$ polymer chain ends, Sigma-Aldrich Chemie GmbH, Schnelldorf, Germany) and a mixture of organic solvents (50 Vol.% ethylene glycol, 50 Vol.% propylene glycol methyl ether) was prepared. The suspension was homogenized by stirring. Suspensions with varying ionomer content were prepared with 15, 20, 25, 30, 35, 40, 45 and 50 wt.% ionomer content in dry layers. For both types of catalyst, this resulted in an I/C ratio in the range of 0.29–1.67. Catalyst layers with various platinum loadings were prepared by screen printing several layers on top of each other with subsequent drying at 110 °C. All catalyst layers were transferred onto a Gore membrane M735.18 containing an anode catalyst layer with a Pt loading of 0.05 mg cm$^{-2}$. The transfer was carried out at a compression of 5 MPa (referred to the printed catalyst area of 20 cm$^2$) and 180 °C for 15 min, producing CCMs with an active area of 12 cm$^2$. As a gas diffusion layer on both the cathode and anode sides, a H23C9 GDL from Freudenberg was used.

The break-in procedure was accomplished by operating the cell at 80 °C and fully humidified gases ($H_2$/Air at anode/cathode). The cell was operated for 1 h at 1.5 A/cm$^2$ before it was cycled for 4 hours between open circuit voltage (OCV) (10 s), 0.6 V (60 s) and 0.4 V (60 s).

### 2.2. In situ Analysis

In this study, the different transport properties and overall performance of catalyst layers, with varying ionomer content, catalyst loading and 2 types of catalyst, were studied in situ using several characterization techniques. The CCMs were assembled in a Baltic FuelCells quickConnect® test cell "high amp" (in Schwerin, Germany) having straight channels and an active area of $3 \times 4$ cm$^2$. The cell was compressed with 5 bara (absolute pressure), corresponding to a compression pressure of 0.8 MPa on the active cell area. All experiments were run at a cell temperature of 80 °C and with various humidification levels using bubbler humidifiers.

In this work, the catalyst layers were characterized using the following measurement protocol:

1. Polarization curves—the polarization curves were recorded at 80 °C cell temperature with fully humidified gases on the anode and cathode (i.e., 80 °C dew point) and an operating pressure of 2.0 bara. A constant flow rate was used with 2.0 nL/min $H_2$ (norm liter per minute) on the anode and 5.0 nL/min air on the cathode. The polarization curves were recorded in potentiostatic mode from 0.2 V to OCV with 0.05 V increments from 0.20–0.75 V and 0.02 V increments from 0.78 - OCV. The holding times at each potential were potential dependent ($U > 0.90$ V: 30 s, $0.70 \text{ V} < U \leq 0.90$ V: 60 s, and $U \leq 0.70$ V: 5 min). A recovery procedure was also followed, where PtO was reduced at a voltage of 0.40 V for 5 s for load points above 0.75.
2. Cyclic voltammetry—the cyclic voltammetry was run between 0.05 and 0.60 V at 100 mV/s with a total of 5 cycles. For the ECSA calculation, an average of the last three cycles is taken. The operating temperature was 80 °C with fully humidified gases and at 1.0 bara.
3. Humidity Sweeps—the humidity sweeps were measured at 80 °C cell temperature and 1.5 bara pressure. The cell was operated at a constant load of 1 A/cm$^2$ with fully humidified gases on the anode while the relative gas humidity on the cathode was varied between 20 % and 120 %. The voltage change was measured after each humidity step was conditioned for 7 min.

## 3. Results and Discussion

### 3.1. Ex Situ Analysis

In order to obtain the desired Pt loading, the catalyst layers were produced by printing various layers on top of each other. It is, hence, important to understand the interaction between these layers and how such a production step affects the interface between them. To do so, SEM images of three catalyst layers were taken and analyzed, as shown in Figure 1. As it can be seen from Figure 1a–c, there are no obvious interfaces visible that would indicate that interfacial effects might influence the performance of the layers. In Figure 1d–f, EDX distributions of carbon, fluorine and platinum are shown, where it can be seen that the ionomer, catalyst and support are homogeneously distributed throughout the different layers, regardless of the number of layers used for the production of the catalyst layer.

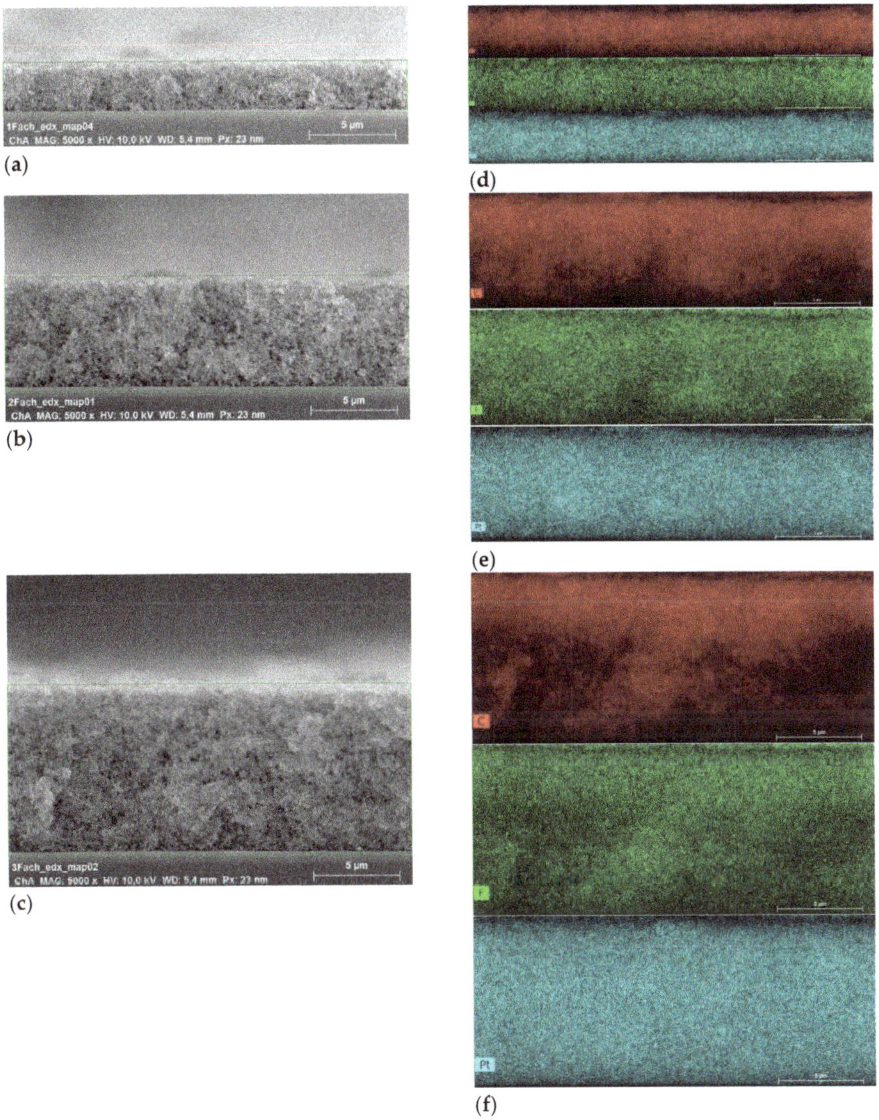


**Figure 1.** SEM images of three catalyst layers with an ionomer content of 30 wt.% and a platinum loading of (**a**) 0.08 mg $cm^{-2}$ (1 Layer), (**b**) 0.180 mg $cm^{-2}$ (2 Layers) and (**c**) 0.271 mg $cm^{-2}$ (3 Layers) with the corresponding carbon, fluorine and platinum distributions given in (**d**) 1 layer, (**e**) 2 layers and (**f**) 3 layers: red corresponds to carbon, green corresponds to fluorine and turquoise to platinum.

The dependency of the thickness on Pt loading and ionomer content was measured using SEM. In Figure 2, the thickness change with Pt loading for catalyst layers produced with 35 wt.% ionomer is given. As it can be seen, the thickness changes linearly with the change of Pt loading, while the ionomer content has no effect on the overall thickness of the catalyst layer. This implies constant porosity throughout the whole thickness of the produced electrodes.

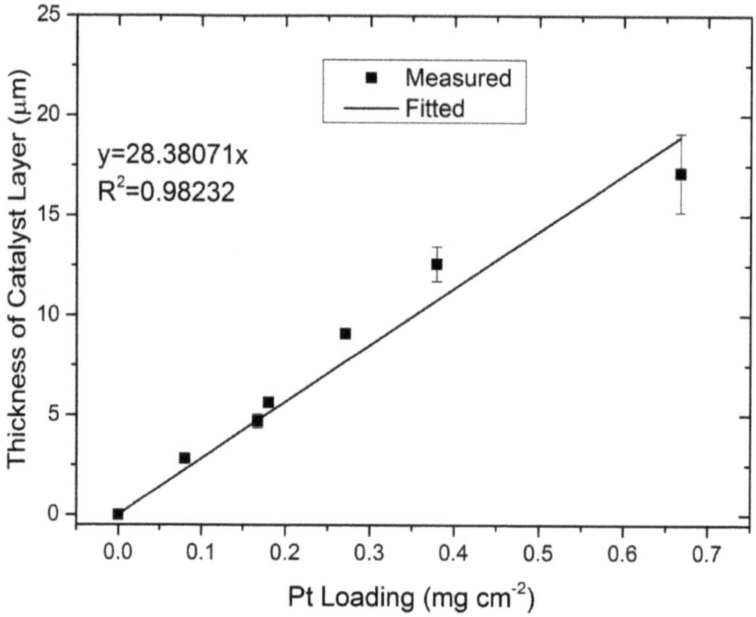


**(a).**

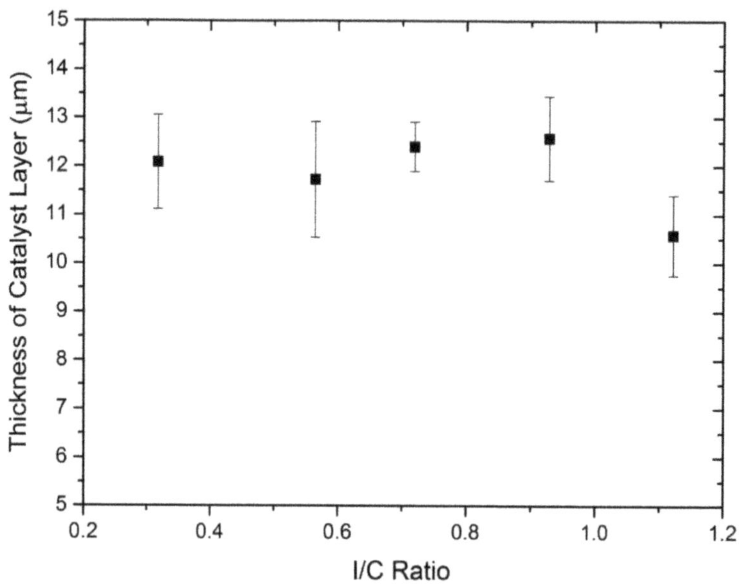


**(b)**

**Figure 2.** Dependence of catalyst layer thickness for the 40 wt% catalyst. (**a**) Measured thickness with an ionomer content of 35 wt.% dependent on the platinum loading, (**b**) effect of ionomer content on the thickness of the catalyst layer with Pt loading of 0.40 mg/cm$^2$.

### 3.2. Reproducibility of Produced CCMs

Prior to investigating the effects of the catalyst layer ingredients on the performance of the cell, it is important to establish the reproducibility of the production and characterization method of the CCMs. In Figure 3a,b, the polarization curves under the wet and dry conditions of various CCMs with different Pt loadings and I/C ratios are given. Figure 3c,d shows the corresponding high frequency resistance (HFR) measurements during the polarization curves. As it can be seen, the production method used in this study results in reproducible catalyst layers and characterization results.

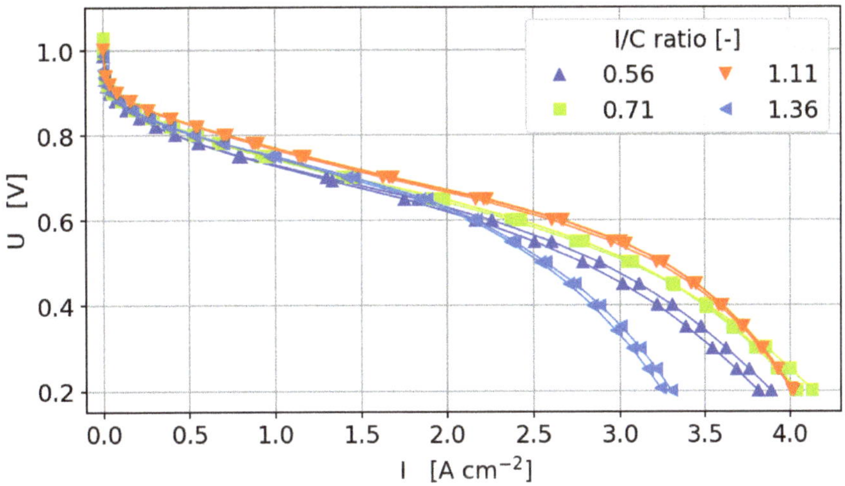


(a)

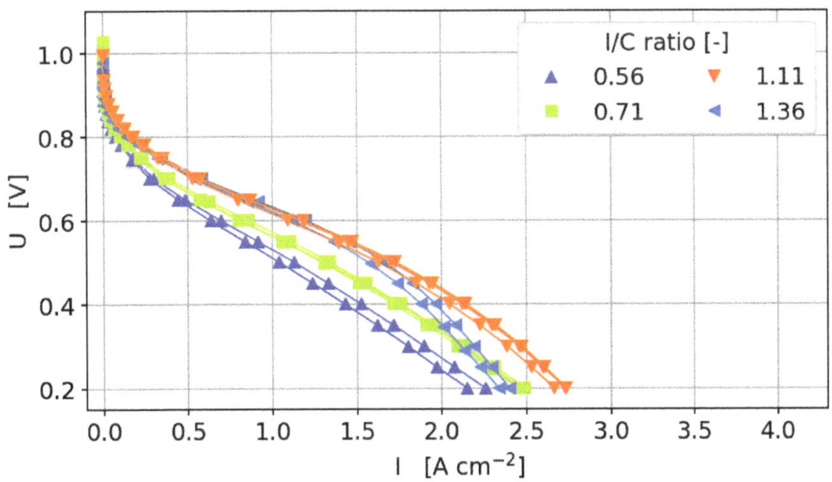


(b)

**Figure 3.** *Cont.*

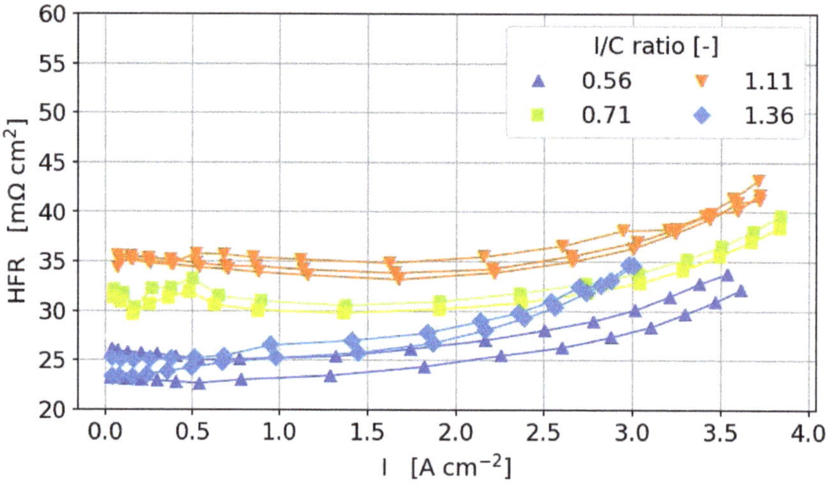


(c)

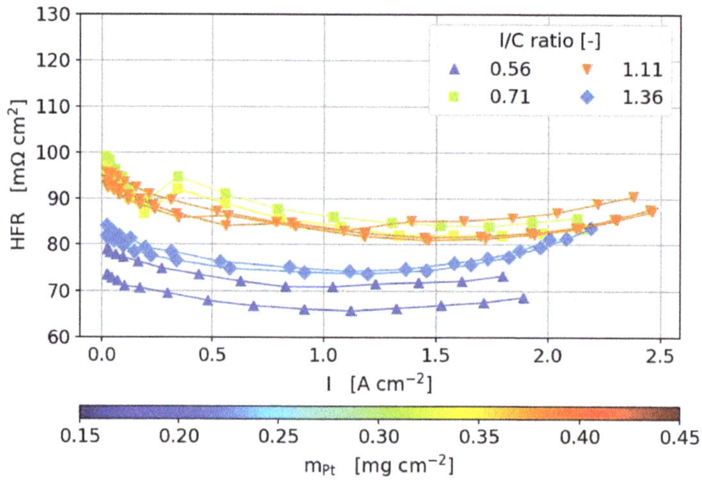


(d)

**Figure 3.** Polarization curves with varying Pt loading and I/C ratio under (**a**) wet conditions, (**b**) dry conditions to illustrate the reproducibility of the production method and characterization technique with their corresponding high frequency resistance under (**c**) wet conditions and (**d**) dry conditions.

### 3.3. *In situ Analysis*

The produced catalyst layers were characterized using the different in situ characterization techniques discussed earlier. In this work, we provide a full data set to understand the effects of various catalyst layer compositions.

#### 3.3.1. Cyclic Voltammetry

The cyclic voltammetry was used in this study in order to estimate the electrochemical surface area (ECSA) for all produced catalyst layers. As shown in Figure 4, the ECSA is plotted against the Pt loading for various I/C ratios. The Pt loading was determined by weighing the catalyst layers after drying on the decal foil with the assumption that all solvent evaporates completely during drying. The ECSA was obtained by integrating the hydrogen adsorption current in the cyclic voltammograms until a fixed potential limit of 90 mV.

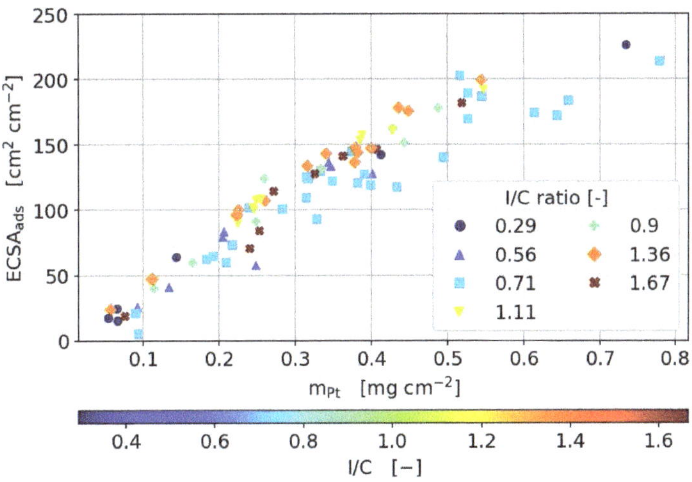


**Figure 4. Electrochemical active surface area**, ECSA, of the different catalyst coated membranes, CCMs, produced with 40wt% Pt/C, dependent on platinum loading and ionomer content.

As it can be seen, a linear dependency of the ECSA on Pt loading is measured for all I/C ratios. This linear dependency implies, (i) the calculation of the Pt loading is adequate at each respective ionomer content and (ii) proper electrical and protonic contact is established for each CL investigated, regardless of how many layers are printed to obtain the desired Pt loading.

### 3.3.2. Polarization Curves

As mentioned earlier, the performance of the produced CCMs was characterized using two polarization curves, a dry curve (dew point 59 °C) and a fully humidified curve (dew point 80 °C). Figure 3a shows the fully humidified polarization curves and HFR data obtained with an I/C ratio of 0.56–1.36 with varying platinum loadings. The HFR is in a low and narrow range between 25 and 35 mOhm cm$^2$, showing that the used production and characterization procedure is stable, and influences by variations of membrane properties can be neglected. This was also found for the other ionomer contents analyzed (data not shown). The influence of the platinum on the current production varies within the different regimes of the polarization curve. In the activation region, there is a clear trend with higher performance for higher catalyst loadings, whereas at higher currents the highest performance is not reached with the highest catalyst loading.

For better visualization of the performance of all 80+ cells discussed in this work, we chose to focus on three regions in the polarization curve as shown in Figures 5 and 6. The three discrete points are extracted from the fully humidified polarization curves: From the activation region (900 mV, Figure 5a), the intermediate current density region (700 mV, Figure 5b) and the mass transport dominated region (300 mV, Figure 5c). Figure 5 shows the current density dependent on I/C ratio (indicated by color and marker type) and platinum loading for the 40 wt% catalyst. Figure 6 shows the same dependencies for the 60 wt% catalyst. In the following section, the general dependencies of both of the 40% catalysts will be discussed first, while the differences to the 60 wt% catalyst will be discussed later in this section.

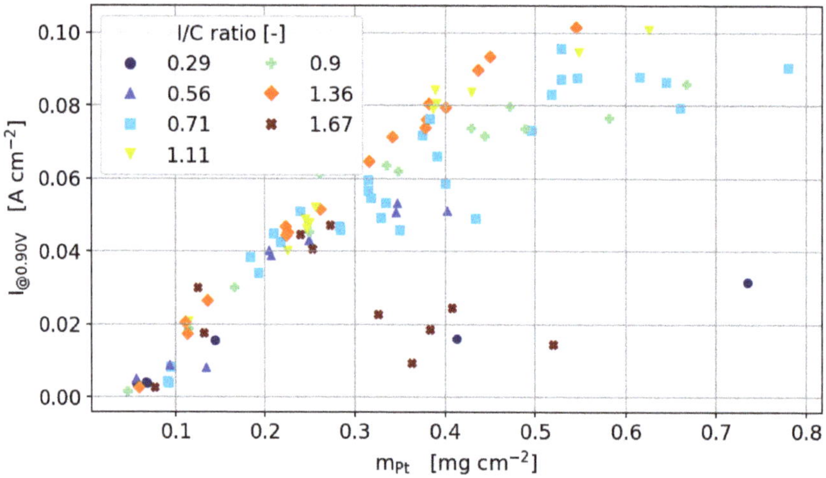


(a)

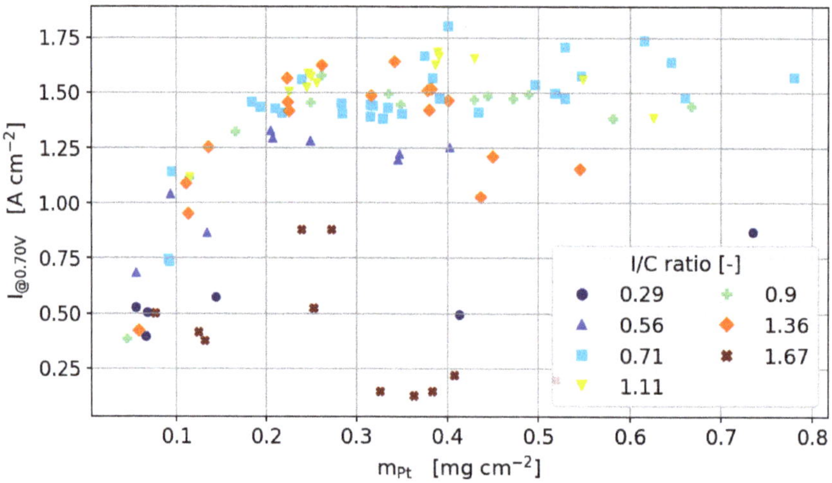


(b)

**Figure 5.** *Cont.*

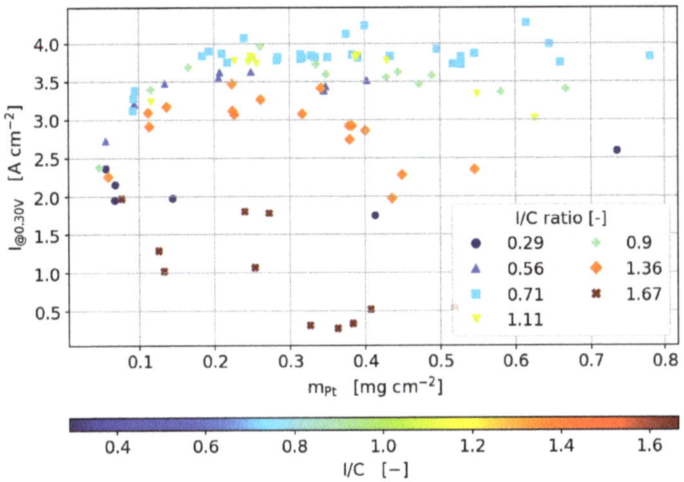


(c)

**Figure 5.** The current density dependency on the Pt loading and I/C ratio at cell voltages of (**a**) 900 mV, (**b**) 700 mV and (**c**) 300 mV for a catalyst of 40 wt% Pt/C.

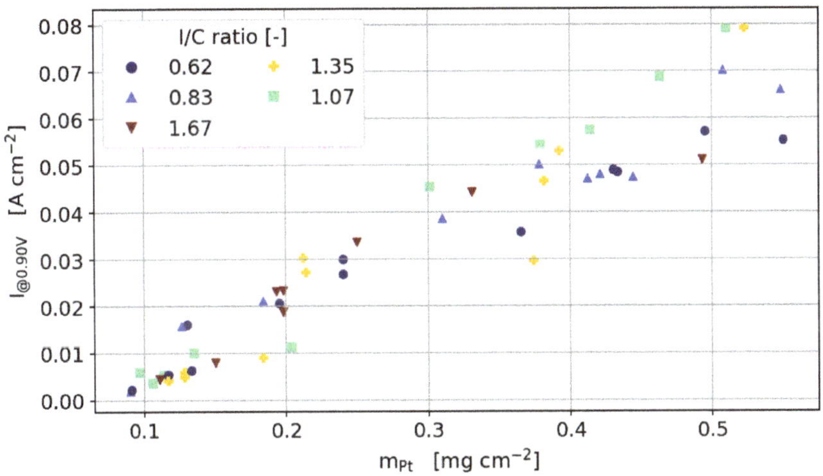


(a)

**Figure 6.** *Cont.*

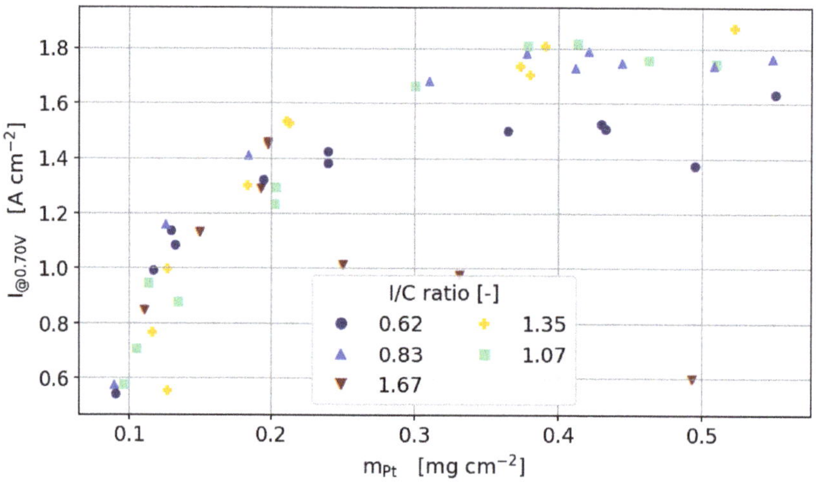


(b)

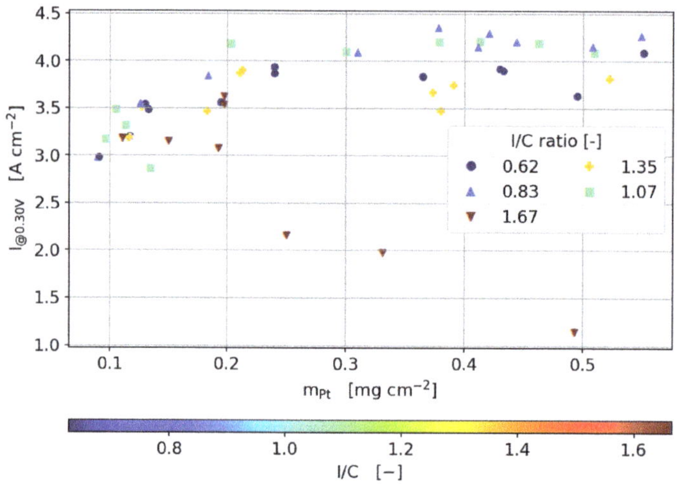


(c)

**Figure 6.** The current density dependency on the Pt loading and ionomer content at cell voltages of (**a**) 900 mV, (**b**) 700 mV and (**c**) 300 mV for a catalyst of 60 wt% Pt/C.

### 3.3.3. 40 wt% Catalyst

In the activation region in Figure 5a, a linear trend is measured for the dependency of the current density on the Pt loading, implying that the oxygen and proton transport effect on the performance is negligible due to the low reaction rates. However, for the catalyst layers with the highest and lowest ionomer content, the voltage losses due to proton transport and oxygen diffusion, respectively, are found to be high enough for a significantly lower current density for thick catalyst layers. The linear dependency measured for I/C ratios of 0.62–1.36 confirms the previous finding that, independent of the number of layers printed on top of each other, the whole catalyst layer thickness is electrochemically active and the layers are connected to both the protonic and electronic phases.

At platinum loadings higher than 0.3 mg $cm^{-2}$, minor differences between the ionomer contents emerge, with an optimum I/C ratio of about 0.71–1.36. The fact that the optimum is found at a relatively high ionomer content indicates that high proton conductivity is more important than oxygen transport at the very low current densities in the activation region. For lower platinum loadings, the performance is rather independent of ionomer content.

At intermediate and high current densities (Figure 5b,c), mass transport and ohmic losses throughout the catalyst layer thickness become more pronounced due to the higher proton flux and oxygen consumption. Since the catalyst layer thickness varies proportionally with the platinum loading, both losses become more relevant for higher platinum loadings. At the higher current densities at 700 mV in Figure 5b and 300 mV in Figure 5c, the current density increases with platinum loading only up to a threshold value. At 700 mV, this threshold is around 0.3 mg $cm^{-2}$ for all ionomer contents. At high ionomer loadings, even a decreasing performance can be observed for increasing platinum loading. Above 0.3 mg $cm^{-2}$, adding further electrochemically active material does not result in a performance increase anymore because diffusive losses become dominant over the performance increase by the increasing the electrochemical active area.

For the low ionomer contents, adding additional material simply results in more inactive material and the current density remains unchanged for higher platinum loadings. This is only possible since for the low ionomer contents, the oxygen diffusion is still high enough to supply the catalyst close to the membrane. Otherwise, adding more material would decrease the cell current by higher diffusion losses due to the increase in the layer's thickness with increased Pt loading.

For I/C ratios higher than 1.07, higher platinum loading results in a performance decrease and a more pronounced platinum loading optimum. Upon the increase in ionomer content over the optimum, the diffusion through the thickness of the catalyst layer seems to limit the cell performance. With increasing catalyst layer thicknesses, (or Pt loading), the oxygen diffusion to the mainly active regions close to the membrane is hindered, which results in moving the reaction zone away from the membrane towards the interface to the GDL. The fact that the current is even decreasing with increasing the platinum loading indicates that despite the high ionomer contents, the proton conductivity is not high enough to enable a proton transport to the reaction zone close to the GDL without significant losses. Increasing the thickness with higher platinum loading, therefore results in ohmic losses that overcompensate the effects by the higher platinum loading.

In Figure 5c, the cell current at 300 mV also shows a distinct optimum for the platinum loading that is also dependent on the ionomer content. Here, differences between the ionomer contents are more pronounced with the optimum at 300 mV around 0.2 mg Pt $cm^{-2}$ compared to 0.3 mg Pt $cm^{-2}$ at 700 mV.

In the fuel cell literature, the discussion on whether the oxygen diffusion to the active sites in the catalyst layer is limited by diffusion through the ionomer film that covers the catalyst particles or by the through-plane diffusion in the pore space of the catalyst layer, is ongoing [21]. Analyzing the findings in the previous section, the results strongly support the hypothesis that the losses are dominated by the through-plane diffusion rather than the film diffusion. Adding thickness to the layer results in decreasing performance for all layers, especially for the higher ionomer contents. This would not be the case if thin-film diffusion would be mainly limiting. If the proton conductivity throughout the layer would prevent the performance to increase, this effect would be more pronounced for lower rather than higher ionomer loadings.

Another indication that limitation by through-plane diffusion rather than its thin-film counterpart is observed through the analysis of catalyst layers with low ionomer contents at low layer thicknesses (low Pt loading). The performance difference between the different ionomer loadings seems to diminish below catalyst loadings smaller than 0.2 $mg_{Pt}$ $cm^{-2}$, resulting in a performance that is independent of ionomer loading for very thin catalyst layers. If the thin film diffusion would limit the cell performance, the differences between ionomer contents would be independent of catalyst layer thickness and also be present for the thin catalyst layers at low platinum loadings.

3.3.4. Humidity Sweeps

In order to investigate the effect of Pt loading and ionomer content on the overall water transport within the cell, relative humidity sweeps were carried out at a current density of 1.0 A $cm^{-2}$. Figure 7 shows the humidity sweeps dependent on the I/C ratio for a Pt loading of 0.25 mg $cm^{-2}$. It is obvious that the catalyst layers with low ionomer content suffer more from drying at low humidification than the catalyst layers with high ionomer content. To investigate this further, the cell voltage at 30% RH and 120% RH, dependent on catalyst loading, is plotted in Figure 8. For the dry conditions, the optimum I/C ratio is found to be at 1.36, while the optimum I/C ratio is found to be in the range of 0.62–1.36 for the wet conditions. Further, it can be seen that an ionomer content above 1.36 might result in flooding of the cell at high humidity levels.

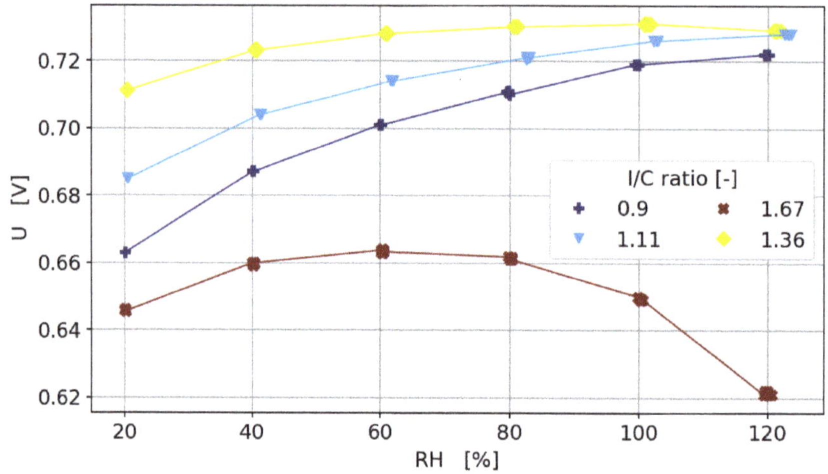


**Figure 7.** Relative humidity sweeps dependent on I/C ratio for Pt/C 40wt.% and a Pt loading of 0.25 mg $cm^{-2}$ at a current density of 1.0 A $cm^{-2}$.

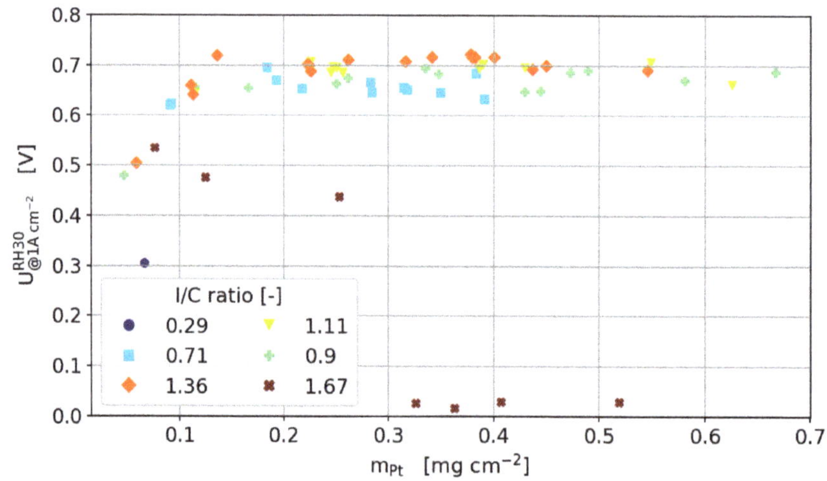


(a)

**Figure 8.** *Cont.*

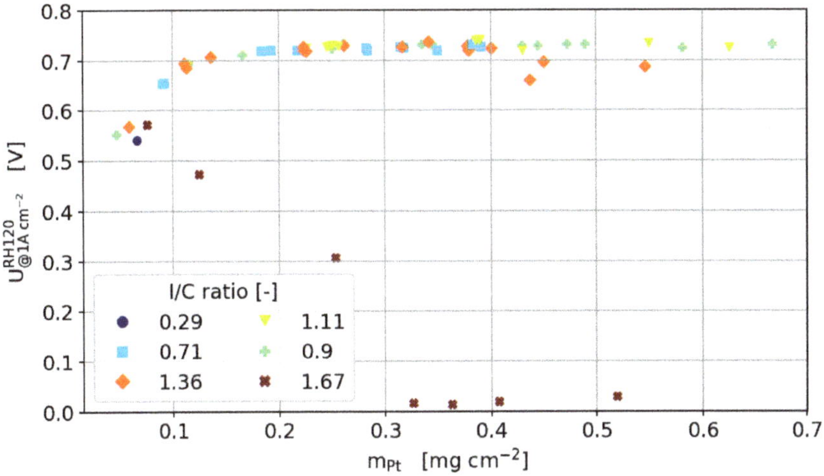


(b)

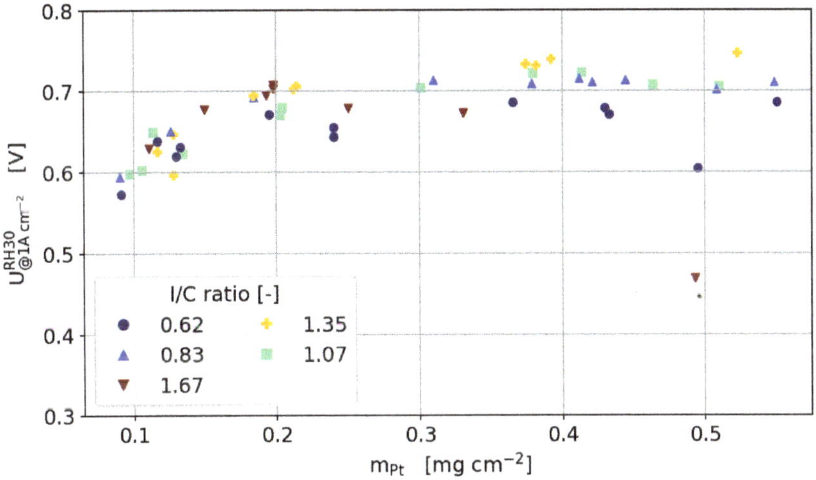


(c)

**Figure 8.** *Cont.*

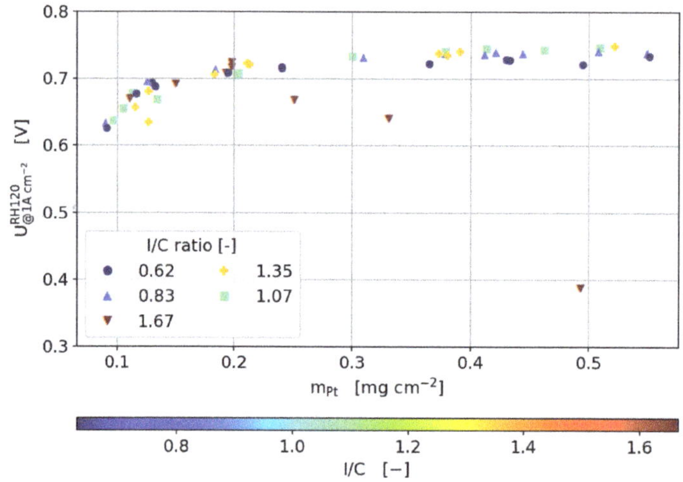


(d)

**Figure 8.** The cell voltage dependent on Pt loading and I/C ratio for a relative humidity of (**a**) 30% and (**b**) 120% for Pt/C of 40wt.% and a relative humidity of (**c**) 30% and (**d**) 120% for Pt/C of 60wt.%. All humidity sweeps are carried out at a current density of 1.0 A cm$^{-2}$.

### 3.3.5. 60 wt % Catalyst

The general dependencies for the optimum ionomer content of the 60 wt% catalyst in terms of ionomer to carbon ratio is the same as for the 40 wt% catalyst. As found by other groups before, the amount of ionomer needed for an optimum compromise between oxygen diffusion and proton transport depends strongly on the carbon surface rather than on the amount of platinum in the catalyst layer. This results in a significantly lower weight percentage of ionomer needed for the 60 wt% catalyst than for the 40 wt% catalyst at the same platinum loading. This is highlighted in Figure 8c,d, where it can be seen that the cells are more tolerant to low humidity.

When considering the change in performance with catalyst layer thickness (or platinum loading), the optimum performance of the 60 wt% catalyst is not as pronounced as that of the 40 wt % catalyst. This obviously results from the thickness difference between the two catalyst types. For the same platinum loading, the 60 wt% catalysts are about 33% thinner than the 40 wt% catalyst since mostly the amount of carbon defines the layer thickness. For higher platinum loadings, the diffusion limitation by the additional thickness is not as pronounced as for the 40 wt% catalyst.

This supports the conclusion of the previous section, where the through-plane diffusion was found to be the main source for the limitations at higher platinum and ionomer loadings. Since the maximum in power output is defined by the interaction between through-plane diffusion and proton conduction, this optimum is shifted towards higher platinum loadings for the thinner catalyst layers with the 60 wt% catalyst. At 300 mV cell voltage, the optimum performance is therefore reached at around 0.25 mg cm$^{-2}$ platinum loading for the 40 wt % catalyst and at around 0.3 mg cm$^{-2}$ for the 60 wt % catalyst.

## 4. Conclusions

A full parametric study of various catalyst layer compositions containing the variation of platinum loading, ionomer content and two types of catalyst material (40 and 60 wt.% Pt/C) was conducted. The following conclusions can be drawn from the work presented in this manuscript.

1. Electrical and protonic contact is established regardless of the printing of layers on topic of each other. This conclusion was established based on SEM images that illustrated a homogeneous distribution of the Pt, C and F. The linear dependency of the ECSA on Pt loading for various I/C ratios without any discontinuity in the slope once again proves the establishment of the electrical and protonic contact between all layers.
2. The optimum ionomer content is not only dependent on the catalyst layer thickness, but also on the load and gas humidity.
3. In the activation region, regardless of which type of catalyst is used, the dependency of the current density on Pt loading is linear as long as the ionomer content is not too low to hinder proton conductivity and is not too high to hinder oxygen diffusion significantly.
4. In the ohmic and mass transport region of the polarization curve, increasing the ionomer content to a value higher than the optimum results in a decrease in current density with higher Pt loading. The diffusion losses outweigh the improvement by the increasing catalytically active area.
5. At high Pt loading and ionomer content, through-plane diffusion losses become more limiting than the proton conductivity.
6. General trends are the same for 40 and 60 wt.% Pt/C catalysts, but the dependency on ionomer content is lower due to reduced catalyst layer thickness.

The impact of this work is in the comprehensive data that it provides. The plan is to use this data set as a basis for our ongoing efforts to investigate the effect of catalyst layer structuring, specifically graded catalyst layers, on its performance. This investigation will be carried out by additionally measuring limiting current density and impedance spectroscopy, and will be further enhanced by accelerated stress tests to understand the dependencies of durability and stability on the structuring of the layer.

**Author Contributions:** Conceptualization, R.A., P.S., K.C., R.S, and N.Z.; Methodology, R.A., P.S., N.Z. and R.S.; Data Curation, R.A., P.S., K.C., R.S., and J.S.; Writing-Original Draft Preparation, R.A., P.S., and N.Z.; Writing-Review & Editing, R.A., K.C., P.S., R.K., and N.Z.; Supervision, R.A. and N.Z.; Project Administration, N.Z. and R.K. All authors have read and agreed to the published version of the manuscript.

**Funding:** This research was funded by the Federal Ministry of Education and Research of Germany, BMBF, under the project "DEKADE" (03SF0544A).

**Conflicts of Interest:** The authors declare no conflict of interest.

## References

1. Zamel, N. The catalyst layer and its dimensionality—A look into its ingredients and how to characterize their effects. *J. Power Sources* **2016**, *309*, 141–159. [CrossRef]
2. Alvar, E.N.; Zhou, B.; Eichhorn, S.H. Composite-supported Pt catalyst and electrosprayed cathode catalyst layer for polymer electrolyte membrane fuel cell. *Int. J. Energy Res.* **2017**, *41*, 1626–1641. [CrossRef]
3. Chan, S.; Jankovic, J.; Susac, D.; Saha, M.S.; Tam, M.; Yang, H. Electrospun carbon nanofiber catalyst layers for polymer electrolyte membrane fuel cells: Structure and performance. *J. Power Sources* **2018**, *392*, 239–250. [CrossRef]
4. *Handbook of Fuel Cells—Fundamentals, Technology and Applications. Novel Catalysts, Catalysts Support and Catalysts Coated Membrane Methods*; Vielstich, W.; Lamm, A.; Gasteiger, H. (Eds.) John Wiley & Sons: Hoboken, NJ, USA, 2003; ISBN 978-0-471-49926-8.
5. O'Hayre, R.; Lee, S.-J.; Cha, S.-W.; Prinz, F. A sharp peak in the performance of sputtered platinum fuel cells at ultra-low platinum loading. *J. Power Sources* **2002**, *109*, 483–493. [CrossRef]
6. Saha, M.S.; Gullá, A.F.; Allen, R.J.; Mukerjee, S. High performance polymer electrolyte fuel cells with ultra-low Pt loading electrodes prepared by dual ion-beam assisted deposition. *Electrochim. Acta* **2006**, *51*, 4680–4692. [CrossRef]
7. Hatzell, K.B.; Dixit, M.B.; Berlinger, S.A.; Weber, A.Z. Understanding inks for porous-electrode formation. *J. Mater. Chem. A* **2017**, *5*, 20527–20533. [CrossRef]

8. Chen, G.-Y.; Wang, C.; Lei, Y.-J.; Zhang, J. Gradient design of Pt/C ratio and Nafion content in cathode catalyst layer of PEMFCs. *Int. J. Hydrog. Energy* **2017**, *42*, 29960–29965. [CrossRef]
9. Yoon, Y.-G.; Yang, T.-H.; Park, G.-G.; Lee, W.-Y.; Kim, C.-S. A multi-layer structured cathode for the PEMFC. *J. Power Sources* **2003**, *118*, 189–192. [CrossRef]
10. Xing, L.; Shi, W.; Das, P.K.; Scott, K. Inhomogeneous Distribution of Platinum and Ionomer in the Porous Cathode to Maximize the Performance of a PEM Fuel Cell. *AIChE J.* **2017**, *63*, 4895–4910. [CrossRef]
11. Herden, S.; Hirschfeld, J.A.; Lohri, C.; Perchthaler, M.; Haase, S. Ionomer equivalent weight structuring in the cathode catalyst layer of automotive fuel cells: Effect on performance, current density distribution and electrochemical impedance spectra. *J. Power Sources* **2017**, *364*, 449–457. [CrossRef]
12. Herden, S.; Riewald, F.; Hirschfeld, J.A.; Perchthaler, M. In-plane structuring of proton exchange membrane fuel cell cathodes: Effect of ionomer equivalent weight structuring on performance and current density distribution. *J. Power Sources* **2017**, *355*, 36–43. [CrossRef]
13. Shahgaldi, S.; Alaefour, I.; Li, X. The impact of short side chain ionomer on polymer electrolyte membrane fuel cell performance and durability. *Appl. Energy* **2018**, *217*, 295–302. [CrossRef]
14. Shahgaldia, S.; Ozdena, A.; Li, X.; Hamdullahpur, F. Cathode catalyst layer design with gradients of ionomer distribution for proton exchange membrane fuel cells. *Energy Convers. Manag.* **2018**, *171*, 1476–1486. [CrossRef]
15. Shahgaldi, S.; Alaefour, I.; Li, X.L. Impact of manufacturing processes on proton exchange membrane fuel cell performance. *Appl. Energy* **2018**, *225*, 1022–1032. [CrossRef]
16. Sassin, M.B.; Garsany, Y.; Gould, B.; Swider-Lyons, K.E. Fabrication Method for Laboratory-Scale High-Performance Membrane Electrode Assemblies for Fuel Cells. *Anal. Chem.* **2017**, *89*, 511–518. [CrossRef] [PubMed]
17. Yu, H.; Roller, J.M.; Mustain, W.E.; Maric, R. Influence of the ionomer/carbon ratio for low-Pt loading catalyst layer prepared by reactive spray deposition technology. *J. Power Sources* **2015**, *283*, 84–94. [CrossRef]
18. Garsany, Y.; Atkinson, R.W., III; Sassin, M.B.; Hjelm, R.M.E. Improving PEMFC Performance Using Short-Side-Chain Low-Equivalent-Weight PFSA Ionomer in the Cathode Catalyst Layer. *J. Electrochem. Soc.* **2018**, *165*, F381–F391. [CrossRef]
19. Ozdena, A.; Shahgaldi, S.; Li, X.; Hamdullahpur, F. The impact of ionomer type on the morphological and microstructural degradations of proton exchange membrane fuel cell electrodes under freeze-thaw cycles. *Appl. Energy* **2019**, *238*, 1048–1059. [CrossRef]
20. Shukla, S.; Stanier, D.; Saha, M.S.; Stumper, J.; Secanell, M. Analysis of Inkjet Printed PEFC Electrodes with Varying Platinum Loading. *J. Electrochem. Soc.* **2016**, *163*, F677–F687. [CrossRef]
21. Salari, S.; Tam, M.; McCague, C.; Stumper, J.; Bahrami, M. The ex-situ and in-situ gas diffusivities of polymer electrolyte membrane fuel cell catalyst layer and contribution of primary pores, secondary pores, ionomer and water to the total oxygen diffusion resistance. *J. Power Sources* **2019**, *449*, 227479. [CrossRef]

**Sample Availability:** Not available.

molecules

MDPI

*Article*

# Development of Porous Pt Electrocatalysts for Oxygen Reduction and Evolution Reactions

**Marika Muto [1], Mayumi Nagayama [2], Kazunari Sasaki [1,2,3,4] and Akari Hayashi [1,2,3,4,5,*]**

1 Department of Hydrogen Energy Systems, Kyushu University, 744 Motooka, Nishi-ku, Fukuoka 819-0395, Japan; mmjb004@gmail.com (M.M.); sasaki@mech.kyushu-u.ac.jp (K.S.)
2 Coevolutionary Research for Sustainable Communities (COI-C2RSC), Kyushu University, 744 Motooka, Nishi-ku, Fukuoka 819-0395, Japan; nagayama.mayumi.630@m.kyushu-u.ac.jp
3 International Research Center for Hydrogen Energy, Kyushu University, 744 Motooka, Nishi-ku, Fukuoka 819-0395, Japan
4 NEXT-FC, Kyushu University, 744 Motooka, Nishi-ku, Fukuoka 819-0395, Japan
5 Q-PIT, Kyushu University, 744 Motooka, Nishi-ku, Fukuoka 819-0395, Japan
* Correspondence: hayashi.akari.500@m.kyushu-u.ac.jp

Academic Editor: Jean St-Pierre
Received: 26 March 2020; Accepted: 18 May 2020; Published: 21 May 2020

check for updates

**Abstract:** Porous Pt electrocatalysts have been developed as an example of carbon-free porous metal catalysts in anticipation of polymer electrolyte membrane (PEM) fuel cells and PEM water electrolyzers through the assembly of the metal precursor and surfactant. In this study, porous Pt was structurally evaluated and found to have a porous structure composed of connected Pt particles. The resulting specific electrochemical surface area (ECSA) of porous Pt was 12.4 $m^2\ g^{-1}$, which was higher than that of commercially available Pt black. Accordingly, porous Pt showed higher oxygen reduction reaction (ORR) and oxygen evolution reaction (OER) activity than Pt black. When the activity was compared to that of a common carbon-supported electrocatalyst, Pt/ketjen black (KB), porous Pt showed a comparable ORR current density (2.5 mA $cm^{-2}$ at 0.9 V for Pt/KB and 2.1 mA $cm^{-2}$ at 0.9 V for porous Pt), and OER current density (6.8 mA $cm^{-2}$ at 1.8 V for Pt/KB and 7.0 mA $cm^{-1}$ at 1.8 V), even though the ECSA of porous Pt was only one-sixth that of Pt/KB. Moreover, it exhibited a higher durability against 1.8 V. In addition, when catalyst layers were spray-printed on the Nafion® membrane, porous Pt displayed more uniform layers in comparison to Pt black, showing an advantage in its usage as a thin layer.

**Keywords:** oxygen reduction; oxygen evolution; PEM fuel cell; PEM water electorolyzer; durability; porous structure; carbon-free

## 1. Introduction

Electrocatalysts composed of noble metal nanoparticles dispersed on carbon supports are commonly used in polymer electrolyte membrane (PEM) fuel cells and PEM water electrolyzers. Such dispersion of metal nanoparticles leads to high catalytic activity, even with small amounts of noble metals. However, carbon supports are sometimes not usable. For example, in the case of PEM water electrolyzers, carbon supports are highly corroded under high potential anodic conditions (over 1.8 V), which are required to obtain a practical current density of 1 A $cm^{-2}$ [1,2]. Even in the PEM fuel cell system, carbon oxidation at the cathode is serious at a locally elevated potential (over 1.5 V) during the start/stop cycling of fuel cell vehicles [3,4]. Additionally, even at the anode catalyst, carbon corrosion has been reported in the situation of fuel starvation, where cell reversal occurs and the anode potential is over 1.5 V [5]. Not only are the carbon supports themselves damaged, but the metal nanoparticles also lose their support, leading to the agglomeration of particles and loss of their catalytic activity [6].

Even though problems related to the durability of carbon supports have been reported, increasing the electroactive surface area without a carbon support is rather challenging. A large quantity of novel metal catalysts with a low surface area should generally be employed without carbon supports, which increases their cost [7]. For that reason, the improvement of electrocatalysts without carbon supports has been extensively studied. In some cases, metal oxide supports have been used, but this still results in a low electronic conductivity in comparison to carbon, even though durability is high [8–14]. For example, although the oxygen reduction reaction (ORR) mass activity of Pt/doped tin oxide at 0.85 V has been reported to be around 600 A $g^{-1}$, which is comparable to that of Pt-deposited carbon, under half-cell measurements in solution the ORR activity at 0.9 V, which is common for standard comparisons, is expected to be low [10]. In other cases, catalysts made of only metal nanoparticles have been studied owing to their high surface area [15–18]. Although nanoparticles show high catalytic activity as powder, they cannot usually maintain their structure in actual devices and mostly result in low performance. However, when these nanoparticles are connected to each other, they can maintain their porous structure, even in actual devices, and a current–voltage performance comparable to that of PEM fuel cells exhibited by Pt supported on carbon at 0.8 V, with a value of 200 mA $cm^{-1}$, has been reported [17]. Among many carbon support-free catalysts, one of the most successful catalysts is 3M's nanostructured thin film (NSTF) catalyst, which is a pure, organic molecular solid in the form of a crystalline whisker coated by metal catalysts. The NSTF catalyst has shown 2–3 times higher mass activity at 0.9 V, with a value of over 600 A $g^{-1}$, and a higher durability than Pt supported on a carbon support [19].

We are rather interested in developing metal-only catalysts and improving their structure as devices. In our prior studies, mesoporous carbon supports were developed through a simple one-pot reaction that involved heating a mixture of a surfactant and carbon precursors for PEM fuel cells, and high performance and durability were achieved in accordance with their mesoporous structure [20–25]. The porous structure was also sustained, even after being built into a device. We are now trying to apply the concept of mesoporous structures to metal-only electrocatalysts with the aim of increasing the surface area and reducing the mass transfer loss of water and gases in the PEM fuel cell and water electrolysis system based on the porous structure. In addition, conductivities superior to those of metal oxide supports are expected for all metallic compositions.

In this paper, porous Pt has been synthesized as one example of a porous metal catalyst by employing a Pt precursor and a surfactant. Step-by-step heat treatments were applied to reduce the Pt precursor and remove the surfactant. Resulting carbon-free porous Pt was structurally evaluated, and its electrochemical activity toward the ORR and oxygen evaluation reaction (OER) was studied in detail. Additionally, a catalyst layer of porous Pt was prepared using a spray printing method and its structure was evaluated.

## 2. Results and Discussion

### 2.1. Heat Treatment and Characterization of Porous Pt

The heat treatment conditions for porous Pt were controlled in order to remove the residual derived from the surfactant and metal precursor. In the case where the sample was calcined at 210 °C for 3 h and 240 °C for 3 h for the thermal reduction of $Pt(acac)_2$ and also heat-treated at 300 °C for 1 h for decomposition of the surfactant under a nitrogen atmosphere, a sticky product was obtained owing to the residual surfactant. Therefore, the heat treatment condition for decomposing the surfactant was altered to 400 °C for 3 h. As a result, a dry powder was obtained. However, based on thermogravimetric (TG) analysis, a weight loss of 62.3% was further observed at around 200 °C (see Supplementary Materials Figure S1a), indicating that the residual surfactant remained in this condition. In order to further remove the residual surfactant completely, additional heat treatment at 200 °C under humidified nitrogen was conducted. Consequently, the weight loss was minimized to 3.6% after calcining for 10 min with this condition (see Figure S1b), resulting in successful removal of the surfactant.

Porous Pt was characterized by nitrogen sorption in comparison to commercial Pt black. Figure 1a,b show nitrogen adsorption/desorption isotherms and corresponding pore size distributions, respectively. The Brunaure–Emmett–Teller (BET) specific surface area of porous Pt was 32.4 $m^2\ g^{-1}$, whereas that of commercially available Pt black was 14.3 $m^2\ g^{-1}$, which was slightly low compared to the value of 25 $m^2\ g^{-1}$ or less given by Sigma-Aldrich (St. Louis, MO, USA). Porous Pt showed mesopores consisting of less than 100 nm in diameter and micropores, while Pt black mostly exhibited small mesopores of less than 30 nm, with relatively fewer micropores. Since Pt black consists of nanoparticles not connected to each other and does not contain pores within the particles, the size of 30 nm reveals the particle size and interparticle pore size achieved by these nanoparticles.

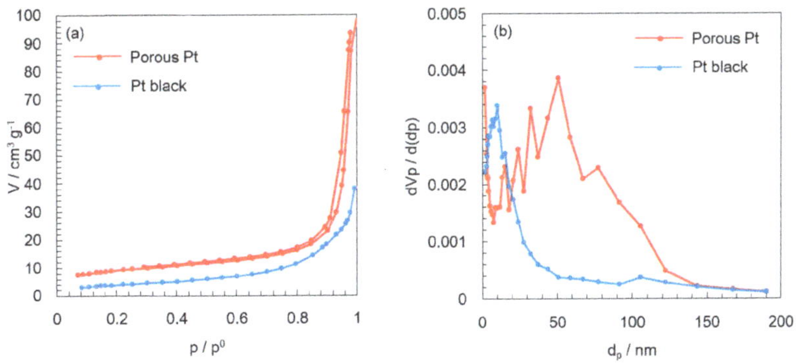


**Figure 1.** (**a**) Nitrogen adsorption/desorption isotherms and (**b**) corresponding pore distributions of porous Pt and Pt black.

In order to view the nanostructure of porous Pt and Pt black, SEM images were observed and are shown in Figure 2a,b, respectively. As seen in Figure 2a, pores consisting of connected Pt particles were confirmed. Even though porous Pt does not have a common ordered porous structure like that of zeolite, it shows pores, displaying a "porous structure" required for device application [17]. The domain size of this porous structure was estimated to range from 500 nm to 2 μm based on SEM images with a low magnification. As seen in SEM images with a higher magnification (Figure S2), the size of the average particles in porous Pt was ca. 20 nm, while Pt black showed much smaller primary particles, which were less than 10 nm. However, in the case of Pt black, much denser aggregates of such Pt particles were observed, as shown in Figure 2b.

**Figure 2.** SEM images of (**a**) porous Pt and (**b**) Pt black.

Pt 4f XPS spectra were recorded to evaluate the surface chemical states. Both porous Pt and Pt black had two peaks at 71.1–71.3 and 74.4–74.6 eV of binding energy, corresponding to $Pt4f_{7/2}$ and $Pt4f_{5/2}$, respectively. Those two peaks indicated that both materials mostly contained the metallic Pt (0) surface [26]. Although oxygen bonding to the Pt surface was also indicated in the XPS spectra, this was mostly likely due to oxygen adsorption that occurred when transferring the sample in the air. This kind of oxygen cannot be easily removed unless in-situ XPS methods, such as EC-XPS [27], are applied.

XRD patterns were also analyzed (Figure S3). Both porous Pt and Pt black showed typical metallic platinum peaks [28], consisting of Pt (111) at 39.8°, Pt (200) at 46.2°, and Pt (220) at 67.4°, which matched the XPS result.

## *2.2. Electrochemical Analyses of Oxygen Reduction and Evolution Reactions*

Cyclic voltammograms of porous Pt and Pt black were recorded and are shown in Figure 3. Pt/ketjent black (KB) was also analyzed and compared as a standard electrocatalyst. The electrochemical surface area (ECSA) of porous Pt was 12.4 $m^2\ g^{-1}$. The ECSA of Pt/KB was 76.5 $m^2\ g^{-1}$, which stayed at the low end of reported values (70–100 $m^2\ g^{-1}$) [29–31], suggesting the formation of a non-uniform thin film. Even though the film formation should be further optimized, with the films made under the same condition, Pt/KB has an ECSA that is six times higher than that of porous Pt. This is reasonable because Pt/KB has highly dispersed 2-nm Pt nanoparticles on the carbon support [32], but porous Pt consists of aggregates of much larger Pt particles. Pt black showed a lower ECSA (4.7 $m^2\ g^{-1}$) than porous Pt. As shown in Figure 2b and Figure S2b, Pt black is composed of dense aggregates, even though primary particles are as small as several nanometers. The structural difference most likely leads to different ECSA. Additionally, cyclic voltammograms of three catalysts, where the current is normalized to the specific surface area of Pt, are shown in Figure S4. The current of Pt/KB becomes smaller than that of porous Pt and Pt black, suggesting that porous Pt and Pt black have higher activity than Pt on KB.

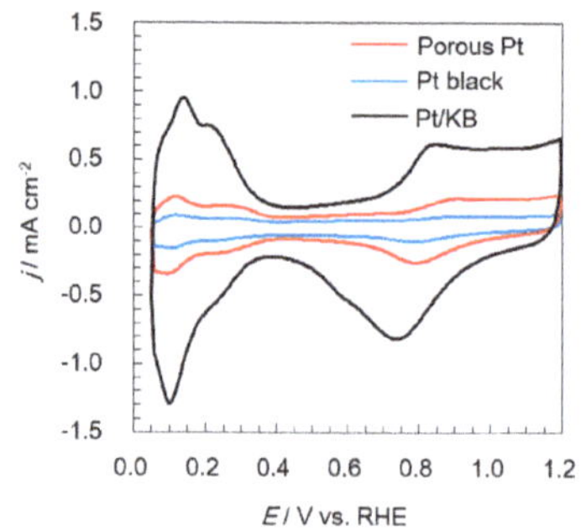


**Figure 3.** Cyclic voltammograms of porous Pt, Pt black, and Pt/ketjen black (KB).

The ORR activities of all electrocatalysts were evaluated at a rotating speed of 1600 rpm and are shown in Figure 4. Pt had equivalent ORR activity to Pt/KB, even though a slightly slow ORR current increase was seen for porous Pt. The values of the current density at 0.9 V were 2.5 and 2.1 mA $cm^{-2}$ for Pt/KB and porous Pt, respectively. The reported ORR activity for Pt/C is even higher, for example 3 mA $cm^{-2}$ [30]. ORR mass activity is commonly calculated through the kinetic current, which is the current at infinite speed extrapolated from a Koutecky–Levich plot, by measuring the ORR polarization curves for different rotation speeds. In this experiment, only one rotational speed was tried. Therefore, another method of calculating the kinetic current using the limiting current at 0.4 V [33] was used. The values of the mass activity at 0.9 V were 230 and 190 A $g^{-1}$ for Pt/KB and porous Pt, respectively. The reported ORR mass activity for Pt/C has the range but is mostly between 200 and 500 A $g^{-1}$ [29,30]. The lower values of ORR current and ORR mass activity in this work is most likely due to the non-uniform thin film acting as a working electrode; the development of uniform thin films is required to precisely discuss ORR activity.

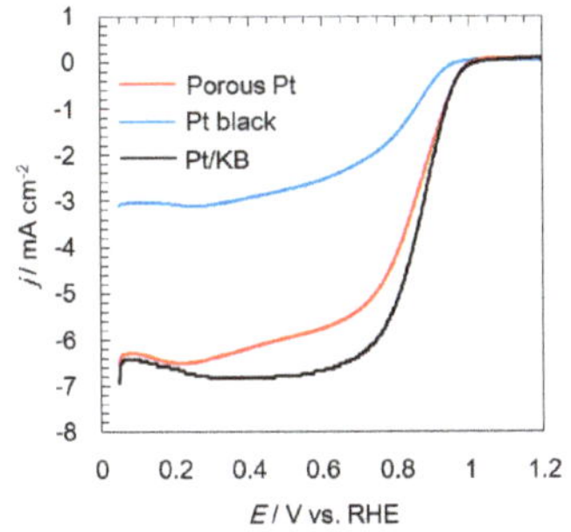


**Figure 4.** Linear sweep voltammograms showing the oxygen reduction reaction (ORR) of porous Pt, Pt black, and Pt/KB at 1600 rpm under oxygen saturation.

The OER activities of three electrocatalysts were evaluated at a rotating speed of 1600 rpm and are shown in Figure 5 with solid lines. Porous Pt had an initial OER current equivalent to that of Pt/KB (6.8 to 7.0 mA $cm^{-2}$ at 1.8 V), while Pt black revealed a lower OER current (3.4 mA $cm^{-2}$ at 1.8 V) than the other two catalysts. When the reported OER activity of Pt catalysts is considered, Pt bulk and nanoparticles show an OER current density of about 4 and 10 mA $cm^{-2}$ at 1.8 V, respectively [34], which reasonably match our results. We believe that the OER activity in the solution half-cell set-up remains constant if enough active surface area is available.

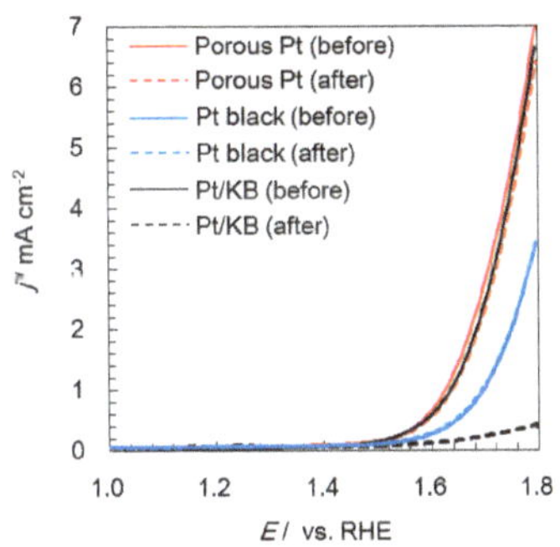


**Figure 5.** Linear sweep voltammograms showing oxygen evolution reaction (OER) of porous Pt, Pt black, and Pt/KB at 1600 rpm under nitrogen saturation. Solid and dotted lines show before and after the durability test, respectively.

## 2.3. *Electrochemical Durability of Catalysts at a High Potential*

In order to evaluate durability against a high potential, the anode potential at PEM water electrolyzers was considered, since it is a more severe condition than the cathode potential at PEM fuel cells. Although the protocol for the durability examination of PEM fuel cells has been well established [4], the protocol for PEM water electrolyzers has not been developed. Therefore, the potential of 1.8 V, which is the voltage required to obtain the practical current density of 1 A $cm^{-2}$, was applied to evaluate the durability in this study. Among the three samples, only Pt/KB was found to continuously lose OER current. Voltammograms of the three samples after the durability test were developed and are indicated with dotted lines in Figure 5. As can be seen in Figure 5, Pt/KB lost most of its OER activity by just applying 1.8 V for 10 min, owing to carbon corrosion that occurred at 1.8 V, while both porous Pt and Pt black maintained the initial OER activity according to the carbon-free condition. A similarly high durability can also be found in the literature for support-free catalysts, even though the condition of durability examination is slightly different [2,17,35]. For PEM fuel cells, current–voltage curves with the cathode of connected metal particles at 80 °C did not change much after 10,000 cycles of the start–stop durability test [17]. For PEM water electrolyzers, the changes in the voltage at 2 A $cm^{-2}$ under 60 °C [2] or 500 mA $cm^{-2}$ under 80 °C [35] were monitored, and the increase was less than 0.02 V during the 300 h test. Therefore, even though a longer test is required to evaluate the durability of porous Pt, an advantage of a carbon-free catalyst was experimentally observed in this study. Regarding the decrease in the current density observed in Figure 5, it was found to be a reversible change. Such reversible loss probably comes from the fact that the unsuccessful detachment of generated oxygen bubbles decreases the mass transfer capability of water on the electrode surface [36].

### *2.4. Evaluation of Catalyst Layers*

Catalyst layers of porous Pt and Pt black were spray-printed on the Nafion® membrane. Pt-loading was fixed to 0.50 mgPt/$cm^2$. The cross section of catalyst layers was observed using Focused Ion Beam (FIB)-SEM and is shown in Figure 6. Although a uniform thin layer of porous Pt with an average thickness of 2.4 µm was observed, as shown in Figure 6a, in the case of Pt black, no continuous layer was seen, as shown in Figure 6b. Even though relatively high catalyst loading is observed for water electrolysis [37,38], the future direction for PEM fuel cells lies in reducing the catalyst amount, for example to 0.10 mgPt/$cm^2$ or less. The formation of thin films will be more difficult when large-scale manufacturing methods of thin layers are considered. A high void volume is also important for enhancing the electrocatalytic activity [39]. Consequently, porous Pt has an advantage in its usage as a thin film, and the performance and durability of the new porous Pt catalyst-based membrane/electrode assembly will be characterized in a future study.

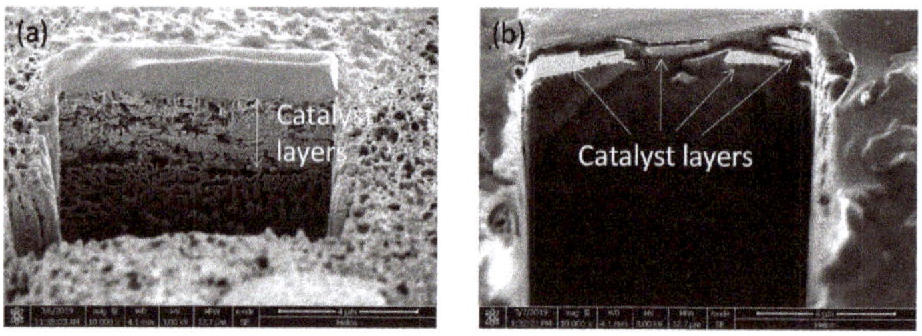


**Figure 6.** Cross section SEM images of (**a**) porous Pt and (**b**) Pt black layers.

## 3. Materials and Methods

### *3.1. Materials*

Platinum (II) acetylacetonate (Pt(acac)$_2$), 5 N hydrochloric acid (HCl), ethanol, 2-propanol, and 0.1 M perchloric acid ($HClO_4$) were all purchased from Wako Pure Chemical Industries Ltd. (Osaka, Japan) Pluronic® F127 was obtained from Sigma-Aldrich. 5% Nafion® dispersion solution and Nafion® 117 were obtained from Dupont (Wilmington, DE, USA). Milli-Q water was used in all cases. Those chemicals were used without any further purification.

For the electrochemical evaluation, Pt black and Pt/KB (TEC10E50E) were used as standard electrocatalysts for the comparison in this study and obtained from Wako Pure Chemical Industries, Ltd. and Tanaka Kikinzoku Kogyo K.K. (Tokyo, Japan), respectively.

### *3.2. Synthesis of Porous Pt*

Pluronic® F127 and Pt(acac)$_2$ were used as a template to form a porous structure and as a metal precursor, respectively. Typically, 0.4275 g of Pluronic® F127 was dissolved in a mixture of water/ethanol/HCl (2.175 g/2.875 g/75 µL). Then, 0.675 g of Pt(acac)$_2$ was added to this solution. This mixture was stirred at 30 °C for 6 h, kept at room temperature for 6 h, and dried at 80 °C for 6 h in the oven. After further heat treatment of the resulting powder, porous Pt materials were obtained. The heat treatment conditions, such as the temperature and atmosphere, were studied to optimize the porous structure and are fully discussed in the "Results and discussion" section.

### 3.3. *Material Characterization*

For material characterization, nitrogen sorption measurements were carried out using BELSORP-mini II-VS (MicrotracBEL Corp. Osaka, Japan). Before the measurement, samples were pre-treated at 200 °C under vacuum for 2 h. The specific surface area was calculated by a BET method using the adsorption area. The Barrett Joyner Hallenda (BJH) method was also applied to estimate the pore size distribution.

Thermogravimetric (TG) analyses were conducted using Thermo Plus Evo2 (Rigaku, Tokyo, Japan). Measurements were carried out in the air from room temperature to 400 °C via raising the temperature by 4 °C/min.

SEM observations were performed using S-5200 (Hitachi High-Tech, Tokyo, Japan), with the accelerating voltage of 30 kV. Studies with X-ray photoelectron spectroscopy (XPS) and X-ray diffraction (XRD) were also conducted using Kratos Axis Ultra (Shimadzu, Kyoto, Japan) and RINT-UltimaIII/PSA (Rigaku), respectively.

### 3.4. *Electrochemical Analyses*

The dispersion containing electrocatalysts (4.3 mg), water (0.35 mL), and 2-propanol (2.56 mL) was mixed using an ultrasonic homogenizer and drop-cast onto a glassy carbon (GC) rod (the diameter: 5 mm, Tokai Fine Carbon Co., Ltd. (Tokyo, Japan)), and dried at room temperature for use as a working electrode. The Pt loading was kept to 17.3 μg $cm^{-2}$ in all cases. In this study, a Nafion® dispersion was not used as a binder in order to simply observe electrochemical characteristics of catalysts without a Nafion® effect. The fact that a binder is not essential for adhesion of the catalyst to the GC and that low-loaded ionomer free catalyst (≤~18 μgPt $cm^{-2}$) adheres with a sufficient strength during measurements for ORR activity evaluation at room temperature has previously been reported [29].

Electrochemical measurements were performed using a common half-cell set-up in the solution. A potentiostat (HZ-7000, Hokuto Denko, Tokyo, Japan) was used for all electrochemical measurements. Besides the working electrodes, Ag|AgCl in a saturated KCl aqueous solution and Pt wire were utilized as reference and counter electrodes, respectively. Measurements were taken in 0.1 M $HClO_4$ under nitrogen atmosphere with the scan rate of 50 mV $s^{-1}$. ORR and OER activities were evaluated at 1600 rpm under the saturation of oxygen and nitrogen, respectively. In this study, all potentials were converted to a reversible hydrogen electrode (RHE), and IR correction was applied for OER evaluation by actually measuring the solution resistance using a potentiostat. Furthermore, in order to observe the durability against the anodic condition of the water electrolysis, the potential of 1.8 V vs. RHE was applied for 10 min, and OER activities before and after applying such potential were evaluated.

### 3.5. *Evaluation of Catalyst Layers*

Catalyst layers were prepared by employing a common spray printing method. The slurry consisting of porous Pt (or Pt black) (81.5 mg), ethanol (1764 μL), 5% Nafion® dispersion solution (919 μL), and MilliQ water (196 μL) was thoroughly stirred by an ultrasonic homogenizer. In this condition, the ratio of ionomer and catalyst was fixed to 0.33. This slurry was spray-printed on Nafion® 117 by a spray printing system (Nordson, Westlake, OH, USA). Catalyst layers were made into a 1 cm × 1 cm square. Pt-loading was fixed to 0.50 mgPt $cm^{-2}$. The cross section of catalyst layers was observed using FIB-SEM (Helios 600, FEI Company, Hillsboro, OR, USA). FIB processing was carried out at an accelerating voltage of 30 kV and a beam current of 0.4 nA.

## 4. Conclusions

A carbon-free porous Pt electrocatalyst was successfully prepared using the assembly of the metal precursor and surfactant under the optimized heat treatment condition. The structure of porous Pt was found to be composed of connected Pt particles with a size of 20 nm, leading to a relatively high BET surface area of 32.4 $m^2$ $g^{-1}$, in comparison to a commercial Pt black electrocatalyst, which has a value of 14.3 $m^2$ $g^{-1}$. The ECSA of porous Pt was 12.4 $m^2$ $g^{-1}$, whereas that of Pt black was 4.7 $m^2$ $g^{-1}$. Accordingly, porous Pt showed higher ORR and OER activity than that of Pt black.

In comparison to a standard electrocatalyst, Pt/KB, an equivalent ORR current density (2.5 mA $cm^{-1}$ at 0.9 V for Pt/KB and 190 mA $cm^{-1}$ at 0.9 V for porous Pt), ORR mass activity (230 A $g^{-1}$ at 0.9 V for Pt/KB and 2.1 A $g^{-1}$ at 0.9 V for porous Pt), and OER current density (6.8 mA $cm^{-1}$ at 1.8 V for Pt/KB and 7.0 mA $cm^{-1}$ at 1.8 V) were obtained, even though the ECSA of porous Pt was only one-sixth that of Pt/KB. However, Pt/KB lost most of its OER activity during the durability test against the anode potential during water electrolysis owing to carbon corrosion, whereas porous Pt maintained its OER activity due to the effect of the carbon-free condition.

In addition, when catalyst layers were spray-printed on the Nafion® membrane, porous Pt showed more uniform layers in comparison to carbon-free Pt black, demonstrating the advantage of its usage as a thin film.

Consequently, porous Pt with high ORR and OER activities and a high durability against a high potential was successfully prepared in this study and showed potential to be used as an electrocatalyst for PEM fuel cells and PEM water electrolyzers.

**Supplementary Materials:** The following are available online. Figure S1: Thermogravimetric (TG) analyses of porous Pt (a) with heat treatment at 400 °C for 3 h under dry nitrogen and (b) with additional heat treatment at 200 °C for 10 min under humidified nitrogen, Figure S2: SEM images of (a) porous Pt and (b) Pt black with a higher magnification, Figure S3: XRD patterns of porous Pt (top, red) and Pt black (bottom, blue), Figure S4: Cyclic voltammograms of porous Pt (red), Pt black (blue), and Pt/KB (black), where the current is normalized to specific surface area of Pt.

**Author Contributions:** Conceptualization, M.M. and A.H.; Methodology, M.M. and A.H.; Validation, M.M., M.N., K.S., and A.H.; Formal Analysis, M.M. and M.N.; Investigation, M.M. and A.H.; Resources, A.K. and K.S.; Writing—Original Draft Preparation, M.A.; Writing—Review and Editing, A.H. and M.N.; Supervision, A.H.; Project Administration, A.H.; Funding Acquisition, A.H. and K.S. All authors have read and agreed to the published version of the manuscript.

**Funding:** This work was partially supported by The Japan Science and Technology Agency (JST) through its "Center of Innovation Science and Technology based Radical Innovation and Entrepreneurship Program (COI Program), Grant Number JPMJCE1318".

**Conflicts of Interest:** The authors declare no conflicts of interest. The funders had no role in the design of the study; in the collection, analyses, or interpretation of data; in the writing of the manuscript; or in the decision to publish the results.

## References

1. Kinoshita, K.; Bett, J.A.S. Potentiodynamic analysis of surface oxides on carbon blacks. *Carbon* **1973**, *11*, 403–411. [CrossRef]
2. Babic, U.; Tarik, M.; Schmidt, T.J.; Gubler, L. Understanding the effects of material properties and operating conditions on component aging in polymer electrolyte water electrolyzers. *J. Power Sources* **2020**, *451*, 227778. [CrossRef]
3. Borup, R.; Meyers, J.; Pivovar, B.; Kim, Y.S.; Mukundan, R.; Garland, N.; Myers, D.; Wilson, M.; Garzon, F.; Wood, D.; et al. Scientific aspects of polymer electrolyte fuel cell durability and degradation. *Chem. Rev.* **2007**, *107*, 3904–3951. [CrossRef]
4. Ohma, A.; Shinohara, K.; Iiyama, A.; Yoshida, T.; Daimaru, A. Membrane and catalyst performance targets for automotive fuel cells by FCCJ membrane, catalyst, MEA. *WG. ECS Trans.* **2011**, *41*, 775.
5. Qin, C.; Wang, J.; Yang, D.; Li, B.; Zhang, C. Proton exchange membrane fuel cell reversal: A review. *Catalysts* **2016**, *6*, 197. [CrossRef]

6. Shao-Horn, Y.; Sheng, W.C.; Chen, S.; Ferreira, P.J.; Holby, E.F.; Morgan, D. Instability of supported platinum nanoparticles in low-Temperature fuel cells. *Top Catal.* **2007**, *46*, 285–305. [CrossRef]
7. Rasten, E.; Hagen, G.; Tunol, R. Electrocatalysis in water electrolysis with solid polymer electrolyte. *Electrochim. Acta* **2003**, *48*, 3945–3952. [CrossRef]
8. Lirong, M.; Sui, S.; Zhai, Y. Investigations on high performance proton exchange membrane water electrolyzer. *Int. J. Hydrog. Energy* **2009**, *34*, 678–684.
9. Huang, S.-Y.; Ganesan, P.; Park, S.; Popov, B.N. Development of a titanium dioxide-Supported platinum catalyst with ultrahigh stability for polymer electrolyte membrane fuel cell applications. *J. Am. Chem. Soc.* **2009**, *131*, 13898–13899. [CrossRef] [PubMed]
10. Kakinuma, K.; Chino, Y.; Senoo, Y.; Uchida, M.; Kamino, T.; Uchida, H.; Deki, S.; Watanabe, M. Characterization of Pt catalysts on Nb-doped and Sb-doped SnO2 support materials with aggregated structure by rotating disk electrode and fuel cell measurements. *Electrochim. Acta* **2013**, *110*, 316–324. [CrossRef]
11. Zhang, Z.; Liu, J.; Gu, J.; Su, L.; Cheng, L. An overview of metal oxide materials as electrocatalysts and supports for polymer electrolyte fuel cells. *Energy Environ. Sci.* **2014**, *7*, 2535–2558. [CrossRef]
12. Yin, M.; Xu, J.; Li, Q.; Jensen, J.O.; Huang, Y.; Cleemann, L.N.; Bjerrum, N.J.; Xing, W. Highly active and stable Pt electrocatalysts promoted by antimony-doped SnO2 supports for oxygen reduction reactions. *Appl. Catal. B* **2014**, *144*, 112–120. [CrossRef]
13. Ioroi, T.; Yasuda, K. PEM-type water electrolysis/fuel cell reversible cell with low PGM catalyst loadings. *ECS Trans.* **2015**, *69*, 919–924. [CrossRef]
14. Karimi, F.; Peppley, B.A. Metal carbide and oxide supports for iridium-Based oxygen evolution reaction electrocatalysts for polymer-Electrolyte-Membrane water electrolysis. *Electrochim. Acta* **2017**, *246*, 654–670. [CrossRef]
15. Sawy, E.N.E.; Birss, V.I. Nano-Porous iridium and iridium oxide thin films formed by high efficiency electrodeposition. *J. Mater. Chem.* **2009**, *19*, 8244–8252. [CrossRef]
16. Nakagawa, T.; Beasley, C.A.; Murray, R.W. Efficient electro-Oxidation of water near its reversible potential by a mesoporous IrOx nanoparticle film. *J. Phys. Chem. C* **2009**, *113*, 12958–12961. [CrossRef]
17. Tamaki, T.; Kuroki, H.; Ogura, S.; Fuchigami, T.; Kitamoto, Y.; Yamaguchi, T. Connected nanoparticle catalysts possessing a porous, hollow capsule structure as carbon-free electrocatalysts for oxygen reduction in polymer electrolyte fuel cells. *Energy Environ. Sci.* **2015**, *8*, 3545–3549. [CrossRef]
18. Da Silva, G.C.; Perini, N.; Ticianelli, E.A. Effect of temperature on the activities and stabilities of hydrothermally prepared IrOx nanocatalyst layers for the oxygen evolution reaction. *Appl. Catal. B* **2017**, *218*, 287–297. [CrossRef]
19. Debe, M.K. Tutorial on the fundamental characteristics and practical properties of nanostructured thin film (NSTF) catalysts. *J. Electrochem. Soc.* **2013**, *160*, F522–F534. [CrossRef]
20. Hayashi, A.; Notsu, H.; Kimijima, K.; Miyamoto, J.; Yagi, I. Preparation of Pt/mesoporous carbon (MC) electrode catalyst and its reactivity toward oxygen reduction. *Electrochim. Acta* **2008**, *53*, 6117–6125. [CrossRef]
21. Hayashi, A.; Kimijima, K.; Miyamoto, J.; Yagi, I. Oxygen transfer and storage processes inside the mesopores of platinum-Deposited mesoporous carbon catalyst thin-Layer electrode. *J. Phys. Chem. C* **2009**, *113*, 12149–12153. [CrossRef]
22. Hayashi, A.; Kimijima, K.; Miyamoto, J.; Yagi, I. Direct observation of well-Dispersed Pt nanoparticles inside the pores of mesoporous carbon through the cross section of Pt/mesoporous carbon particles. *Chem. Lett.* **2009**, *38*, 346–347. [CrossRef]
23. Minamida, Y.; Noda, Z.; Hyashi, A.; Sasaki, K. Development of MEAs with Pt/Mesoporous carbon as a cathode catalyst. *ECS Trans.* **2014**, *64*, 137–144. [CrossRef]
24. Sonoda, Y.; Hayashi, A.; Minamida, Y.; Matsuda, J.; Akiba, E. Nanostructure control of porous carbon materials through changing acidity with a soft-Template method. *Chem. Lett.* **2015**, *44*, 503–505. [CrossRef]
25. Fu, B.; Minamida, Y.; Noda, Z.; Sasaki, K.; Hayashi, A. Development of MEAs by controlling carbon structures in cathode layers. *ECS Trans.* **2016**, *75*, 827–835. [CrossRef]
26. Yue, B.; Ma, Y.; Tao, H.; Yu, L.; Jian, G.; Wang, X.; Wang, X.; Lu, Y.; Hu, Z. CNx nanotubes as catalyst support to immobilize platinum nanoparticles for methanol oxidation. *J. Mater. Chem.* **2008**, *18*, 1747–1750. [CrossRef]

27. Wakisaka, M.; Mitsui, S.; Hirose, Y.; Kawashima, K.; Uchida, H.; Watanabe, M. Electronic structures of Pt-Co and Pt-Ru alloys for CO-tolerant anode catalysts in polymer electrolyte fuel cells studied by EC-XPS. *J. Phys. Chem. B* **2006**, *110*, 23489–23496. [CrossRef] [PubMed]
28. Zhang, Y.; Zhang, H.; Ma, Y.; Cheng, J.; Zhong, H.; Song, S.; Ma, H. A novel bifunctional electrocatalyst for unitized regenerative fuel cell. *J. Power Sources* **2010**, *195*, 142–145. [CrossRef]
29. Shinozaki, K.; Zack, J.W.; Pylypenko, S.; Pivovar, B.S.; Kocha, S.S. Oxygen reduction reaction measurements on platinum electrocatalysts utilizing rotating disk electrode technique II. Influence of ink formulation, catalyst layer uniformity and thickness. *J. Electrochem. Soc.* **2015**, *162*, F1384–F1396. [CrossRef]
30. Garsany, Y.; Ge, J.; St-Pierre, J.; Rocheleau, R.; Swider-Lyons, K.E. Analytical procedure for accurate comparison of rotating disk electrode results for the oxygen reduction activity of Pt/C. *J. Electrochem. Soc.* **2014**, *161*, F628–F640. [CrossRef]
31. Sievers, G.W.; Jensen, A.W.; Brüser, V.; Arenz, M.; Escudero-Escribano, M. Sputtered platinum thin-films for oxygen reduction in gas diffusion electrodes: A Model System for Studies under Realistic Reaction Conditions. *Surfaces* **2019**, *2*, 25. [CrossRef]
32. Zhao, X.; Hayashi, A.; Noda, Z.; Kimijima, K.; Yagi, I.; Sasaki, K. Evaluation of change in nanostructure through the heat treatment of carbon materials and their durability for the start/stop operation of polymer electrolyte fuel cells. *Electrochim. Acta* **2013**, *97*, 33–41. [CrossRef]
33. Garsany, Y.; Baturina, O.A.; Swider-Lyons, K.E.; Kocha, S.S. Experimental Methods for Quantifying the Activity of Platinum Electrocatalysts for the Oxygen Reduction Reaction. *Anal. Chem.* **2010**, *82*, 6321–6328. [CrossRef] [PubMed]
34. Reier, T.; Oezaslan, M.; Strasser, P. Electrocatalytic oxygen evolution reaction (OER) on Ru, Ir, and Pt catalysts: A comparative study of nanoparticles and bulk materials. *ACS Catal* **2012**, *2*, 1765–1772. [CrossRef]
35. Li, G.; Yu, H.; Song, W.; Dou, M.; Li, Y.; Shao, Z.; Yi, B. A hard-Template method for the preparation of IrO2, and its performance in a solid-Polymer-Electrolyte water electrolyzer. *ChemSusChem* **2012**, *5*, 858–861. [CrossRef] [PubMed]
36. Dastafkan, K.; Li, Y.; Zeng, Y.; Han, L.; Zhao, C. Enhanced surface wettability and innate activity of an iron borate catalyst for efficient oxygen evolution and gas bubble detachment. *J. Mater. Chem. A* **2019**, 15252–15261. [CrossRef]
37. Siracusano, S.; Baglio, V.; Di Blasi, A.; Briguglio, N.; Stassi, A.; Ornelas, R.; Trifoni, E.; Antonucci, V.; Arico, A.S. Electrochemical characterization of single cell and short stack PEM electrolyzers based on a nanosized IrO2 anode electrocatalyst. *Int. J. Hydrog. Energy* **2010**, *35*, 5558–5568. [CrossRef]
38. Liu, C.; Carmo, M.; Bender, G.; Everwand, A.; Lickert, T.; Young, J.L.; Smolinka, T.; Stolten, D.; Lehnert, W. Performance enhancement of PEM electrolyzers through iridium-Coated titanium porous transport layers. *Electrochem. Commun.* **2018**, *97*, 96–99. [CrossRef]
39. Lee, M.; Uchida, M.; Tryk, D.A.; Uchida, H.; Watanabe, M. The effectiveness of platinum/carbon electrocatalysts: Dependence on catalyst layer thickness and Pt alloy catalytic effects. *Electrochim. Acta* **2011**, *56*, 4783–4790. [CrossRef]


**Sample Availability:** Samples of the compounds porous Pt is available from the authors.

molecules

MDPI

Article

# Transport and Electrochemical Interface Properties of Ionomers in Low-Pt Loading Catalyst Layers: Effect of Ionomer Equivalent Weight and Relative Humidity

**Sushmit Poojary, Muhammad Naoshad Islam, Udit N. Shrivastava, Edward P. L. Roberts and Kunal Karan ***

Department of Chemical and Petroleum Engineering, University of Calgary, Calgary, AB T2N 1N4, Canada; sushmit.poojary@ucalgary.ca (S.P.); muhammadnaoshad.isla@ucalgary.ca (M.N.I.); udit.shrivastava@ucalgary.ca (U.N.S.); edward.roberts@ucalgary.ca (E.P.L.R.)

* Correspondence: kkaran@ucalgary.ca; Fax: +1-403-220-4854

Academic Editor: Jean St-Pierre

Received: 21 May 2020; Accepted: 18 July 2020; Published: 26 July 2020

check for updates

**Abstract:** Catalyst layer (CL) ionomers control several transport and interfacial phenomena including long-range transport of protons, local transport of oxygen to Pt catalyst, effective utilization of Pt catalyst, electrochemical reaction kinetics and double-layer capacitance. In this work, the variation of these properties, as a function of humidity, for CLs made with two ionomers differing in side-chain length and equivalent weight, Nafion-1100 and Aquivion-825, was investigated. This is the first study to examine humidity-dependent oxygen reduction reaction (ORR) kinetics in-situ for CLs with different ionomers. A significant finding is the observation of higher ORR kinetic activity ($A/cm^2_{Pt}$) for the Aquivion-825 CL than for the Nafion-1100 CL. This is attributed to differences in the interfacial protonic concentrations at Pt/ionomer interface in the two CLs. The differences in Pt/ionomer interface is also noted in a higher local oxygen transport resistance for Aquivion-825 CLs compared to Nafion-1100 CLs, consistent with stronger interaction between ionomer and Pt for ionomer with more acid groups. Similar dependency on Pt utilization (ratio of electrochemically active area at any relative humidity (RH) to that at 100% RH) as a function of RH is observed for the two CLs. As expected, strong influence of humidity on proton conduction is observed. Amongst the two, the CL with high equivalent weight ionomer (Nafion-1100) exhibits higher conduction.

**Keywords:** catalyst layer; polymer electrolyte fuel cell; oxygen transport resistance; oxygen reduction reaction kinetics; platinum ionomer interface; ionomer thin film

## 1. Introduction

Ion-containing polymer or ionomer is a key material in polymer electrolyte fuel cells (PEFCs). Until a decade ago, the focus of studies on ionomeric materials and their fuel cell properties were largely limited to their application as the polymer electrolyte membrane separating the anode and the cathode. However, ionomer is also one of the critical material constituents of the cathode and the anode catalyst layers. Catalyst layers of the polymer electrolyte fuel cells are complex, nanoporous, nanocomposite of ionomer, and Pt/C catalyst with co-continuous phases [1–3]; see Figure 1 [3]. In a typical Pt/C-based catalyst layer, a 4–10 nm thin film of acid ionomer covers the aggregates of Pt/C catalyst [4]. The catalyst layer ionomer is often described as a binder [5], which does not capture the multiple crucial functions it serves: (i) as an ion-conducting material phase that ensures transport of protons over 10–20 micron thick catalyst layers, (ii) as an acidic medium that together with Pt catalyst forms the electrochemically active interface where the oxygen reduction reaction (cathode) or hydrogen oxidation reactions (anode) occur, and (iii) as a material phase that controls the transport of reactants ($O_2$ or $H_2$) and products to/from the active Pt sites.

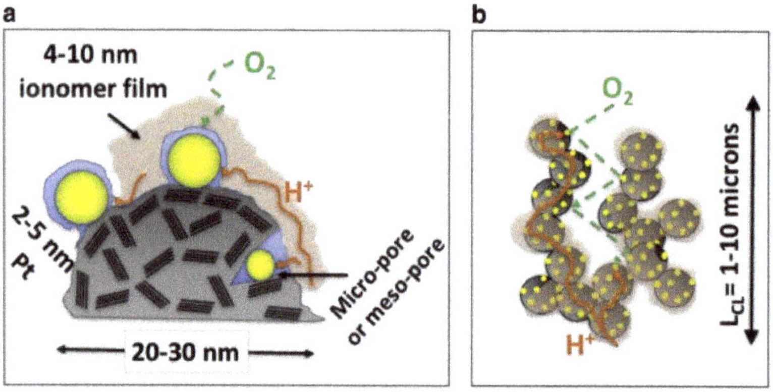


**Figure 1.** Depiction of microstructure of and key processes in a polymer electrolyte fuel cell (PEFC) catalyst layer. (**a**) Local transport of oxygen and protons to the surface of Pt catalyst particles (yellow) on the surface and in the micropores of carbon support coated with ionomer thin film (**b**) Long-range transport of proton and oxygen transport through the thickness of the catalyst layer. Reproduced with permission from Karan [3], Current Opinions in Electrochemistry; published by Elsevier, 2017.

The ionomer in a catalyst layer is essential for facilitating proton transport to achieve high in-operando electrochemical surface area (ECSA) utilization [6]. Pt catalysts not in direct contact with ionomer (see Figure 1a) may not be accessible to protons under low relative humidity (RH) conditions. Recently, it has been identified that the chemical structure of ionomer dictates the chemical environment of the electrochemically active interface including poisoning by sulfonic group and ultimately affecting the oxygen reduction reaction (ORR) activity [7].

Oxygen transport through the ionomer thin film covering the catalyst is known to be a limiting factor for low-Pt loading catalysts [8–11]. It is now well established that confinement effects and interactions with substrates (e.g., Pt/C) strongly influence the structure and properties of the ionomer such that it differs significantly from that of the free-standing membrane [12–20]. The proton conduction, the nature of the electrochemical interface, and the local transport of reactants in the catalyst layer ionomer all depend on its hydration state (water content). In addition to bulk water content, the interfacial water content also becomes important for various transport processes and electrochemical phenomena in catalyst layer ionomers. Both bulk and interfacial water content depend on the relative humidity (RH). Exposure to humid environment stimulates the evolution of hydrophilic domains in the bulk ionomer and water sorption at the Pt-ionomer interface [15]. Hydrophilic domains are necessary for proton transport [21] and also assist in oxygen permeation through the thin film [11,22]. Hydration can lead to reorganization of surface, bulk, and buried interfacial structure of ionomers [23,24]. Although discussed sparingly in the literature, it can be expected that interfacial water will influence (a) the extent of acid dissociation, i.e., local pH, and thereby the electrochemical reaction kinetics, (b) the conduction of protons on the catalyst surface, and (c) the transport resistance of oxygen to the Pt catalyst. Moreover, increase in RH enhances the accessibility of Pt catalyst residing in the internal pores of the support and thus affects ECSA [25]. The chemical structure of ionomer and the chemical nature of substrate surface affect the size of the hydrophilic domains in the bulk and at the interface [14,26].

A handful of studies in the literature reports the effect of ionomer type on fuel cell performance [5,27,28]. Even fewer studies discuss the ionomer-dependent relevant catalyst layer electrochemical properties such as ECSA, double-layer capacitance, kinetics, and the local oxygen transport resistance [29–32] and no single study discusses all of these properties for the same catalyst layer. Incorporation of higher ion exchange capacity (IEC) ionomers improve ECSA and specific activity [29] but at the expense of higher local oxygen transport resistance [31]. However, only a few studies have examined the effect of RH on the aforementioned characteristics—ECSA, ORR kinetics, double-layer capacitance, oxygen transport resistance—of the catalyst layer. The literature lacks a systematic account of the combined effect of ionomer structure and RH on the catalyst layer electrochemical properties. In addition, the discussion of the catalyst layer electrochemistry in the context of Pt/ionomer interface is absent. Overall, humidity-dependent probing of catalyst layer properties can provide significant insight into how ionomer molecular structure influence both transport properties and Pt/ionomer interfacial characteristics.

Here, the RH-dependent bulk and interfacial properties of catalyst layer ionomer are reported for long side chain (LSC) Nafion ionomer with an equivalent weight (EW) of ca. 1100 (Naf-1100) and short side chain (SSC) Aquivion ionomer, EW ca. 825 (Aq-825) as a function of relative humidity. Scheme 1 presents the molecular structure of both ionomers. The method of catalyst layer fabrication and details of experimental measurements are provided in the Materials and Method section.

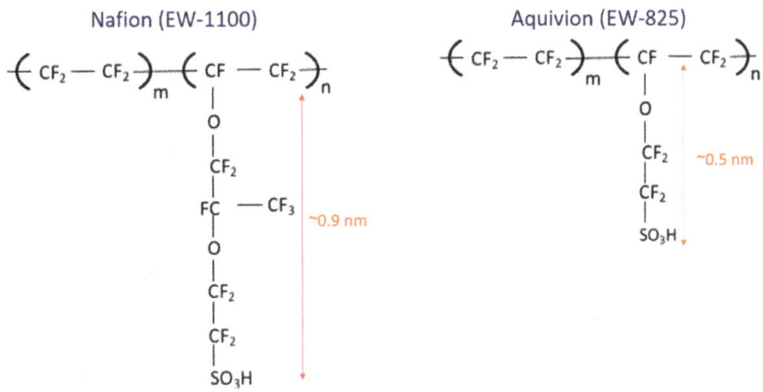


**Scheme 1.** Chemical structure of Nafion and Aquivion ionomers.

## 2. Results

The RH-dependent electrochemical and transport properties for low-Pt loading catalyst layers (nominal Pt loading of 0.04 mg/cm$^{-2}$) prepared with 10 wt% Pt/C catalyst and ionomer:carbon ratio of 0.8 for the two different ionomers are reported. Electrochemical properties studied include the double-layer capacitance, electrochemically active surface area, and ORR kinetics, while the transport properties studied are the proton conduction and oxygen transport resistance.

Double-layer capacitance, electrochemically active surface area, and protonic conductivity: The electrochemically active surface area obtained from H-adsorption peak integration from cyclic voltammetry (CV) scan was normalized with respect to the electrode area and denoted as the roughness factor (RF). Figure 2a compares the RH-dependent RF and double-layer capacitance ($C_{dl}$) for the catalyst layers Aq-825 and Naf-1100. RF for both ionomers increases with RH, similar to that reported in other studies [8,25], whereas $C_{dl}$ almost remains constant for Aq-825 but increases with RH for Naf-1100. $C_{dl}$ mainly arises from the interfacial charges at the ionomer/carbon and ionomer/Pt interface. Differences in the catalyst layer microstructure or interfacial characteristics can result in the differences in $C_{dl}$ for two catalyst layers. If the ionomer coverage on the carbon for the two catalyst layers is different, it would result in different $C_{dl}$. If the charge concentration at the ionomer/carbon and ionomer/Pt interface is different, it could also result in different $C_{dl}$. The higher magnitude of $C_{dl}$ for Aq-825 catalyst layer than Naf-1100 catalyst layer points toward either a higher coverage of ionomer in Aq-825 catalyst layer or high interfacial charge concentration than that in Naf-1100 catalyst layer. Although the ionomer to carbon ratio was kept the same for the two catalyst layers and similar coverages are expected, we do not have direct evidence of the microstructural similarity. Thus, the origin of differences in magnitude of $C_{dl}$ as well as its RH-dependence remains unresolved. Only one previous study that has reported $C_{dl}$ of catalyst layers with different ionomers [31] could be found. In that study, for catalyst layers made with Nafion and Aquivion ionomers with similar ionomer loading (30 wt%), the $C_{dl}$ was reported to be 19 mF cm$^2$. However, it was also reported that a catalyst layer with lower ionomer content (10 wt% Aquivion ionomer) had higher $C_{dl}$ than that for high ionomer content catalyst layer, which is counterintuitive. To the best of our knowledge, no studies have reported a comparison of RH-dependent $C_{dl}$ of CLs with different ionomers. Thus, the differences in RH-dependent $C_{dl}$ observed for two ionomers in this study cannot be cross-checked with results from other studies. Moreover, it is not so straightforward to delineate the effect of RH on double-layer capacitance. The double-layer capacitance is known to be pH-dependent and complicated by the contributions from charges in the Helmholtz and diffuse double layer, even in simpler liquid electrolyte/bare metal electrode systems [33]. Double-layer capacitance in such systems usually increases with decreasing pH. However, the system studied here has a solid electrolyte (ionomer), wherein the water content is controlled by RH. The interfacial water will control the interfacial pH, while the bulk water (in the ionomer phase) will control the connectivity of the proton-conducting water channel. Thus, while the interfacial water controls the charge distribution across the solid/polymer interface, the bulk water controls the accessibility of the protons to the interface during potential scanning. Interestingly, for Aq-825, in dry conditions, the CL double-layer capacitance was 1.33 times greater than that of the Naf-1100 CL, which is equal to the ratio of their ion exchange capacities (IECs). At a given RH, the RF for Aq-825 catalyst layer is greater than that of the Naf-1100 catalyst layer. The electrochemical surface area (ECSA) at 100% RH for Aq-825 catalyst layer is 78 m$^2$/g$_{Pt}$, while that for Naf-1100 catalyst layer is 56 m$^2$/g$_{Pt}$, indicating higher Pt accessibility in Aq-825 catalyst layer. A previous study also reported a higher Pt utilization for catalyst layer prepared with high IEC (EW 980) Aquivion ionomer than that prepared from low IEC (EW 1100) Nafion ionomer [29]. Both higher IEC and, possibly, higher coverage of Pt/C by ionomer in Aq-825 catalyst layer than in Naf-1100 catalyst layer must contribute to the higher $C_{dl}$ as well as RF of Aq-825 catalyst layer. The RH-dependent Pt utilization was calculated by dividing ECSA at any RH by the ECSA at 100%RH. The Pt utilization (see inset of Figure 2a) of the two catalyst layers at any given RH was very similar, indicating that the type of ionomer does not affect the RH-dependent Pt utilization. For 10 wt% Pt/Vulcan catalyst, Padgett et al. [25] reported that Pt utilization was dependent on RH. From tomography performed on 10 wt% Pt particles, they found that Pt particles located in the internal pores contributed about 20–30% to the total surface area. However, they did not report the double-layer capacitance, and as such a correlation between ECSA accessibility and double-layer capacitance is not available from their study. The observed increase in RF with RH could be due to the increased accessibility of protons to the Pt in the internal pores.

Figure 2b shows the variation of the protonic conductivity of Aq-825 and Naf-1100 catalyst layers as a function of humidity. For comparison, the protonic conductivity of Nafion membrane has also been included. The protonic resistances for the catalyst layers and the membrane were determined from analyses of the electrochemical impedance spectra (EIS) at each RH. The inset of Figure 2b shows an example of the impedance spectra at 30% RH in $H_2/N_2$. Similar impedance spectra were used at each RH to estimate the high-frequency resistance or HFR (marked by point A in inset of Figure 2b) and the catalyst layer protonic resistance $R_{CL}$. The high-frequency resistance is the sum of the electrical and membrane protonic resistances, and Equation (1) below was used to calculate the membrane protonic conductivity.

$$\sigma_{mem} = \frac{t_{mem}}{HFR - R_e} \tag{1}$$

where $\sigma_{mem}$ is the membrane conductivity, $R_e$ is area-specific electrical resistance (10 m$\Omega$ cm$^2$) determined from an ex-situ resistance measurement, and $t_{mem}$ is the thickness of the membrane. In a typical EIS of a fuel cell catalyst layer, the point along the 45° line at which spectra transition to a completely capacitive behavior, for example, point B in the inset, the real component of spectra at this point is summation of HFR and a third of $R_{CL}$ [34]. Using Equation (2) (see below), the catalyst layer conductivity is obtained [35].

$$\sigma_{CL} = \frac{t_{CL}}{R_{CL}\varepsilon^n} \tag{2}$$

where $\sigma_{CL}$ is catalyst layer conductivity, $\varepsilon$ is ionomer volume fraction, the Bruggeman exponent $n$ is 1.5 [34], and $t_{CL}$ is thickness of the catalyst layer.

Expectedly, protonic conductivity of catalyst layer and membrane are strong functions of RH. At 30% RH, both Aq-825 and Naf-1100 catalyst layer conductivities are almost an order of magnitude smaller than the Nafion membrane conductivity, and at any given RH, Naf-1100 catalyst layer conductivity is smaller than the Aq-825 conductivity. In a catalyst layer, ionomer exists in thin film form (depiction shown in Figure 1) and it is known that at comparable RH, ionomer thin film exhibits much lower protonic conductivity than the bulk membrane [13]. The IEC of Aq-825 ionomer (IEC ~1.2) is higher than that for Naf-1100 ionomer (IEC ~0.9). The difference in conductivity can be attributed to the intrinsic effect of difference in acid content of the ionomers and to the extrinsic effect of how ionomer is spatially distributed in the catalyst layer. The former effect is well known for bulk membranes, while the latter effect (differences in microstructure) is complicated. For example, lower coverage of ionomer would imply thicker ionomer films that have higher conductivity but the connectivity and tortuosity may also be higher.

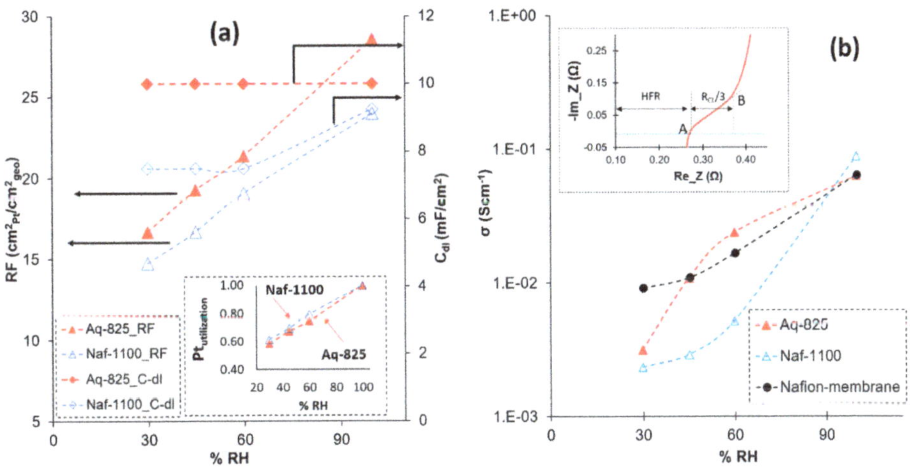


**Figure 2.** Comparison among relative humidity (RH)-dependent properties of Naf-1100 and Aq-825 derived from cyclic voltammetry (CV) and electrochemical impedance spectra (EIS) in $H_2/N_2$ at 70 °C and 100% RH (**a**) roughness factor and double-layer capacitance (**b**) protonic conductivity.

### 2.1. Oxygen Reduction Reaction (ORR) Kinetics

Little is understood about the influence of ionomer EW and side chain on the ORR kinetics. Ionomer EW or IEC is a measure of its acidic strength. From early studies of ORR on Pt electrodes in liquid electrolyte, Damjanovic and Brusic proposed the following kinetic expression highlighting the dependency of ORR kinetic current ($i_{ORR}$) on proton activity/concentration [36]:

$$i_{ORR} = k\, P_{O_2}{}^{n} \left[H^+\right]^m \exp\!\left(-\frac{\alpha F}{RT}\eta_{ORR}\right) \tag{3}$$

where $k$ is the electrochemical rate constant akin to exchange current density, $n$ and $m$ are reaction order stated to be 1 and 1.5 [36], $\alpha$ is transfer coefficient, $p_{O_2}$ is oxygen partial pressure, $\eta_{ORR}$ is ORR overpotential, $F$ is Faraday constant, $R$ is universal gas constant, and $T$ is temperature. It must be noted that in the original work [33], the formal potential rather than overpotential was used.

The interfacial proton concentration in the catalyst layer, i.e., proton concentration at the Pt/ionomer interface, would depend on the abundance of sulfonic acid groups at or near the Pt/ionomer interface as well as the interfacial water content, both of which are not directly accessible in fuel cell experiments. From our neutron reflectometry (NR) study of different ionomers on planar Pt, it is known that the amounts of water present at the Pt/ionomer interface varies with RH and at 100% RH the water content correlates to the ionomer side-chain length [12,14]. However, the abundance of sulfonic group is not known. CO desorption electrochemistry applied to estimate Pt-sulfonic group interactions could offer insight into this but was not applied in the current study [37]. Regardless, it can be expected that the proton concentration at the Pt/ionomer interface would vary with RH and could be different for different ionomers. Accordingly, proton concentration-dependent ORR kinetics (Equation (3)) can be expected to occur. A comparison of RH-dependent ORR kinetic behavior of Aq-825 and of Naf-1100 catalyst layers is presented in Figure 3. By neglecting oxygen transport resistance and hydrogen oxidation reaction overpotential, the $\eta_{ORR}$ was estimated using Equation (4) below and plotted against the log of specific current density ($i_s$, A $cm^{-2}{}_{Pt}$), see Figure 3(a1–a4), clearly following the Tafel behavior.

$$\eta_{ORR} = OCV - E_{cell} - i\ (HFR + R_{CL}/3) \quad (4)$$

where $OCV$ is open-circuit voltage, $i$ is current density, and $E_{cell}$ is cell voltage.

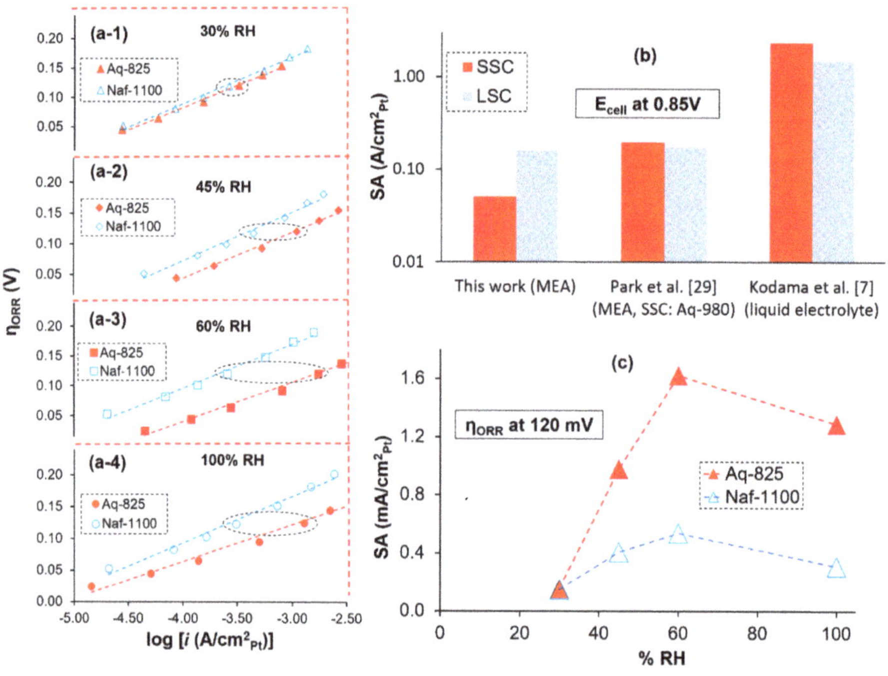


**Figure 3.** Comparison among kinetic properties of Naf-1100 and Aq-825 in $H_2$/air (**a**) activation overpotential against current density in Tafel region at different RHs (**b**) specific activity at 0.85 V from this work, park et al. [29], Kodama et al. [7] in liquid electrolyte (**c**) specific activity at activation overpotential of 120 mV and at different RHs.

For most electrochemical reactions, two kinetic parameters are often considered—the Tafel slope, which is related to the transfer coefficient ($\alpha$), and the exchange current density. From Figure 3a, it can be noted that the slopes of the plot, which is the Tafel slope, are similar for both catalyst layers at each RH. The Tafel slopes were found to be 66–70 mV/decade. It is also obvious from the data in Figure 3a that at each RH, ORR current in the Aq-825 catalyst layer is superior to that in the Naf-1100 catalyst layer. Since Tafel slope (or $\alpha$) are similar, considering the Damjanovic and Brusic formulation of ORR kinetics [36], it can be deduced that higher current density (expressed on a per $cm^2$ of Pt basis) for Aq-825 catalyst layer compared to that for Naf-1100 catalyst layer would be a result of higher interfacial concentration of protons at Pt/ionomer interface for Aq-825 catalyst layer. The higher IEC of Aq-825 and its shorter side chain can be expected to create a higher abundance of sulfonic groups at the interface and, thereby, in a higher interfacial proton concentration. To the best of our knowledge, there has been no prior study comparing RH-dependent ORR kinetics for catalyst layers with different ionomers.

In literature, the kinetic performance of a catalyst layer is generally evaluated by defining specific activity or mass activity at a voltage in kinetic or Tafel region of the polarization curve. In Figure 3b, the specific current density, i.e., specific activities (SA) for ORR at 0.85 V for catalyst layers made with short and long side chain (SSC and LSC) ionomers in this work are compared along with results from other studies for SA determined in membrane electrode assembly (MEA) [29] and in liquid electrolyte [7]. In MEAs, regardless of the type of ionomer used in fabricating the catalyst layer, SA at 0.85 V in MEA can be as much as ten times lower than that in liquid electrolyte (rotating disc electrode or RDE setup) [7]. In the present work, SA at 0.85 V in 100% RH for Aq-825 catalyst layer is lower than that for Naf-1100 catalyst layer. Since OCV for the Aq-825 catalyst layer at 100% RH is 40 mV lower than the Naf-1100 catalyst layer, as per Equation (4), $\eta_{ORR}$ for both cells are different at same $E_{cell}$ and, hence, comparison of SA at same voltage may not be valid. Therefore, SA in Figure 3c is compared at a similar $\eta_{ORR}$ = 120 mV marked by the dotted ellipses in Figure 3a. Except at 30% RH, at each RH, SA for Aq-825 catalyst layer is 2–5 times greater than Naf-1100 catalyst layer. For both catalyst layers, SA increased with RH. From 30% RH to 60% RH, SA for Aq-825 SA catalyst layer increased ten folds whereas for Naf-1100 SA increased by three times. Then, for both catalyst layers between 60% and 100%, RH SA decreases—trend is similar to a previous study [29].

### 2.2. Oxygen Transport Resistance

Limiting current technique explained elsewhere [38] was employed to determine the oxygen transport resistance ($R_{O_2,T}$, s/cm) via Equation (5) below.

$$R_{O_2,T} = \frac{4FC_{O2,ch}}{i_{lim}} = \underbrace{[R_{O_2,GDL} + R^{M}_{O_2,MPL}]}_{\text{Pressure dependent}} + \underbrace{[R^{Kn}_{O_2,MPL} + R_{O_2,CL}]}_{\text{Pressure independent}} \quad (5)$$

where $C_{O2,ch}$ is gas concentration at the channel, $i_{lim}$ is limiting current density, $F$ is Faraday constant. $R_{O_2,T}$ is a combination of the pressure-dependent and -independent terms. In gas diffusion layer (GDL), pore size is in order of 1–10 μm [39] in microporous layer (MPL) pore size varies from ~50 nm to 1 μm [39,40], and in catalyst layer pore sizes are below 100 nm [41]. Therefore, oxygen transport through the gas diffusion layer and some of the pores of MPL occurs via molecular diffusion and the associated transport resistances are denoted as $R_{O_2,GDL}$ and $R^{M}_{O_2,MPL}$, respectively. Transport through smaller pores in MPL occurs via pressure-independent Knudsen diffusion. The associated oxygen transport resistance is denoted as $R^{Kn}_{O_2,MPL}$. In addition to the GDL and MPL, catalyst layer offers additional pressure-independent resistance $R_{O_2,CL}$, which includes Knudsen oxygen diffusion resistance as well as oxygen permeation resistance through the thin ionomer film coating of the Pt particles.

The inset in Figure 4 shows an example of the linear increase in $R_{O_2,T}$ with the pressure. The slope of this line is inversely proportional to the molecular diffusion resistance in GDL and MPL, the pressure-dependent terms, and intercept represents the pressure-independent terms. From our internal study, we estimated $R^{Kn}_{O_2,MPL}$ is 0.1 s/cm, which is very small compared to the intercept of

$R_{O_2,T} \sim 1.5$ s/cm. The catalyst layer thickness is determined to be in the 10–12 μm range. Considering the Vulcan carbon support, most of the Pt catalysts are expected to be on the surface of the carbon particle. Thus, Knudsen diffusion within the micropores of carbon support as expected for high surface area carbon can be neglected for the present study wherein Vulcan carbon support has been used. A mix of Knudsen and molecular diffusion through the macro-pores of catalyst layer is expected. In our analyses, the gas phase $O_2$ transport resistance is considered to be significantly smaller than local transport resistance. Hence, the magnitude of intercept in the inset mainly corresponds to the $R_{O_2,CL}$. Accordingly, the average local transport resistance to Pt/ionomer interface ($R_{O_2,Pt}$) can be approximated via Equation (6) as [9]:

$$R_{O_2,Pt} = R_{O_2,CL} \cdot RF \tag{6}$$

$R_{O_2,Pt}$ in both catalyst layers decrease significantly with RH. Between 30% and 80% RH, Aq-825 exhibits 30% reduction in $R_{O_2,Pt}$ whereas Naf-1100 shows 50% reduction in $R_{O_2,Pt}$. The magnitude of $R_{O_2,Pt}$ for Naf-1100 catalyst layer is similar to the reported values in the literature [9,22]. At each RH, $R_{O_2,Pt}$ for Naf-1100 catalyst layer is lower than that for Aq-825 catalyst layer, e.g., at 80% RH $R_{O_2,Pt}$ of Aq-825 catalyst layer is close to 40% higher than $R_{O_2,Pt}$ for Naf-1100 catalyst layer. Ono et al. also reported higher local $R_{O_2,Pt}$ for higher IEC ionomers than lower IEC ionomers [31]. RH dependency of $R_{O_2,Pt}$follows the trend reported by the Toyota group in a study that is the only known direct measurement of oxygen transport resistance of ionomer on Pt [22]. Hydrated ionomer promotes oxygen transport through the ionomer, while the side-chain interactions with Pt likely influences the ionomer thin film morphology, especially the interfacial structure.

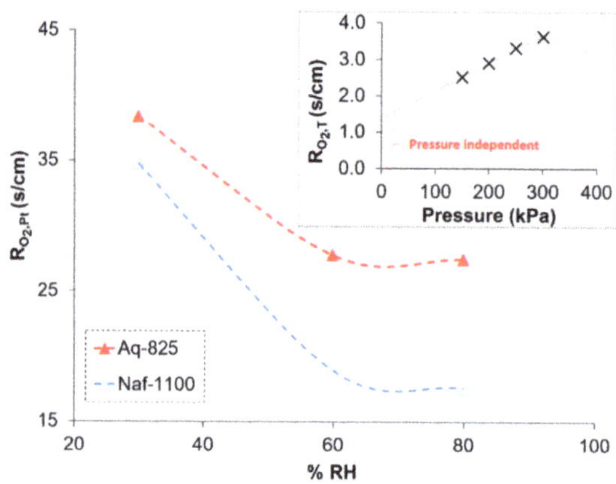


**Figure 4.** RH-dependent local ionomer transport resistance for both Naf-1100 and Aq-825 catalyst layers estimated at 70 °C. Inset shows total transport resistance as a function of pressure.

## 3. Discussion

In a majority of the prior studies, a key stated motivation of preparing fuel cell catalyst layers with higher IEC ionomer such as Aq-825 is to achieve higher protonic conductance within the catalyst layer (see e.g., Park et al. [29]). However, interfacial processes, whether ORR kinetics or local $R_{O_2,Pt}$, can also be affected by the ionomer. The higher conductivity of catalyst layer prepared with high IEC ionomer (Aq-825) than that of catalyst layer prepared with low IEC ionomer (Naf-1100) is evident in our study (Figure 2b) and other works [31]. Additionally, our work indicates that the ionomer in a catalyst layer also affects the ORR electrochemical kinetics, an aspect previously not investigated in depth in other studies. Consistent with previous findings, a significant influence of CL ionomer on microstructure-dependent properties (i) Pt utilization (Figure 2a) and (ii) local oxygen transport

(Figure 4) is also noted. Our study elucidates the RH dependency of these properties, which has been investigated to a limited extent.

A key point we would like to emphasize is that the interfacial properties are significantly affected by the nature of Pt-ionomer interface. In a previous neutron reflectometry (NR) study from our group [12,14], we examined the temperature- and RH-dependent bulk and interfacial water distribution in 15 nm ionomer films on planar Pt substrates for different ionomers including 3M EW-725 (SSC) and Nafion-1100 (LSC). Based on these findings, Figure 5 depicts the Pt/Aq-825 and Pt/Naf-1100 interfaces [12,14,42]. Although 3M-725 is a different ionomer from Aq-825, both of them have same backbone and similar side chain length, the only difference is spacing between the ionic groups. At 97% RH, Pt/Aq-825 interface may have only a monolayer of water (3 Å) separating Pt surface from the polytetrafluoroethylene (PTFE) backbone. In same condition, two monolayers thick water (6 Å) is formed at Pt/ionomer interface. In both cases, sulfonic groups are shown to interact with Pt surface (based on IR [43]), and hydrophobic back is separated by hydrophilic domains (based on GISAXS [15]).

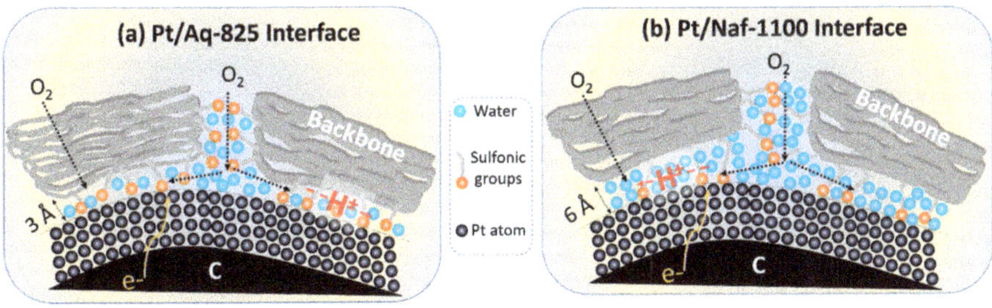


**Figure 5.** Depiction of Pt/ionomer interface based on neutron reflectometry (NR) at 97% RH (**a**) Pt/Aq-825 interface (**b**) Pt/Naf-1100 interface.

## 3.1. Pt Utilization

At each RH investigated in our study, Aq-825 catalyst layer exhibited 1.4 times higher ECSA than Naf-1100 catalyst layer. Pt utilization is essentially a quantification of accessibility of protons to the Pt catalysts. Ionomer coverage and poisoning effect are two factors that can affect the Pt utilization.

Naf-1100 has a longer side chain with two ether groups, while Aquivion-825 has a shorter side chain with only one ether group. The combination of RDE and surface-enhanced infrared absorption spectroscopy provides evidence of absorption of the oxygen atom of flexible LSC ether group on Pt atom. Such absorption of oxygen atom of SSC ether group on Pt atom is absent. Thus, a higher blockage of Pt sites is observed in catalyst layer made with LSC ionomer (Nafion) compared to that made with SSC ionomer (Aq-825), resulting in a lower ECSA for Naf-1100 catalyst layer [7]. In absence of any microstructural characterization, we cannot ascertain to what extent ionomer coverage differences contribute to the Pt utilization differences.

## 3.2. ORR Kinetics

As discussed earlier, the differences in the Pt-area normalized kinetic current for the two catalyst layers at a given overpotential is attributed to the interfacial proton concentration. As discussed above, the interfacial water and the sulfonic acid abundance at the interface both will affect the interfacial proton concentration. At low RH (30% RH), the ORR kinetic currents for Aq-825 and Naf-1100 catalyst layers are similar. At this RH, the interfacial water content is expected to be similar for both CLs. It would then appear that there is very little difference in sulfonic acid abundance at the Pt/ionomer interface for the two CLs. At higher RH, significant differences are observed. As depicted in Figure 4 above, there will be higher amount of water at the Pt/ionomer interface in the Naf-1100 CLs. This would be tantamount to diluting the sulfonic acid by different amounts of water, effectively lowering the interfacial proton concentration. Using Equation (7) below, ratio of proton concentration is estimated to be ~3, and the kinetic current of Aq-825 is nearly 4 times greater than Naf-1100.

$$C_{H^+,interface} \propto \frac{IEC}{\phi_{w,interface}} \tag{7}$$

where $C_{H^+,interface}$ is the interfacial proton concentration and $\phi_{w,interface}$ is the interfacial water volume fraction.

### 3.3. Oxygen Transport Resistance

The oxygen transport to the Pt in a catalyst layer comprises of gas-phase transport (mix of Knudsen and molecular diffusion depending on the local pore dimension), transfer from gas-phase to the ionomer phase, diffusion through the ionomer film, and then additional interfacial resistance. For high surface area carbon and agglomerated Pt/C structure, diffusion through the micropores within carbon and pores inside the agglomerate structure, respectively, would also have to be considered. At low Pt loadings, ionomer thin films in the catalyst layer thought to be large contributors to the local oxygen transport resistance [9,11]. The local ionomer thin film resistance for both catalyst layers significantly decreases with RH similar to reported by an ex-situ oxygen transport study on Nafion thin film on Pt [22]. Within the ionomer thin films, oxygen has two transport pathways: (a) through the free volume within the hydrophobic matrix, and (b) via the water-filled hydrophilic domains. Oxygen has relatively high solubility in the hydrophobic Teflon-like matrix but restrictive diffusion. On the other hand, oxygen can be solvated in the water phase of hydrophilic domains and be diffused with ease. As RH increases, both the bulk and interfacial water content increases, thereby making the oxygen transport more facile [11,22]. At same RH, water volume fraction in bulk of Aq-825 ionomer film is expected to be higher than that Naf-1100 ionomer film. However, as depicted in Figure 5, the interfacial water layer thickness for Naf-1100 is almost two times the thickness of Aq-825 interfacial layer. The local oxygen transport resistance ($R_{O_2,Pt}$) at 80% RH for the Aq-825 catalyst layer is 1.6 times higher than that for the Naf-1100 catalyst layer. If the oxygen transport through the bulk portion of the ionomer film rather than near interface region was the dominant resistance, Aq-825 catalyst layer would not exhibit higher $R_{O2}$ than Naf-1100 catalyst layer. It is hypothesized that the higher interfacial water content at the Pt/ionomer interface in the Naf-1100 catalyst layer promotes faster oxygen transport than that in the Pt/Aq-825 catalyst layer with low interfacial water content.

## 4. Materials and Methods

### 4.1. Catalyst Ink Preparation

The catalyst inks were prepared using commercially available 10.2 wt% Pt/C (Tanaka) electrocatalyst and Nafion (EW 1100) purchased from Ion power Inc. (New Castle, DE, USA) and Aquivion (EW 825) purchased from Sigma Aldrich (St. Louis, MO, USA). The ionomer to carbon (I/C) and solid to liquid (S/L) ratio was maintained at 0.8 and 0.1, respectively. The ionomer stock dispersion was first diluted using a mixture of deionized (DI) water and isopropanol, and the resulting mixture was probe-sonicated for 2 min using an ice jacket to break up the ionomer aggregates. Then, 285 mg of Pt/C catalyst was added into the diluted ionomer dispersion. Then, the mixture was bath sonicated for 15 min followed by three hours of magnetic stirring and 24 h of ball milling.

### 4.2. Membrane Electrode Assembly (MEA) Preparation

The catalyst layer decals were prepared by coating the catalyst inks on ethylene tetrafluoroethylene (ETFE) sheet using a micrometer adjustable film casting doctor blade (EQ-Se-KTQ-250, MTI corporation, Richmond, CA, USA) and air dried for 24 h. The doctor blade was adjusted to a thickness of 100 μm for coating of both catalyst layer decals. Decal transfer method was used to prepare an MEA. Nafion membrane (25 μm, NRE-211, DuPont, Wilmington, DE, USA) was hot-pressed in between the anode

(0.2 $mg_{Pt}/cm^2$) and the cathode decals at 150 °C and 2 MPa pressure for 3 min [44,45]. The MEAs had an anode and cathode active area of 1.44 and 1 $cm^2$, respectively. The larger anode area was maintained to ensure redundant supply of reactant (protons) from the anode. The thickness, platinum loading, and ionomer loading of the cathode CLs are listed in Table 1. The changes in local and interfacial transport properties become very insignificant and remain obscured at high cathode Pt loading as it scales inversely with Pt ECSA, thus CLs with ultra-low Pt loading were prepared in this study.

**Table 1.** Summary of the cathode catalyst layer properties.

| Catalyst Layer Type | CL Thickness (μm) | Pt Loading ($mg_{Pt}/cm^2_{geo}$) | Ionomer Loading (wt%) | Ionomer Loading ($mg/cm^2_{geo}$) |
|---|---|---|---|---|
| Nafion | 11 ± 1 | 0.043 | 42.07 | 0.34 |
| Aquivion | 10 ± 1 | 0.036 | 42.04 | 0.28 |

### *4.3. Fuel Cell Assembly and Testing*

A small scale and differential flow field cell was assembled by sandwiching the hot-pressed MEA between 160 μm thick Toray gas diffusion layers coated with microporous layer (MPL) (TGP H-60, Toray, Japan). The differential cell was used to avoid any discrepancies arising from the gradient in RH and reactant gas concentrations along the active area. The single cell was tested using a fuel cell test station (100W, Greenlight Innovation, Burnaby, Canada) coupled with two potentiostats (Biologic SP-200, Seyssinet-pariset, France and Ivium Vertex, Eindhoven, The Netherlands).

Electrochemical impedance spectroscopy (EIS), cyclic voltammetry (CV), and linear sweep voltammetry (LSV) tests were performed under $H_2/N_2$ for determining high-frequency resistance (HFR) for the electrodes, ECSA, and $H_2$ crossover, respectively. The CV and LSV tests were performed at 200 and 5 mV/s, respectively. The protonic resistance of ionomer in the cathode CL was determined by performing EIS at 0.4 V under $H_2/N_2$ by sweeping frequencies in the range of 1 Hz to 1 MHz with an amplitude of 10 mV. Pt electrochemical surface area (ECSA) was determined by integrating the hydrogen underpotential deposition ($H_{upd/desorption}$) region while subtracting capacitive currents. The cell was initially conditioned (see Table 2 for the conditioning protocol details).

**Table 2.** Summary of the conditioning protocol used in this study.

| Step # | Test | $T_{cell}$ (°C) | $T_{hum}$ (°C) | Flow Rate An/Ca (NLPM) | Reactant Gas An/Ca | Back Pressure (kPag) | Time (hrs) |
|---|---|---|---|---|---|---|---|
| 1 | $H_2$ pumping | 30 | 45 | 0.5/0.5 | $H_2/H_2$ | 0 | 0.5 |
| 2 | Flooding | 60 | 70 | 0.05/0.1 | $H_2/N_2$ | 50 | 8–12 |
| 3 | 0.6 V hold | 70 | 70 | 0.3/0.5 | $H_2/O_2$ | 200 | 12 |
| 4 | Potential cycling (0.6–0.8 V) | 70 | 70 | 0.3/0.5 | $H_2/O_2$ | 50 | Holding 5 min at each potential |

The cell performance was characterized by collecting potentiostatic polarization curves with in-situ high-frequency resistance (HFR) using an Ivium Vertex potentiostat (Ivium, Eindhoven, Netherlands). The potentiostatic polarization curve was obtained by holding the voltage for 3 min at each voltage at a resolution of 0.1 V. The limiting current study was performed by varying $O_2$ concentrations (1–24% $O_2$:$N_2$) [38,46]. The detail of all the testing conditions is outlined in Table 3.

**Table 3.** Summary of membrane electrode assembly (MEA) testing conditions for all the tests performed in this study.

| Test | $T_{cell}$ (°C) | RH (%) | Flow Rate An/Ca (NLPM) | Reactant Gas An/Ca | Back Pressure (kPag) |
|---|---|---|---|---|---|
| Diagnostics (CV, LSV, EIS) | | 30, 45, 60, and 100 | 0.1/0.2 | $H_2/N_2$ | 50 |
| Performance | 70 | | 0.3/0.5 | $H_2$/Air | |
| Limiting current | | 30, 60, and 80 | 0.2/0.5 | $H_2/O_2$ (1–24%):$N_2$ | 50, 100, 150, and 200 |

## 5. Conclusions

In this work, the effect of ionomer side chain length (or EW) on the electrochemical interfacial properties and mass transport properties including long-range proton transport, local $O_2$-transport, Pt utilization, double-layer capacitance, and ORR reaction kinetics was investigated at varying RH. In summary, the CL prepared with shorter side chain (Aq-825) exhibited higher ECSA, higher CL ionic conductivity, higher CL double-layer capacitance, and higher CL local $O_2$-transport resistance compared to the CL prepared with longer side chain (Naf-1100). The differences in these properties can be explained on the basis of differences in the EW and side chain. However, the differences in catalyst layer microstructure such the ionomer coverage and connectivity or the pore size can also be responsible. A systematic study combining microstructural characterization and catalyst layer properties is needed to ascertain the origin of the observed differences in catalyst layer properties. At 120 mV ORR overpotential, the Aq-825 CL showed 2-5 times higher ORR activity compared to the Naf-1100 CL at any given RH, except at 30% RH, which can be ascribed to the higher interfacial concentration of protons at the Pt/ionomer interface for Aq-825 CL. Our findings also indicate that at each RH, Naf-1100 CL illustrated lower local oxygen transport resistance ($R_{O_2,Pt}$) compared to Aq-825 CL, for instance, Aq-825 CL showed 40% higher $R_{O_2,Pt}$ at 80% RH. For both CLs, $R_{O_2,Pt}$ decreased with increasing RH as higher interfacial water content at the Pt/ionomer interface in the CL promotes faster oxygen transport.

**Author Contributions:** Conceptualization, K.K.; methodology, U.N.S. and M.N.I.; formal analysis, S.P., M.N.I., and U.N.S.; investigation, S.P.; resources, K.K. and E.P.L.R.; data curation, S.P.; writing—original draft preparation, U.N.S., S.P. and M.N.I.; writing—review and editing, U.N.S., S.P., M.N.I., K.K. and E.P.L.R.; supervision, K.K.; project administration, K.K.; funding acquisition, K.K. and E.P.L.R. All authors have read and agreed to the published version of the manuscript.

**Funding:** This research was funded by Alberta Innovates, grant number AI SRP G2018000904, Natural Science and Engineering Research Council of Canada, (NSERC) CREATE program, grant number 495455-2017, and partially funded by KK's and ER's NSERC Discovery Grants.

**Conflicts of Interest:** The authors declare no conflict of interest.

## References

1. Uchida, M. PEFC catalyst layers: Effect of support microstructure on both distributions of Pt and ionomer and cell performance and durability. *Curr. Opin. Electrochem.* **2020**. [CrossRef]
2. Holdcroft, S. Fuel cell catalyst layers: A polymer science perspective. *Chem. Mater.* **2014**, *26*, 381–393. [CrossRef]
3. Karan, K. PEFC catalyst layer: Recent advances in materials, microstructural characterization, and modeling. *Curr. Opin. Electrochem.* **2017**, *5*, 27–35. [CrossRef]
4. Lopez-Haro, M.; Guétaz, L.; Printemps, T.; Morin, A.; Escribano, S.; Jouneau, P.-H.; Bayle-Guillemaud, P.; Chandezon, F.; Gebel, G. Three-dimensional analysis of Nafion layers in fuel cell electrodes. *Nat. Commun.* **2014**, *5*, 1–6. [CrossRef] [PubMed]

5. Balogun, E.O.; Hussain, N.; Chamier, J.; Barendse, P. Performance and durability studies of perfluorosulfonic acid ionomers as binders in PEMFC catalyst layers using Electrochemical Impedance Spectroscopy. *Int. J. Hydrog. Energy* **2019**, *44*, 32219–32230. [CrossRef]
6. Neyerlin, K.; Gu, W.; Jorne, J.; Clark, A.; Gasteiger, H.A. Cathode catalyst utilization for the ORR in a PEMFC analytical model and experimental validation. *J. Electrochem. Soc.* **2007**, *154*, 279–287. [CrossRef]
7. Kodama, K.; Motobayashi, K.; Shinohara, A.; Hasegawa, N.; Kudo, K.; Jinnouchi, R.; Osawa, M.; Morimoto, Y. Effect of the Side-Chain Structure of Perfluoro-Sulfonic Acid Ionomers on the Oxygen Reduction Reaction on the Surface of Pt. *Acs Catal.* **2018**, *8*, 694–700. [CrossRef]
8. Fukuyama, Y.; Shiomi, T.; Kotaka, T.; Tabuchi, Y. The impact of platinum reduction on oxygen transport in proton exchange membrane fuel cells. *Electrochim. Acta* **2014**, *117*, 367–378. [CrossRef]
9. Greszler, T.A.; Caulk, D.; Sinha, P. The impact of platinum loading on oxygen transport resistance. *J. Electrochem. Soc.* **2012**, *159*, 831–840. [CrossRef]
10. Gittleman, C.S.; Kongkanand, A.; Masten, D.; Gu, W. Materials research and development focus areas for low cost automotive proton-exchange membrane fuel cells. *Curr. Opin. Electrochem.* **2019**, *18*, 81–89. [CrossRef]
11. Weber, A.Z.; Kusoglu, A. Unexplained transport resistances for low-loaded fuel-cell catalyst layers. *J. Mater. Chem. A* **2014**, *2*, 17207–17211. [CrossRef]
12. Karan, K.; Shrivastava, U.N.; Fritzsche, H. Neutron Reflectometry Investigation of Near Ionomer/Catalyst Interface Structure in Polymer Electrolyte Based Energy Devices. *Phys. Can.* **2018**, *74*, 45–48.
13. Paul, D.K.; McCreery, R.; Karan, K. Proton transport property in supported Nafion nanothin films by electrochemical impedance spectroscopy. *J. Electrochem. Soc.* **2014**, *161*, 1395–1402. [CrossRef]
14. Shrivastava, U.N.; Fritzsche, H.; Karan, K. Interfacial and Bulk Water in Ultrathin Films of Nafion, 3M PFSA, and 3M PFIA Ionomers on a Polycrystalline Platinum Surface. *Macromolecules* **2018**, *51*, 9839–9849. [CrossRef]
15. Shrivastava, U.N.; Suetsugu, K.; Nagano, S.; Fritzsche, H.; Nagao, Y.; Karan, K. Cross-correlated humidity-dependent structural evolution of Nafion thin films confined on a platinum substrate. *Soft Matter* **2020**, *16*, 1190–1200. [CrossRef] [PubMed]
16. Karan, K. Interesting Facets of Surface, Interfacial, and Bulk Characteristics of Perfluorinated Ionomer Films. *Langmuir* **2019**, *35*, 13489–13520. [CrossRef]
17. Kusoglu, A.; Dursch, T.J.; Weber, A.Z. Nanostructure/Swelling Relationships of Bulk and Thin-Film PFSA Ionomers. *Adv. Funct. Mater.* **2016**, *26*, 4961–4975. [CrossRef]
18. Kusoglu, A.; Kushner, D.; Paul, D.K.; Karan, K.; Hickner, M.A.; Weber, A.Z. Impact of substrate and processing on confinement of Nafion thin films. *Adv. Funct. Mater.* **2014**, *24*, 4763–4774. [CrossRef]
19. DeCaluwe, S.C.; Kienzle, P.A.; Bhargava, P.; Baker, A.M.; Dura, J.A. Phase segregation of sulfonate groups in Nafion interface lamellae, quantified via neutron reflectometry fitting techniques for multi-layered structures. *Soft Matter* **2014**, *10*, 5763–5776. [CrossRef]
20. DeCaluwe, S.C.; Baker, A.M.; Bhargava, P.; Fischer, J.E.; Dura, J.A. Structure-property relationships at Nafion thin-film interfaces: Thickness effects on hydration and anisotropic ion transport. *Nano Energy* **2018**, *46*, 91–100. [CrossRef]
21. Kusoglu, A.; Weber, A.Z. New insights into perfluorinated sulfonic-acid ionomers. *Chem. Rev.* **2017**, *117*, 987–1104. [CrossRef] [PubMed]
22. Kudo, K.; Jinnouchi, R.; Morimoto, Y. Humidity and temperature dependences of oxygen transport resistance of Nafion thin film on platinum electrode. *Electrochim. Acta* **2016**, *209*, 682–690. [CrossRef]
23. Kabir, S.; Myers, D.J.; Kariuki, N.; Park, J.; Wang, G.; Baker, A.; Macauley, N.; Mukundan, R.; More, K.L.; Neyerlin, K.C. Elucidating the Dynamic Nature of Fuel Cell Electrodes as a Function of Conditioning: An ex Situ Material Characterization and in Situ Electrochemical Diagnostic Study. *Acs Appl. Mater. Interfaces* **2019**, *11*, 45016–45030. [CrossRef] [PubMed]
24. Paul, D.K.; Karan, K. Conductivity and wettability changes of ultrathin Nafion films subjected to thermal annealing and liquid water exposure. *J. Phys. Chem. C* **2014**, *118*, 1828–1835. [CrossRef]
25. Padgett, E.; Andrejevic, N.; Liu, Z.; Kongkanand, A.; Gu, W.; Moriyama, K.; Jiang, Y.; Kumaraguru, S.; Moylan, T.E.; Kukreja, R. Connecting fuel cell catalyst nanostructure and accessibility using quantitative cryo-STEM tomography. *J. Electrochem. Soc.* **2018**, *165*, 173–180. [CrossRef]
26. Dura, J.A.; Murthi, V.S.; Hartman, M.; Satija, S.K.; Majkrzak, C.F. Multilamellar interface structures in Nafion. *Macromolecules* **2009**, *42*, 4769–4774. [CrossRef]

27. Xu, H.; Kunz, H.R.; Bonville, L.J.; Fenton, J.M. Improving PEMFC Performance Using Low Equivalent Weight PFSA Ionomers and Pt-Co/C Catalyst in the Cathode. *J. Electrochem. Soc.* **2007**, *154*, 271–278. [CrossRef]
28. Lei, C.; Bessarabov, D.; Ye, S.; Xie, Z.; Holdcroft, S.; Navessin, T. Low equivalent weight short-side-chain perfluorosulfonic acid ionomers in fuel cell cathode catalyst layers. *J. Power Sources* **2011**, *196*, 6168–6176. [CrossRef]
29. Park, Y.-C.; Kakinuma, K.; Uchida, H.; Watanabe, M.; Uchida, M. Effects of short-side-chain perfluorosulfonic acid ionomers as binders on the performance of low Pt loading fuel cell cathodes. *J. Power Sources* **2015**, *275*, 384–391. [CrossRef]
30. Garsany, Y.; Atkinson, R.W.; Sassin, M.B.; Hjelm, R.M.; Gould, B.D.; Swider-Lyons, K.E. Improving PEMFC performance using short-side-chain low-equivalent-weight PFSA ionomer in the cathode catalyst layer. *J. Electrochem. Soc.* **2018**, *165*, 381–391. [CrossRef]
31. Ono, Y.; Ohma, A.; Shinohara, K.; Fushinobu, K. Influence of equivalent weight of ionomer on local oxygen transport resistance in cathode catalyst layers. *J. Electrochem. Soc.* **2013**, *160*, 779–787. [CrossRef]
32. Jomori, S.; Komatsubara, K.; Nonoyama, N.; Kato, M.; Yoshida, T. An experimental study of the effects of operational history on activity changes in a PEMFC. *J. Electrochem. Soc.* **2013**, *160*. [CrossRef]
33. Smith, C.P.; White, H.S. Theory of the interfacial potential distribution and reversible voltammetric response of electrodes coated with electroactive molecular films. *Anal. Chem.* **1992**, *64*, 2398–2405. [CrossRef] [PubMed]
34. Makharia, R.; Mathias, M.F.; Baker, D.R. Measurement of catalyst layer electrolyte resistance in PEFCs using electrochemical impedance spectroscopy. *J. Electrochem. Soc.* **2005**, *152*, 970–977. [CrossRef]
35. Liu, Y.; Murphy, M.W.; Baker, D.R.; Gu, W.; Ji, C.; Jorne, J.; Gasteiger, H.A. Proton conduction and oxygen reduction kinetics in PEM fuel cell cathodes: Effects of ionomer-to-carbon ratio and relative humidity. *J. Electrochem. Soc.* **2009**, *156*, 970–980. [CrossRef]
36. Damjanovic, A.; Brusic, V. Electrode kinetics of oxygen reduction on oxide-free platinum electrodes. *Electrochim. Acta* **1967**, *12*, 615–628. [CrossRef]
37. Garrick, T.R.; Moylan, T.E.; Yarlagadda, V.; Kongkanand, A. Characterizing electrolyte and platinum interface in PEM fuel cells using CO displacement. *J. Electrochem. Soc.* **2017**, *164*, 60–64. [CrossRef]
38. Baker, D.R.; Caulk, D.A.; Neyerlin, K.C.; Murphy, M.W. Measurement of oxygen transport resistance in PEM fuel cells by limiting current methods. *J. Electrochem. Soc.* **2009**, *156*, 991–1003. [CrossRef]
39. Ostadi, H.; Rama, P.; Liu, Y.; Chen, R.; Zhang, X.; Jiang, K. 3D reconstruction of a gas diffusion layer and a microporous layer. *J. Membr. Sci.* **2010**, *351*, 69–74. [CrossRef]
40. Chun, J.H.; Park, K.T.; Jo, D.H.; Lee, J.Y.; Kim, S.G.; Lee, E.S.; Jyoung, J.-Y.; Kim, S.H. Determination of the pore size distribution of micro porous layer in PEMFC using pore forming agents under various drying conditions. *Int. J. Hydrog. Energy* **2010**, *35*, 11148–11153. [CrossRef]
41. Sabharwal, M.; Pant, L.; Putz, A.; Susac, D.; Jankovic, J.; Secanell, M. Analysis of catalyst layer microstructures: From imaging to performance. *Fuel Cells* **2016**, *16*, 734–753. [CrossRef]
42. Kurihara, Y.; Mabuchi, T.; Tokumasu, T. Molecular analysis of structural effect of ionomer on oxygen permeation properties in PEFC. *J. Electrochem. Soc.* **2017**, *164*, 628–637. [CrossRef]
43. Yagi, I.; Inokuma, K.; Kimijima, K.i.; Notsu, H. Molecular structure of buried perfluorosulfonated ionomer/Pt interface probed by vibrational sum frequency generation spectroscopy. *J. Phys. Chem. C* **2014**, *118*, 26182–26190. [CrossRef]
44. Shrivastava, U.N.; Tajiri, K. Sources of current density distribution in the land-channel direction of a pemfc. *J. Electrochem. Soc.* **2016**, *163*, 1072–1083. [CrossRef]
45. Shrivastava, U.N.; Tajiri, K.; Chase, M. Current density and ohmic resistance distribution in the land-channel direction of a proton exchange membrane fuel cell. *J. Power Sources* **2015**, *299*, 189–194. [CrossRef]
46. Caulk, D.A.; Baker, D.R. Heat and water transport in hydrophobic diffusion media of PEM fuel cells. *J. Electrochem. Soc.* **2010**, *157*, 1237–1244. [CrossRef]

molecules

MDPI

Article

# Effect of GDM Pairing on PEMFC Performance in Flow-Through and Dead-Ended Anode Mode

**Yannick Garsany [1,*], Cornelius H. Bancroft [2], Robert W. Atkinson III [1], Keith Bethune [3], Benjamin D. Gould [2] and Karen E. Swider-Lyons [2]**

1 EXCET, Inc., Springfield, VA 22151, USA; robert.atkinson.ctr@nrl.navy.mil
2 US Naval Research Laboratory, Washington, DC 20375, USA; cornelius.bancroft@nrl.navy.mil (C.H.B.); benjamin.gould@nrl.navy.mil (B.D.G.); karen.lyons@nrl.navy.mil (K.E.S.-L.)
3 Hawaii Natural Energy Institute, University of Hawaii, Honolulu, HI 96822, USA; bethune@hawaii.edu
* Correspondence: yannick.garsany.ctr@nrl.navy.mil; Tel.: +1-202-404-1825

Received: 8 February 2020; Accepted: 21 March 2020; Published: 24 March 2020

check for updates

**Abstract:** Asymmetric gas diffusion media (GDM) pairing, which feature distinct GDM at the anode and cathode of the proton electrolyte membrane fuel cell (PEMFC), enhance water management compared to symmetric pairing of GDM (anode and cathode GDM are identical). An asymmetric pairing of Freudenberg GDM (H24C3 at anode and H23C2 at cathode) reduces ohmic resistances by up to 40% and oxygen transport resistances by 14% en route to 25% higher current density in dry gas flows. The asymmetric GDM pairing effectively hydrates the membrane electrode assembly (MEA) while minimizing liquid water saturation in the cathode compared to a commonly used symmetric GDM pairing of SGL 29BC at the anode and cathode. Superior water management observed with asymmetric GDM in flow-through mode is also realized in dead-ended anode (DEA) mode. Compared to the symmetric GDM pairing, the asymmetric GDM pairing with Freudenberg GDM increases cell voltage at all current densities, extends and stabilizes steady-state voltage behavior, slows voltage decay, and vastly reduces the frequency of anode purge events. These results support that the asymmetric Freudenberg GDM combination renders the PEMFC less prone to anode water saturation and performance loss from the anticipated increase in water back-diffusion during DEA mode operation.

**Keywords:** PEMFCs; asymmetric & symmetric GDM; Freudenberg; SGL 29BC; dead-ended anode (DEA) mode

## 1. Introduction

Proton exchange membrane fuel cells (PEMFCs) are clean electrochemical power sources for use in a broad array of applications [1]. System costs are still the major challenge to widespread commercialization of PEMFCs. One approach to reduce cost and system complexity is to operate the PEMFC in dead-ended anode (DEA) mode. Operating a PEMFC in DEA mode simplifies the balance of plant requirements by removing the hydrogen ejector/blower, gas humidifier, mass flow meter, and redundant piping [2–4].

During the DEA operation of a PEMFC, hydrogen is supplied to the inlet of the DEA PEMFC system and a normally closed solenoid valve blocks the outlet. Using this simple set-up reduces the system cost and increases the PEMFCs hydrogen utilization [5,6]. There are two major failure modes for the voltage decay in PEMFCs operated in DEA mode: dilution of the anode fuel concentration via $N_2$ crossover and excessive accumulation of back diffused liquid water from the cathode. Both of these failure modes can be exacerbated by improper materials selection inside the PEMFC. Liquid water generated at the cathode can back diffuse through the membrane and accumulate inside the anode gas diffusion media (GDM) and flow channels [7–12], blocking the gas transport pathway. When air is supplied as the oxidant, $N_2$ can diffuse through the membrane due to a pressure concentration gradient [7,12–15], resulting in local fuel (i.e., $H_2$) starvation and performance loss.

A typical PEMFC is composed of bipolar plates (BPPs) and a membrane electrode assembly (MEA), which is in a five layer structure comprising a proton exchange membrane (PEM) at the center, two coated catalyst layers (CLs) and two GDM [16]. The GDM are essential components of the PEMFCs, and play multiple functions during the PEMFC operation such as transportation of reactant gases to the CLs, removing produced water and heat from the MEA to the BPPs, and conducting electrons from the CLs to the BPPs [17–19]. GDM typically consist of porous carbon matrices (cloths, papers, or nonwovens) and are comprised by two regions. The fibrous gas diffusion layer (GDL) substrate has larger pores or voids and serves as a relatively robust substrate that gives the MEA its mechanical integrity. The GDL is comprised of carbon for electrical conductivity and is commonly treated with fluoropolymers (i.e., PTFE) for hydrophobicity. A carbon particle-based, hydrophobic, microporous layer (MPL) is coated on the GDL to improve thermal and electrical contact with the CLs, and maintain humidification of the adjacent CLs and the PEM while avoiding flooding of its porous backing (the GDL substrate), which would compromise the reactant gas supply to the CLs [20–22].

To date, the vast majority of the PEMFC open literature focuses on testing the PEMFCs using a symmetric GDM pairing, i.e., the same GDM is used on the anode and cathode side of the MEA. However, when selecting GDM for PEMFCs, it is critical to consider the anode and cathode GDM properties in tandem as each electrode plays a role in global cell water management. Careful pairing of the two GDM in concert can have a tremendous impact on the PEMFC performance. Recently, Schweiss reported that PEMFCs tested using an asymmetric GDM pairing, with a distinct anode GDM that improves water retention and a porous cathode GDM that promotes high oxidant diffusivity, are less sensitive to relative humidity and increases PEMFC current density compared to all tested symmetric GDM configurations (same GDM used on the anode and cathode) [23]. We have previously shown in open-cathode fuel cells that an asymmetric GDM pairing featuring higher porosity in the anode GDM than the cathode significantly improves hydration and power production [24,25].

In this work, we compare an asymmetric GDM pairing utilizing Freudenberg GDM (H24C3 at anode, H23C2 at cathode) to a symmetric GDM pairing frequently used in the open literature containing SGL 29BC at both the anode and cathode, to highlight the impact of the GDM water management on fuel cell operation in flow-through mode and DEA mode in a range of cathode inlet relative humidity.

## 2. Materials and Methods

### 2.1. PEMFC Performance in Regular Flow-Through Mode

Two distinct anode and cathode GDM pairings were tested in this study: symmetric and asymmetric. The symmetric GDM pairing was comprised of SGL 29BC carbon paper (SIGRACET, SGL Carbon Inc.) at both the anode and cathode, which consisted of 5% PTFE loading in the GDL and 23% PTFE loading in the MPL. The asymmetric GDM pairing investigated in this study utilized H23C2 (Freudenberg FCCT SE.) carbon felt on the cathode side, which was not PTFE-treated in the GDL and had 40% PTFE loading in the MPL. The anode side contained H24C3 (Freudenberg FCCT SE & Co.), which was PTFE-treated in the GDL and MPL.

Fuel cells were tested in a 25 $cm^2$ active area single cell hardware (Fuel Cell Technologies) with individually designed graphite flow fields (Poco Graphite), which comprised mirror symmetric flow patterns for the anode and the cathode. The flow fields consisted of 33 parallel channels with a channel and land width of 1 mm and a channel depth of 0.9 mm, which were arranged in triple parallel serpentine.

Membrane electrode assemblies (MEAs) were prepared by combining the Freudenberg GDM with a Primea® MESGA catalyst coated membrane (CCM, W. L. Gore & Associates A510.4/M710.18/C510.4, containing sub-gaskets) with Pt loading of 0.40 $mg_{Pt}$ $cm^{-2}$ on the anode side and 0.40 $mg_{Pt}$ $cm^{-2}$ on the cathode side, respectively. The square active electrode area was 25 $cm^2$, while the GDM were 27.04 $cm^{-2}$. When perfectly aligned, the GDM extended ~0.2 cm beyond the electrode active area on all sides.

All Freudenberg GDM were compressed by 28% of the initial uncompressed thickness, which we found to be optimal in our prior work [26]. The SGL 29BC GDM were compressed to 14% of the initial uncompressed GDM thickness, the optimal value for these materials in these operating conditions [26,27]. In order to compress the MEA to achieve the desired GDM compression, the thicknesses of the CCM and GDM were measured with a digimatic micrometer (Mitutoyo, Model MDC-1" PX) at 9 evenly spaced locations over each component area. These measurements were averaged to calculate the component thickness and used to determine the required gasket thickness. PTFE Skived Tape (Enflo) were placed on the anode and cathode side of the membrane to enable the desired GDM compression [16,26,27]. The final fuel cell assembly was sealed with 8 bolts torqued to 10 Nm per bolt in a star pattern.

Once assembled, the performance of single cells was tested using Scribner 850e Fuel Cell Test Systems from Scribner Associates, Inc. All experiments were conducted at 65 °C and ambient pressure (1 atm), unless otherwise noted. Humidifiers were filled with 18 MΩ cm water from a Barnstead Nanopure System. The inlet dew point for both gases was set at 50.3 °C and 37 °C, corresponding to 50% and 25% inlet relative humidity (RH), respectively. Ultrapure gases, $H_2$ (HY UHP300) or Air (UZ 300), purchased from Air-liquide, were supplied to the anode and cathode, respectively, under stoichiometric flow conditions 2|2 for $H_2$|air, unless otherwise noted. All experiments started by pre-conditioning the PEMFC with the following sequence: the cell voltage was first held at 0.60 V in $H_2$|air for 2 h, followed by 20 cycles that alternated between 0.70 V and 0.40 V with each voltage held for 10 min. The cell and gas temperature used in the "break-in" procedure were used to collect the current-voltage (I–V) polarization curves. I–V polarization curves were recorded at increments of 25 mV from open circuit voltage (OCV) to 0.40 V with hold times of 1 min/point. Cell internal resistance was measured at current densities above 100 $mAcm^{-2}$ using the current interrupt technique with the load box and the Fuel Cell V.3.2 software (Scribner Associates Inc.).

The total $O_2$ mass transport resistance was derived from limiting current measurements at different total pressures and varied $O_2$ concentrations [28–30]. High stoichiometric flow rates of the reactant gases were used to maintain uniform gas conditions in the flow channel. A 1.0 slpm flow of $H_2$ was used at the anode, while 1.5 slpm mixtures of $O_2/N_2$ of varying oxidant concentrations were used at the cathode with $O_2$ flow exceeding a stoichiometric ratio of 10 at all testing conditions. The cell temperature was maintained at 80 °C with inlet gases at 62% RH during measurements. Dry $O_2$ mole fractions (1.0%, 1.5%, 2.0%, and 2.5%) in $N_2$ were regulated from an ultra-high purity air tank equipped with a mass-flow controller (mks, Model M100B01322CS1BV). The anode and cathode total pressures (110 $kPa_{abs}$, 150 $kPa_{abs}$, 200 $kPa_{abs}$, and 300 $kPa_{abs}$) were varied to isolate pressure-dependent and pressure-independent $O_2$ transport resistances. The cell voltage was scanned from 0.3 V to 0.06 V in 0.03 V steps, held 2 min at each voltage, and the limiting current was measured in each gas mixture and pressure. The analysis to calculate oxygen transport resistances followed the procedure outlined in the Results and Discussion section.

### 2.2. PEMFC Performance in Dead-Ended Anode (DEA) Mode

Figure 1 presents a schematic of the experimental set-up used to test the PEMFCs in a dead-ended anode configuration.

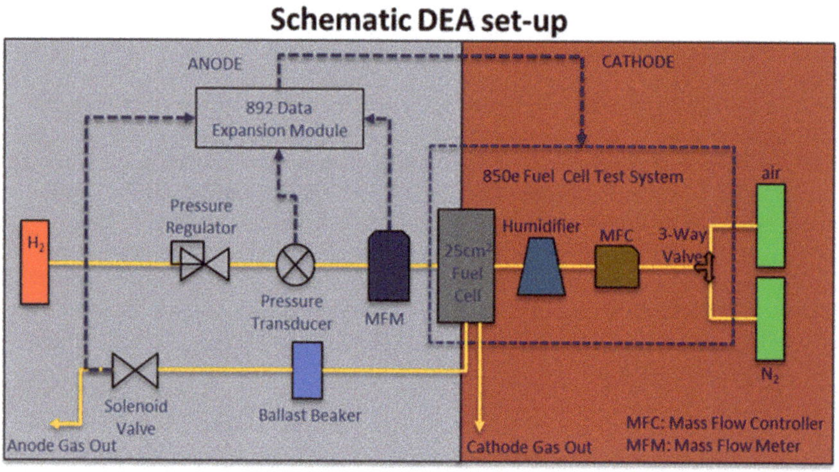


**Figure 1.** Experimental set-up schematic to test a single cell PEMFC in dead-ended anode (DEA) configuration.

In DEA mode, the anode compartment was fed with dry $H_2$ from a gas cylinder. The hydrogen inlet pressure was controlled using a pressure regulator (Harris Mechanical Regulator) and monitored using a pressure transducer (Omega, part# PX359-015CG5V). The inlet $H_2$ flow rate was measured using a mass-flow meter (Masterflex Differential Pressure Flowmeter, Item# EW-32908-69, Cole Palmer). Signals for the pressure transducer and mass flow rate were recorded using an 892e Data Expansion Module connected to the 850e Fuel Cell Test System (Scribner Associates, Inc.). A normally closed solenoid valve (P/N: 009-0631-900, Parker), controlled and activated by the 850e Fuel Cell Test System, was installed at the outlet of the anode to accomplish the DEA operation. On the cathode side, the 850e Fuel Cell Test System was used to control air flow rate (i.e., stoichiometric flow rate of 2) and air relative humidity (RH) supplied under ambient pressure. The humidifier was filled with 18 MΩ cm water from a Barnstead Nanopure System. The cell operating temperature was set to 65 °C. The inlet dew point for the air was set at 50.3 and 37 °C, corresponding to 50% and 25% $RH_{inlet}$, respectively. Ultrapure air (UZ 300) and $H_2$ (HY UHP 300) purchased from Air-liquide was supplied to the cathode and anode side, respectively.

To measure I-V polarization curves in DEA mode, we adapted a purging scheme from references [31–33] that entailed a 1 s duration purge every 60 s. The PEMFCs were operated at constant current (i.e., galvanostatic) mode with selected current densities of 200, 400, 600, 800, 1000, 1200, and 1400 mA $cm^{-2}$. The anode compartment was fed with dry hydrogen. The anode $H_2$ inlet pressure was set to 2 psi. After the load current was applied to the PEMFC, the solenoid valve was opened for 1 min to purge impurities inside the anode compartment and then closed for DEA mode. Humidified air was supplied to the cathode compartment under ambient pressure with a stoichiometric flow rate of 2 for all the current densities. The cell voltage was recorded and monitored by the Scribner 850e Fuel Cell Test System. All selected current densities were held for 3 min. During this time frame, the Scribner 850e Fuel Cell Test System sent a signal to open the solenoid every minute to purge the anode compartment for a set purging duration of 1 s and then the solenoid valve was closed again. Following this purge schedule, cell voltages did not fluctuate significantly during the DEA mode galvanostatic measurement, confirming that the PEMFCs had achieved steady-state equilibrium.

The temporal evolution of the cell voltages was also investigated. The anode compartment was fed with dry $H_2$. The anode $H_2$ inlet pressure was set to 2 psi. Humidified air was supplied to the cathode compartment under ambient pressure with a stoichiometric flow rate of 2 for all the current densities. The PEMFCs were operated at two current densities, i.e., 800 mA $cm^{-2}$ and 1200 mA $cm^{-2}$.

Initially, the PEMFC was operated at a selected current density in the flow-through mode. After the voltage was stabilized, the solenoid valve was closed. When the cell voltage dropped by 0.10 V, the solenoid valve was opened for 1 s. This purging cycle was repeated over a 50 min minimum period for each operating condition.

## 3. Results and Discussion

### *3.1. PEMFC Performance in Regular Flow-Through Mode*

The PEMFC with the asymmetric GDM pairing has lower Ohmic resistances and significantly greater current and power densities at all operating conditions tested in flow-through mode, which flows $H_2$ continuously at a fixed stoichiometric value relative to the air flow. Figure 2 shows the typical I–V polarization and power density curves (A, B, and C) with the associated Ohmic resistances (D, E, and F) measured for a PEMFC containing a symmetric pairing of SGL 29BC GDM on the anode and cathode side (black circle) to a PEMFC containing an asymmetric pairing of Freudenberg GDM on the anode (i.e., H24C3) and cathode (i.e., H23C2) side (red triangle) at a cell working temperature of 65 °C, fed with ambient pressure air, in $H_2$|air at stoichiometric flow of 2/2 humidified at 100%, 50%, and 25% $RH_{inlet}$, respectively.

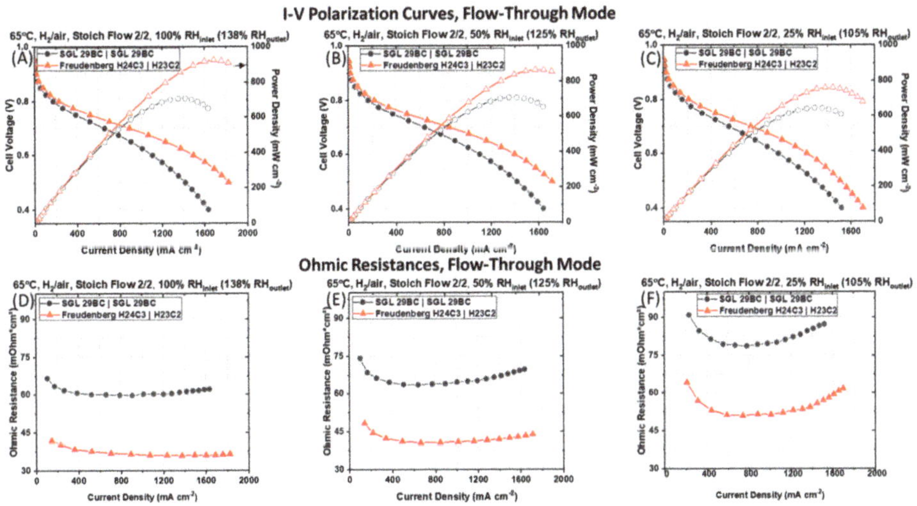


**Figure 2.** I–V polarization and power densities curves (**A–C**) with the associated Ohmic resistances (**D–F**) measured for a PEMFC containing a typical symmetric pairing of SGL 29BC GDM on the anode and cathode side and a PEMFC containing an asymmetrical pairing of Freudenberg GDM [i.e., H24C3 (anode) | H23C2 (cathode)] recorded at a cell temperature of 65 °C, ambient pressure, in $H_2$|air environment at stoichiometric flow rates of 2|2 at (**A**,**D**) 100%, (**B**,**E**) 50%, and (**C**,**F**) 25% $RH_{inlet}$, respectively, from left to right columns.

In the kinetic region of the polarization curve (i.e., operating cell voltage ≥ 0.80 V), there is a smaller effect of mass transport and current densities are very similar for both PEMFCs at all operating conditions. The differences in performance between the PEMFCs containing the symmetric SGL 29BC GDM pairing and the PEMFC containing the asymmetric Freudenberg GDM pairing are more pronounced at the lower cell operating voltage (i.e., operating cell voltage ≤ 0.60 V). Higher current densities are systematically measured for the PEMFC containing the asymmetric Freudenberg GDM pairing (i.e., H24C3 (anode) | H23C2 (cathode)) at lower cell voltage requiring a higher consumption of $O_2$, $H^+$, and rejection of $H_2O$, compared to the PEMFCs containing the symmetric SGL 29BC GDM pairing at all operating conditions.

The current density measured at an operating cell voltage of 0.60 V increases from 1101 to 1394 mA cm$^{-2}$ at 50% $RH_{inlet}$ and from 976 to 1217 mA cm$^{-2}$ at 25% $RH_{inlet}$ when the symmetric GDM pairing of SGL 29BC is replaced by the asymmetric pairing of Freudenberg GDM. The benefit is even greater in more humidified gas flows, when the presence of liquid water is more certain, as we observe a 36% increase in the measured current density at 100% $RH_{inlet}$ in the PEMFC with the asymmetric Freudenberg GDM pairing (i.e., from 1107 to 1491 mA cm$^{-2}$). The peak power density increases by 31.5% at 100% $RH_{inlet,}$ 22% at 50% $RH_{inlet}$, and 18% at 25% $RH_{inlet,}$ respectively, when the symmetric SGL 29BC GDM pairing is replaced by the asymmetric Freudenberg GDM pairing.

As shown in Figure 2D–F, the Ohmic resistances measured at a current density of 1000 mA cm$^{-2}$ for the PEMFC containing the symmetrical SGL 29BC GDM pairing were 60 mOhm cm$^2$ at 100% $RH_{inlet,}$ 64 mOhm cm$^2$ at 50% $RH_{inlet}$, and 80 mOhm cm$^2$ at 25% $RH_{inlet}$, respectively, compared to 36 mOhm cm$^2$ at 100% $RH_{inlet,}$ 42 mOhm cm$^2$ at 50% $RH_{inlet}$, and 51 mOhm cm$^2$ at 25% $RH_{inlet}$ for the PEMFC containing the asymmetric Freudenberg GDM pairing. The lower values of the cell Ohmic resistance for the PEMFC containing the asymmetric Freudenberg GDM pairing suggests that the cell is less sensitive to the adverse effects of dehydration in dry operating conditions. We describe these effects extensively in our prior work [26]. In short, Freudenberg GDM have a significantly lower compressibility, maintain a relatively large void volume at high levels of compressive stress, and have smoother MPL surfaces. These traits are expected to minimize cell contact resistances without sacrificing high gas transport.

The difference in the polarization behavior is not simply a function of the reduction of the cell Ohmic resistance for the PEMFC containing the asymmetric Freudenberg GDM pairng, as this GDM pairing still maintains a higher current density after *iR*-correcting the cell voltages, as shown in Figure 3. The *iR*-corrected polarization curves, which account for differences in cell Ohmic resistances, reveal that the PEMFC containing the symmetric SGL 29BC GDM pairing maintain lower current densities at all *iR*-corrected cell voltages compared to the PEMFC containing the asymmetric Freudenberg GDM pairing. This indicates that in addition to the differences in Ohmic resistance, there are additional sources of resistances in the PEMFCs with different GDM pairing.

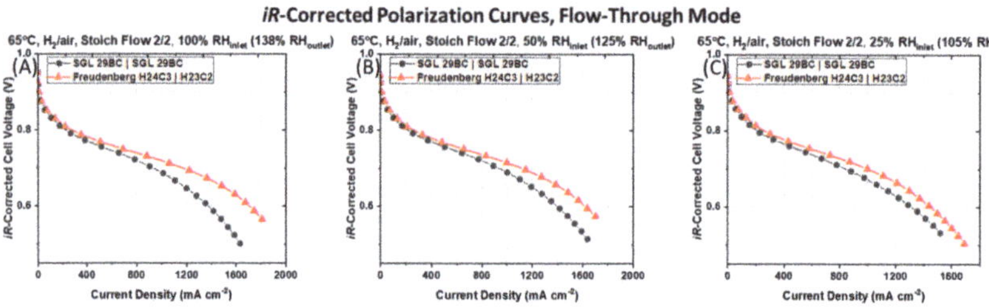


**Figure 3.** *iR*-corrected I–V polarization curves measured for a PEMFC containing a typical symmetric pairing of SGL 29BC GDM on the anode and cathode side and a PEMFC containing an asymmetric pairing of Freudenberg GDM (i.e., H24C3 (anode) | H23C2 (cathode)) recorded at a cell temperature of 65 °C, ambient pressure, in $H_2$|air environment at stoichiometric flow rates of 2|2 at (**A**) 100% $RH_{inlet}$, (**B**) 50% $RH_{inlet}$, and (**C**) 25% $RH_{inlet}$, respectively.

To quantify the total oxygen transport resistance ($R_{total\text{-}O_2}$), $O_2$ limiting current measurements were performed. This experiment resolves where the liquid water is saturating to occlude gas transport. The $R_{total\text{-}O_2}$ was calculated using Equation (1) [28] from the slope of the plot between the limiting current and dry $O_2$ mole fraction at different total pressures as described in the experimental section.

$$R_{Total-O_2} = \frac{4FC_{O_2}}{i_{lim}} = \frac{4F}{i_{lim}} \times \frac{P_{abs} - P_{H_2O}}{R \times T} \times x_{O_{2-dry}} \quad (1)$$

In Equation (1), $C_{O2}$ is the gas channel $O_2$ concentration, $i_{lim}$ is the measured limiting current density (A $cm^{-2}{}_{geometric}$), F is the Faraday constant (96485 C $mol^{-1}$), R is the universal gas constant (8.3145 J $mol^{-1}$ $K^{-1}$), T is the cell temperature (K), $P_{abs}$ is the absolute gas pressure, and $P_{H_2O}$ is the partial pressure of water. The $R_{total\text{-}O_2}$ can be also be described by Equation (2) [30]:

$$R_{Total-O_2} = R_{O_2}^{P-dep} + R_{O_2}^{P-ind} \tag{2}$$

where $R_{O_2}{}^{P\text{-}dep}$ is the pressure dependent $O_2$ bulk diffusion resistance and $R_{O_2}{}^{P\text{-}ind}$ is the pressure independent oxygen transport resistance. $R_{O_2}{}^{P\text{-}dep}$ describes Fickian intermolecular gas diffusion through larger pores (>100 nm diameter), while $R_{O_2}{}^{P\text{-}ind}$ comprises Knudsen diffusion in smaller pores of the microporous layer and the catalyst layers (<100 nm diameter) as well as diffusion through the ionomer film covering the Pt particles.

The asymmetric Freudenberg GDM pairing reduces both the pressure-dependent and pressure-independent $O_2$ transport resistances, suggesting a decrease in liquid water saturation in the cathode. The calculated $R_{total\text{-}O_2}$ are plotted in Figure 4A as a function of the total gas pressure. We observe that both PEMFCs tested have a decrease in $R_{total\text{-}O_2}$ with a decrease of the total gas pressure. Figure 4A shows the effect of the GDM pairing on the $R_{total\text{-}O_2}$, which is the sum of the $R_{O_2}{}^{P\text{-}dep}$ (solid bars) and $R_{O_2}{}^{P\text{-}ind}$ (shaded bars) calculated for the 150 $kPa_{abs}$ back-pressure data set. The symmetric anode and cathode GDM pairing (SGL 29BC) yields a $R_{O_2}{}^{P\text{-}ind}$ that is ~3× larger and $R_{O_2}{}^{P\text{-}dep}$ that is ~1.1× larger than the asymmetric Freudenberg GDM pairing.

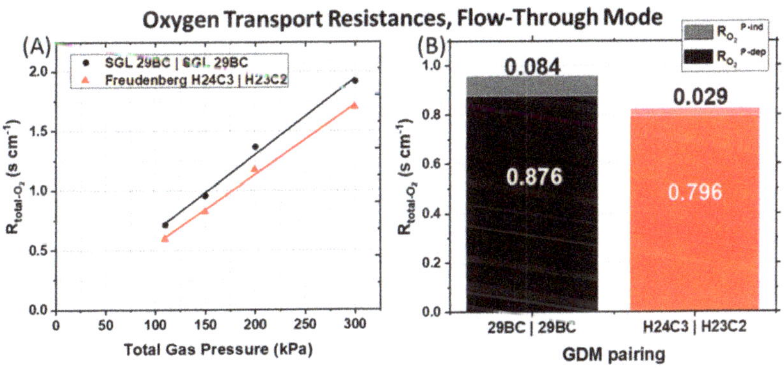


**Figure 4.** (**A**) Total $O_2$ transport resistance ($R_{total\text{-}O_2}$) calculated from Equation (1) as a function of total gas pressure. (**B**) Total $O_2$ transport resistance calculated for the 150 $kPa_{abs}$ back-pressure data set ($R_{total\text{-}O_2}$ = sum of the solid and shaded bars) which can be separated into pressure dependent ($R_{O_2}{}^{P\text{-}dep}$) and pressure independent ($R_{O_2}{}^{P\text{-}ind}$) terms as described by Equation (2).

This indicates that in the PEMFC with the symmetric SGL 29BC GDM pairing, oxygen diffusion is hindered by poor gas transport in the two pore size regimes: the fine pores in the cathode CL also including the ionomer or internal water ($R_{O_2}{}^{P\text{-}ind}$), as well as in larger pores such as those in the MPL, GDL, and the gas channel ($R_{O_2}{}^{P\text{-}dep}$). The exact form of water is uncertain, but possibilities include continuous or discontinuous film and droplets. The higher $R_{O_2}{}^{P\text{-}ind}$ can arise from either more water saturation in the cathode catalyst layer or from a very poorly hydrated cathode CL ionomer that impedes oxygen diffusion through the ionomer film to the active sites [28,34,35]. The accompanying high $R_{O_2}{}^{P\text{-}dep}$ suggests a greater oxygen diffusion resistance through the larger GDL or MPL pores or the gas channels, which is likely the result of the higher tortuosity in the GDL of SGL 29BC, which Zenyuk et al. [36] have reported to be several times greater than the tortuosity of the GDL region of the Freudenberg H23C2 used at the cathode in the asymmetric GDM pairing.

Compared to the symmetric SGL 29BC GDM pairing, the asymmetric Freudenberg GDM pairing overall displays superior power, presumably due to better management of the water, reactants and products. The latter GDM configuration can effectively remove significant amounts of liquid water for high humidity and high power operation as well as maintain membrane hydration during dry operation. Considering that the PEMFC with asymmetric Freudenberg GDM pairing has lower Ohmic resistances (Figure 2D–F) and lower oxygen transport resistances (Figure 4), we conclude that this asymmetric Freudenberg GDM pairing promotes more effective water management in flow-through mode.

### 3.2. *PEMFC Performance in Dead-Ended Anode (DEA) Mode*

The PEMFC with the asymmetric GDM pairing promotes more effective water management in flow-through mode (Section 2.1), and we postulate that this GDM selection will also enhance PEMFC performance in DEA mode and resolve voltage decay due to excessive accumulation of back-diffused liquid water from the cathode and dilution of the anode fuel concentration via $N_2$ crossover. During the DEA operation of a PEMFC, when air is supplied as the oxidant, water produced at the cathode back-diffuses across the membrane to the dry anode and accumulates in the anode flow channel [7–12]. This accumulation of water at the anode blocks the gas transport pathway. Additionally, nitrogen may be transported through the membrane from cathode to anode due to pressure and concentration gradients [7,12–15], resulting in local fuel (i.e., $H_2$) starvation and performance loss. Cell voltage decay in this study will have a stronger sensitivity to the former failure mode, anode water accumulation from insufficient water management, because these PEMFCs only differ in GDM selection. We assume that the $N_2$ crossover rate will not change for the two PEMFCs tested because they are comprised of identical catalyst coated membranes (CCMs).

The advantage of using asymmetric GDM observed in flow-through mode is maintained when the PEMFCs are operated in DEA mode. Polarization curves for the two PEMFCs with different GDM pairings are shown with open symbols in Figure 5.

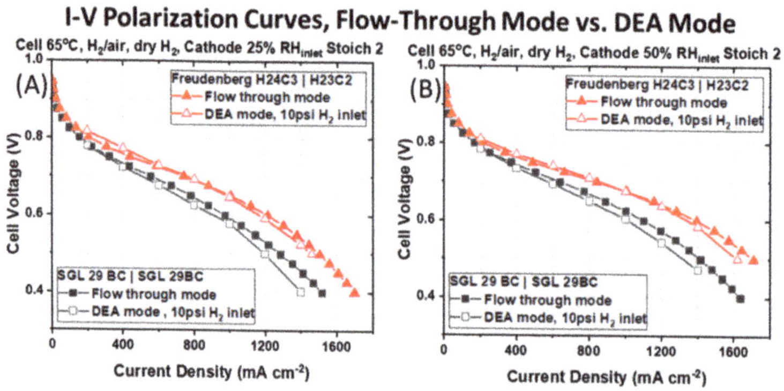


**Figure 5.** Galvanostatic I–V polarization curves measured for a PEMFC containing a typical symmetric pairing of SGL 29BC GDM on the anode and cathode side and a PEMFC containing an asymmetric pairing of Freudenberg GDM. Polarization curves are measured in DEA mode (open symbols) and flow-through mode (closed symbols) at (**A**) 25% $RH_{inlet}$ cathode and (**B**) 50% $RH_{inlet}$ cathode. In all cases, the cell temperature is 65 °C, in $H_2$|air, with air supplied to the cathode at stoichiometric ratio of 2 at atmospheric pressure. During DEA mode, dry $H_2$ is supplied to the anode at 2 psi. In flow-through mode, the inlet RH of the $H_2$ at the anode matches the cathode RH and $H_2$ flows at a stoichiometric ratio of 2.

Owing to the pressure and concentration gradients between anode and cathode during DEA mode, we anticipate higher amounts of water back-diffusion across the membrane, and greater liquid water saturation at the anode. That the asymmetric GDM pairing maintains higher cell voltages and power compared to the PEMFC with symmetric GDM during DEA mode supports that an asymmetric GDM pairing imparts superior water management. At 800 mA cm$^{-2}$ and 25% $RH_{inlet}$ at the cathode [Figure. 5(A)], the PEMFC with asymmetric GDM (Freudenberg H24C3 | H23C2, red curve) pairing produces 0.67 V compared to 0.61 V for the PEMFC with symmetric GDM (SGL 29BC). Similarly, a considerable increase in voltage (and power) is measured at 1200 mA cm$^{-2}$, 0.57 to 0.48 V, when an asymmetric GDM pairing is used in place of a symmetric GDM pairing commonly reported to be used in the open literature. The same trends are observed with a more humidified cathode gas stream in Figure 5B. There are marginal decreases in cell voltages during operation in DEA mode compared to flow-through mode (~42 mV at 1200 mA cm$^{-2}$) for both PEMFCs under study. These are attributed to the use of dry hydrogen on the anode side during testing in DEA mode, while in flow-through mode, the anode $RH_{inlet}$ matches that at the cathode.

The time evolution of the cell voltage is significantly different for the PEMFC containing the symmetric SGL 29BC GDM pairing and the PEMFC containing the asymmetric Freudenberg GDM pairing. Figure 6 shows the time evolution of the cell voltage obtained in DEA mode at two applied current densities of (A) 800 mA cm$^{-2}$ and (B) 1200 mA cm$^{-2}$ for a $H_2$ inlet pressure of 2 psi and with 25% $RH_{inlet}$ air flowing at the cathode. The purge interval is defined as the time between each valve opening event. Examples of these are illustrated at the bottom of Figure 6A,B for each pairing of GDM by either the black or red arrows. The criterion for triggering a valve-opening event was a voltage decay of 0.10 V during constant current operation. All of these observations support that water accumulates in the anode faster for a PEMFC with a symmetric GDM pairing. First, the time between purge events is much greater for the asymmetric GDM pairing. It requires more time at a given current density for the anode to saturate with water and cell voltage decay to be observed. The asymmetric GDM spends a longer time at steady-state prior to voltage decay, and its periods of voltage decay are more gradual. Further, the applied current density has a major impact on the purge interval for each of the PEMFCs under study. A reduction of the purge interval with increasing current density, from Figure 6A to Figure 6B, highlights the inferior water management of the symmetric GDM pairing (black circles) compared to the asymmetric GDM pairing (red triangles). A longer purge duration is desirable in a fuel cell system because it increases $H_2$ utilization and reduces valve wear.

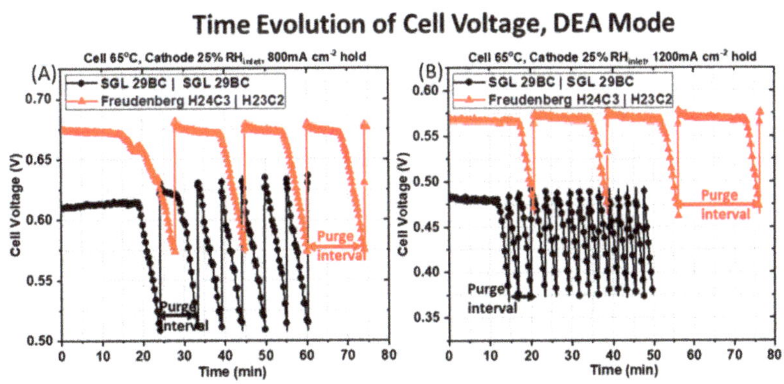


**Figure 6.** Time evolution of the cell voltage obtained at applied current densities of (**A**) 800 mA cm$^{-2}$ and (**B**) 1200 mA cm$^{-2}$ for an $H_2$ inlet pressure of 2 psi and 25% $RH_{inlet}$ cathode. The purge interval is defined as the time between each valve opening event; examples of a purge interval are depicted by the black and red arrows at the bottom of the figures.

As shown in Figure 6A for the PEMFC containing the symmetric SGL 29BC GDM pairing (black circles) at the applied current density of 800 mA cm$^{-2}$ the cell output voltage is maintained at a constant value of ~0.61 V initially from 0 min to 20 min 37 s. This behavior represents an absence of liquid water in the GDL and the channels at the beginning of the experiment after dry $H_2$ was purged through the anode compartment and prior to water production at the cathode. The cell operating voltage drops rapidly after 20 min 37 s, likely due to liquid water saturation in the anode GDM or flow fields from back-diffused water. After 3 min 21 s of steady voltage decline, the first purge event happens and the solenoid is opened to remove water/impurities accumulated in the anode. After the 1 s gas purging duration, the cell operating voltage rises and recovers rapidly to its original level, implying that the 1 s purging duration is sufficient to discharge the accumulated water at this applied current density of at 800 mA $cm^{-2}$. The cell output voltage is stable again for ~7 min, and then steadily declines for about 3 min 21 s, when the next purge event happens. As the experiment progresses, the gap between each purge intervals becomes shorter. There is a mean value of 2 min 24 s ± 21 s for the constant voltage hold period (steady-state, prior to the rapid voltage decay) and a mean value of 2 min 58 s ± 18 s for the regions of rapid voltage decline. This sums to a mean purge interval of 5 min 22 s ± 22 s for the total of five purge events for the last 24 min of the experiment.

For the same operating condition, the response obtained for the PEMFC containing the asymmetric Freudenberg GDM pairing is drastically different, as shown in Figure 6A. The average purge time for the PEMFC with asymmetric GDM is extended by ~10 min compared to the PEMFC with symmetric GDM that are more prone to anode water saturation. The cell output voltage is maintained at a constant value of 0.67 V initially from 0 min to 17 min 46 s, again probably representing an absence of liquid water in the GDL and the channels during this period after dry hydrogen was purged through the anode compartment prior to current generation. The cell operating voltage begins to drop smoothly after 17 min 46 s, likely due to the initial saturation by back-diffused water. The cell voltage decays relatively slowly, ~10 min of steady cell output decline, before the first purge event happens. As the experiment progresses, the gap between each purge intervals remains constant, with mean value of 10 min 45 s ± 1 min 27 s for the constant voltage hold period and a mean value of 4 min 42 s ± 10 s for the steady voltage decline period. This corresponds to mean purge interval of 15 min 27 s ± 1 min 33 s for a total of three purge events for the last 47 min of the experiment.

As the applied current density increases from 800 to 1200 mA $cm^{-2}$, the water generated in the cathode becomes larger, and there is an expected increase in the water back-diffused from the cathode to the anode due to the difference in water vapor concentration. Therefore, the accumulated water in the anode compartment becomes larger as the current density increases. This is reflected in the cell voltage response shown in Figure 6B for the PEMFC containing the symmetric SGL 29 BC GDM pairing at 1200 mA $cm^{-2}$ and 2 psi $H_2$ inlet pressure. The cell voltage is maintained at a constant value of ~ 0.48 V initially from 0 min to 13 min 25s, again probably representing an absence of liquid water in the GDL and the channels during this period, where dry hydrogen is purged through the anode compartment. The cell operating voltage begins to drop smoothly after 13 min 25s, likely due to the initial saturation of back diffusive water vapor. After ~1 min of steady cell output decline, the first purge event happens, the solenoid is opened to remove water/impurities accumulated in the anode compartment. After the 1 s gas purging duration, the cell operating voltage rose and recovered rapidly to its original level, implying that the 1 s purging duration is sufficient to discharge the accumulated water at this applied current density of at 1200 mA $cm^{-2}$. However, as the experiment progresses, the gap between each purge interval is very stable and brief, with a mean value of 1 min 30 s ± 26 s for the constant voltage hold period and a mean value of 1 min ± 3 s for the steady voltage decline period. This corresponds to a mean purge interval of 2 min 33 s± 28 s for a total of 13 purge events for the last 33 min of the experiment. This is a significant increase in purge frequency compared to the lower current density operation.

The PEMFC containing the asymmetric Freudenberg GDM pairing has a different behavior, as shown in Figure 6B. Unlike with the symmetric GDM pairing, which saw a reduction in purge interval at higher current density, the average purge interval marginally increases at higher current density when asymmetric GDM are used. The cell voltage is maintained at a constant value of 0.57 V initially from 0 min to 17 min 34 s, again probably representing an absence of liquid water in the GDM and the channels during this period, where dry hydrogen is purged through the anode compartment.

The cell operating voltage begins to drop smoothly after 17 min 34 s, likely due to the initial saturation of back-diffused water vapor. After ~3 min 10 s of steady cell output decline, the first purge event happens, the solenoid is opened to remove water/impurities accumulated in the anode compartment. As the experiment progresses, the gap between each purge interval remains stable, with a mean value of 15 min 14 s ± 1 min 40 s for the constant voltage hold period and a mean value of 3 min 16 s ± 20 s for the steady voltage decline period. This corresponds to a mean purge interval of 18 min 29 s ± 1 min 21 s for a total of three purge events for the last 55 min of the experiment.

Figure 7 compares the variation of the mean purge interval observed for the data in Figure 6 as a function of the fixed current density for both PEMFCs under study when testing in steady state DEA mode. There is a longer time before the first purge is required due to the initial absence of liquid water in the GDM and the channels prior to the start of the DEA mode operation, as dry hydrogen is purged through the anode compartment. It should be noted that the first purge interval is not included in this calculation because the PEMFC has not yet reached an equilibrium state in hydration. For reference, the blue line shows the purge interval used during polarization characterization in Figure 5.

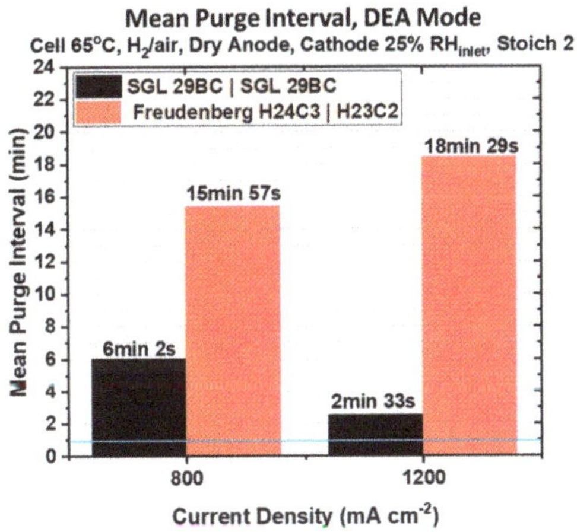


**Figure 7.** Variation of the mean purge interval as a function of the applied current density. The horizontal blue line at 1 min marks the purge duration used to collect the polarization curves reported in Figure 5.

Superior water management imparted by an asymmetric GDM pairing will conserve $H_2$ fuel by vastly reducing anode purging frequency, another advantage of using the Freudenberg asymmetric GDM pairing for a PEMFC in DEA mode. At the applied current density of 800 mA $cm^{-2}$, the mean purge interval is 2.7× longer for the PEMFC containing the asymmetric Freudenberg GDM pairing. As shown in Figure 6A, only four purging events are necessary over an 80 min period compared to seven over a 60 min period for the PEMFC containing the symmetric SGL 29BC pairing. As the applied current density increases from 800 to 1200 mA $cm^{-2}$, the water generated in the cathode increases and we expect a greater back-diffusion driving force from the cathode to the anode due to the greater difference in water vapor concentration between the two electrodes. Therefore, we expect greater water accumulation in the anode compartment as the current density increases. It can be seen from Figure 7, that the PEMFC containing the symmetric SGL 29BC pairing is the most affected by the increase in the current density. The mean purge interval is more than 50% shorter, and a total of fourteen purge events are necessary over a 50 min period compared to only four in an 80 min period for the PEMFC containing the Freudenberg asymmetric GDM pairing, as shown in Figure 6B. The PEMFC constructed with the asymmetric Freudenberg GDM pairing requires even less frequent anode purging at higher current densities. We attribute this to an expected difference in thermal conductivity for the two cathode GDM used in these PEMFCs. Fibrous GDL substrates from Freudenberg have been shown to have a lower thermal conductivity than those GDL substrates developed by SGL [37,38]. With a lower thermal conductivity in the cathode GDL substrate, the cathode catalyst layer temperature would increase, facilitating product water evaporation for rejection from the cathode.

## 4. Conclusions

PEMFCs with an asymmetric Freudenberg GDM pairing that feature different GDM at the anode and the cathode have superior I–V polarization curves in flow-through mode as compared to PEMFCs with a typical symmetric SGL 29BC GDM pairing (i.e., the same GDM is used on the anode and cathode side of the MEA). Compared to the symmetric SGL 29BC GDM pairing, the asymmetric Freudenberg GDM pairing overall displays superior water management. The PEMFC with the asymmetric GDM pairing has lower Ohmic resistances and significantly greater current and power densities at all operating conditions tested in flow-through mode. Additionally, the asymmetric GDM pairing reduces oxygen transport resistances. This result coupled with the significantly lower ohmic resistance suggests that the asymmetric GDM pairing hydrates the membrane and catalyst layers without excess liquid water saturation at the cathode.

The advantage of using asymmetric GDM observed in flow-through mode is maintained when the PEMFCs are operated in DEA mode, supporting that an asymmetric GDM pairing imparts superior water management, even when the driving force for water back-diffusion to the anode is enhanced. This is manifested during DEA mode in several ways. First, the cell voltage is consistently greater at all current densities at both 25% and 50% cathode inlet RH. Second, the asymmetric GDM pairing significantly increases the mean time required between anode purge events. It requires significantly more time at a given current density for the anode to saturate with water and cell voltage decay to be observed.

This promises $H_2$ fuel conservation for a PEMFC stack comprised of asymmetric GDM pairings. Additionally, the asymmetric GDM spends longer time at steady-state prior to voltage decays, and its periods of voltage decay are more gradual. The applied current density has a unique impact on the purge interval for each of the PEMFCs under study. At the applied current density of 800 mA cm$^{-2}$, the mean purge interval is 2.7× longer for the PEMFC containing the asymmetric Freudenberg GDM pairing. The PEMFC containing the symmetric SGL 29BC pairing is the most affected by the increase in the current density. After increasing the current density from 800 to 1200 mA cm$^{-2}$, the mean purge interval is more than 50% shorter, and a total of fourteen purge events are necessary over a 50 min period compared to only four in an 80 min period for the PEMFC containing the Freudenberg asymmetric GDM pairing at the higher current density. The PEMFC constructed with the asymmetric Freudenberg GDM pairing requires even less frequent anode purging at higher current densities. The PEMFC with symmetric GDM pairing must be purged 7.4× more often at this current density.

**Author Contributions:** Conceptualization, Y.G., C.H.B., K.B., B.D.G., and K.E.S.-L.; Methodology, Y.G., C.H.B., B.D.G., and K.B.; Validation, Y.G., R.W.A., K.B., B.D.G., and K.E.S.-L.; Investigation, Y.G. and C.H.B.; Formal analysis, Y.G.; Writing—Original Draft Preparation, Y.G.; Writing—Review & Editing, Y.G., R.W.A., B.G., and K.E.S-L. All authors have read and agreed to the published version of the manuscript.

**Funding:** The authors are grateful to the Office of Naval Research for the financial support of this project.

**Conflicts of Interest:** The authors declare no conflict of interest. The funders had no role in the design of the study; in the collection, analyses, or interpretation of data; in the writing of the manuscript, or in the decision to publish the results.

## References

1. Litster, S.; McLean, G. PEM fuel cell electrodes. *J. Power Sources* **2004**, *130*, 61–76. [CrossRef]
2. Brunner, D.A.; Marcks, S.; Bajpai, M.; Prasad, A.K.; Advani, S.G. Design and characterization of an electronically controlled variable flow rate ejector for fuel cell applications. *Int. J. Hydrogen Energy* **2012**, *37*, 4457–4466. [CrossRef]

3. Kim, B.J.; Kim, M.S. Studies on the cathode humidification by exhaust gas recirculation for PEM fuel cell. *Int. J. Hydrogen Energy* **2012**, *37*, 4290–4299. [CrossRef]
4. Pei, P.; Chen, H. Main factors affecting the lifetime of Proton Exchange Membrane fuel cells in vehicle applications: A review. *Appl. Energ* **2014**, *125*, 60–75. [CrossRef]
5. Han, I.-S.; Jeong, J.; Shin, H.K. PEM fuel-cell stack design for improved fuel utilization. *Int. J. Hydrogen Energy* **2013**, *38*, 11996–12006. [CrossRef]
6. Nikiforow, K.; Karimäki, H.; Keränen, T.M.; Ihonen, J. Optimization study of purge cycle in proton exchange membrane fuel cell system. *J. Power Sources* **2013**, *238*, 336–344. [CrossRef]
7. Abbou, S.; Dillet, J.; Maranzana, G.; Didierjean, S.; Lottin, O. Local potential evolutions during proton exchange membrane fuel cell operation with dead-ended anode—Part I: Impact of water diffusion and nitrogen crossover. *J. Power Sources* **2017**, *340*, 337–346. [CrossRef]
8. Chevalier, S.; Ge, N.; Lee, J.; Antonacci, P.; Yip, R.; George, M.G.; Liu, H.; Banerjee, R.; Fazeli, M.; Bazylak, A. In situ analysis of voltage degradation in a polymer electrolyte membrane fuel cell with a dead-ended anode. *Electrochem. Commun.* **2015**, *59*, 16–19. [CrossRef]
9. Choi, J.W.; Hwang, Y.-S.; Cha, S.W.; Kim, M.S. Experimental study on enhancing the fuel efficiency of an anodic dead-end mode polymer electrolyte membrane fuel cell by oscillating the hydrogen. *Int. J. Hydrogen Energy* **2010**, *35*, 12469–12479. [CrossRef]
10. Lee, Y.; Kim, B.; Kim, Y. An experimental study on water transport through the membrane of a PEFC operating in the dead-end mode. *Int. J. Hydrogen Energy* **2009**, *34*, 7768–7779. [CrossRef]
11. Siegel, J.B.; McKay, D.A.; Stefanopoulou, A.G.; Hussey, D.S.; Jacobson, D.L. Measurement of Liquid Water Accumulation in a PEMFC with Dead-Ended Anode. *J. Electrochem. Soc.* **2008**, *155*, B1168. [CrossRef]
12. Abbou, S.; Dillet, J.; Maranzana, G.; Didierjean, S.; Lottin, O. Local potential evolutions during proton exchange membrane fuel cell operation with dead-ended anode—Part II: Aging mitigation strategies based on water management and nitrogen crossover. *J. Power Sources* **2017**, *340*, 419–427. [CrossRef]
13. Chen, Y.S.; Yang, C.W.; Lee, J.Y. Implementation and evaluation for anode purging of a fuel cell based on nitrogen concentration. *Appl. Energy* **2014**, *113*, 1519–1524. [CrossRef]
14. Manokaran, A.; Pushpavanam, S.; Sridhar, P.; Pitchumani, S. Experimental analysis of spatio-temporal behavior of anodic dead-end mode operated polymer electrolyte fuel cell. *J. Power Sources* **2011**, *196*, 9931–9938. [CrossRef]
15. Yu, J.L.; Jiang, Z.W.; Hou, M.; Liang, D.; Xiao, Y.; Dou, M.L.; Shao, Z.G.; Yi, B.L. Analysis of the behavior and degradation in proton exchange membrane fuel cells with a dead-ended anode. *J. Power Sources* **2014**, *246*, 90–94. [CrossRef]
16. Sassin, M.B.; Garsany, Y.; Gould, B.D.; Swider-Lyons, K.E. Fabrication Method for Laboratory-Scale High-Performance Membrane Electrode Assemblies for Fuel Cells. *Anal. Chem.* **2017**, *89*, 511–518. [CrossRef]
17. Wang, Y.; Chen, K.S.; Mishler, J.; Cho, S.C.; Adroher, X.C. A review of polymer electrolyte membrane fuel cells: Technology, applications, and needs on fundamental research. *Appl. Energy* **2011**, *88*, 981–1007. [CrossRef]
18. Millichamp, J.; Mason, T.J.; Neville, T.P.; Rajalakshmi, N.; Jervis, R.; Shearing, P.R.; Brett, D.J.L. Mechanisms and effects of mechanical compression and dimensional change in polymer electrolyte fuel cells—A review. *J. Power Sources* **2015**, *284*, 305–320. [CrossRef]
19. Jayakumar, A.; Sethu, S.P.; Ramos, M.; Robertson, J.; Al-Jumaily, A. A technical review on gas diffusion, mechanism and medium of PEM fuel cell. *Ionics* **2015**, *21*, 1–18. [CrossRef]
20. Gostick, J.T.; Ioannidis, M.A.; Fowler, M.W.; Pritzker, M.D. On the role of the microporous layer in PEMFC operation. *Electrochem. Commun.* **2009**, *11*, 576–579. [CrossRef]
21. Simon, C.; Kartouzian, D.; Müller, D.; Wilhelm, F.; Gasteiger, H.A. Impact of Microporous Layer Pore Properties on Liquid Water Transport in PEM Fuel Cells: Carbon Black Type and Perforation. *J. Electrochem. Soc.* **2017**, *164*, F1697–F1711. [CrossRef]
22. Weber, A.Z.; Newman, J. Effects of Microporous Layers in Polymer Electrolyte Fuel Cells. *J. Electrochem. Soc.* **2005**, *152*, A677. [CrossRef]
23. Schweiss, R. Benefits of Membrane Electrode Assemblies with Asymmetrical GDL Configurations for PEM Fuel Cells. *Fuel Cells* **2016**, *16*, 100–106. [CrossRef]
24. Atkinson, R.W., III; Hazard, M.W.; Rodgers, J.A.; Stroman, R.O.; Gould, B.D. Influence of Gas Diffusion Media Compression on Open-Cathode Fuel Cells. *J. Electrochem. Soc.* **2019**, *166*, F926–F934. [CrossRef]
25. Atkinson, R.W., III; Rodgers, J.A.; Hazard, M.W.; Stroman, R.O.; Gould, B.D. Influence of Cathode Gas Diffusion Media Porosity on Open-Cathode Fuel Cells. *J. Electrochem. Soc.* **2018**, *165*, F1002–F1011. [CrossRef]

26. Atkinson, R.W., III; Garsany, Y.; Gould, B.D.; Swider-Lyons, K.E.; Zenyuk, I.V. The Role of Compressive Stress on Gas Diffusion Media Morphology and Fuel Cell Performance. *ACS Appl. Energy Mater.* **2018**, *1*, 191–201. [CrossRef]
27. Sassin, M.B.; Garsany, Y.; Gould, B.D.; Swider-Lyons, K. Impact of Compressive Stress on MEA Pore Structure and Its Consequence on PEMFC Performance. *J. Electrochem. Soc.* **2016**, *163*, F808–F815. [CrossRef]
28. Baker, D.R.; Caulk, D.A.; Neyerlin, K.C.; Murphy, M.W. Measurement of Oxygen Transport Resistance in PEM Fuel Cells by Limiting Current Methods. *J. Electrochem. Soc.* **2009**, *156*, B991. [CrossRef]
29. Greszler, T.A.; Caulk, D.; Sinha, P. The Impact of Platinum Loading on Oxygen Transport Resistance. *J. Electrochem. Soc.* **2012**, *159*, F831–F840. [CrossRef]
30. Owejan, J.P.; Owejan, J.E.; Gu, W. Impact of Platinum Loading and Catalyst Layer Structure on PEMFC Performance. *J. Electrochem. Soc.* **2013**, *160*, F824–F833. [CrossRef]
31. Chen, B.; Cai, Y.; Tu, Z.; Chan, S.H.; Wang, J.; Yu, Y. Gas purging effect on the degradation characteristic of a proton exchange membrane fuel cell with dead-ended mode operation I. With different electrolytes. *Energy* **2017**, *141*, 40–49. [CrossRef]
32. Chen, B.; Tu, Z.; Chan, S.H. Performance degradation and recovery characteristics during gas purging in a proton exchange membrane fuel cell with a dead-ended anode. *Appl. Therm. Eng.* **2018**, *129*, 968–978. [CrossRef]
33. Lin, Y.F.; Chen, Y.S. Experimental study on the optimal purge duration of a proton exchange membrane fuel cell with a dead-ended anode. *J. Power Sources* **2017**, *340*, 176–182. [CrossRef]
34. Novitski, D.; Holdcroft, S. Determination of O2 Mass Transport at the Pt | PFSA Ionomer Interface under Reduced Relative Humidity. *ACS Appl Mater Interfaces* **2015**, *7*, 27314–27323. [CrossRef]
35. Orfanidi, A.; Madkikar, P.; El-Sayed, H.A.; Harzer, G.S.; Kratky, T.; Gasteiger, H.A. The Key to High Performance Low Pt Loaded Electrodes. *J. Electrochem. Soc.* **2017**, *164*, F418–F426. [CrossRef]
36. Zenyuk, I.V.; Parkinson, D.Y.; Connolly, L.G.; Weber, A.Z. Gas-diffusion-layer structural properties under compression via X-ray tomography. *J. Power Sources* **2016**, *328*, 364–376. [CrossRef]
37. Bock, R.; Hamre, B.; Onsrud, M.; Karoliussen, H.; Seland, F.; Burheim, O.S. The Influence of Argon, Air and Hydrogen Gas on Thermal Conductivity of Gas Diffusion Layers and Temperature Gradients in PEMFCs. *ECS Trans.* **2019**, *92*, 223–245. [CrossRef]
38. Bock, R.; Shum, A.D.; Xiao, X.; Karoliussen, H.; Seland, F.; Zenyuk, I.V.; Burheim, O.S. Thermal Conductivity and Compaction of GDL-MPL Interfacial Composite Material. *J. Electrochem. Soc.* **2018**, *165*, F514–F525. [CrossRef]

molecules

*Article*

# Optimization of Membrane Electrode Assembly of PEM Fuel Cell by Response Surface Method

**Rohit K. S. S. Vuppala [1,2], Benitta A. Chaedir [1], Lishuai Jiang [3], Lianjun Chen [3], Muhammad Aziz [4] and Agus P. Sasmito [1,*]**

1 Department of Mining and Materials Engineering, McGill University, 3450 University, Frank Dawson Adams Bldg., Montreal, QC H3A2A7, Canada
2 Mechanical Engineering Department, National Institute of Technology, Tiruchirapalli 620 015, India
3 State Key Laboratory of Mining Disaster Prevention and Control, Shandong University of Science and Technology, Qingdao 266590, China
4 Institute of Industrial Science, The University of Tokyo, 4-6-1 Komaba, Meguro-ku, Tokyo 153-8505, Japan
* Correspondence: agus.sasmito@mcgill.ca; Tel.: +1-514-398-3788

Received: 10 July 2019; Accepted: 21 August 2019; Published: 26 August 2019

**Abstract:** The membrane electrode assembly (MEA) plays an important role in the proton exchange membrane fuel cell (PEMFC) performance. Typically, the structure comprises of a polymer electrolyte membrane sandwiched by agglomerate catalyst layers at the anode and cathode. Optimization of various parameters in the design of MEA is, thus, essential for reducing cost and material usage, while improving cell performance. In this paper, optimization of MEA is performed using a validated two-phase PEMFC numerical model. Key MEA parameters affecting the performance of a single PEMFC are determined from sensitivity analysis and are optimized using the response surface method (RSM). The optimization is carried out at two different operating voltages. The results show that membrane thickness and membrane protonic conductivity coefficient are the most significant parameters influencing cell performance. Notably, at higher voltage (0.8 V per cell), the current density can be improved by up to 40% while, at a lower voltage (0.6 V per cell), the current density may be doubled. The results presented can be of importance for fuel cell engineers to improve the stack performance and expedite the commercialization.

**Keywords:** PEM fuel cell; membrane electrode assembly (MEA); response surface method; computational fuel cell dynamics

## 1. Introduction

As a clean energy device, a proton exchange membrane (PEM) fuel cell is a promising power-generating technology that has received increasing attention over the last decade. Fuel cell is an electrochemical device that converts chemical energy into electrical energy. Due to its high energy-conversion efficiency and zero-emission potential, the fuel cell is considered as the best power-generating device, especially in transportation applications. Among the different kinds of fuel cells, PEM fuel cell (PEMFC) offers desirable features, such as low operational temperature and high-power density, which makes it the most promising alternative technology for power production.

In order for the fuel cell technology to be competitive with conventional power systems, some challenges associated with it, including reliability, longevity, and cost, must be overcome. A better understanding of the system is required to achieve the ideal price-performance balance. Studies have been carried out to characterize the behavior of the PEMFC system as affected by different parameters. Wang et al. [1] conducted parametric experiments to study the effect of various operating parameters on the performance of a single PEMFC and used the results to validate the 3D model they developed. The parameters studied were pressure, fuel cell temperature, and anode and cathode relative humidity. It was observed that, generally, increasing the fuel cell temperature and pressure increases its performance, except when the temperature is higher than the gas stream humidification temperatures, especially at a low current density region. Cathode humidification temperature was found to have no significant impact on fuel cell performance, while increasing anode humidification temperature increases performance at the low current density region. These results are in accordance with the results obtained by Ferng et al. [2] and Amirinejad et al. [3] who concluded that operation at higher pressure and elevated temperature can improve the electrode kinetic performance and increase the ionic conductivity in membrane and electrodes, which results in high power density in the PEMFC system. Santarelli and Torchio [4] experimentally analyzed the performance of a PEMFC by varying cell temperature, anode and cathode flow temperatures in both saturation and dry conditions, and reactant pressure. The results showed that, in addition to cell temperature, an increase in reactant saturation temperature also leads to a better performance and the best improvements due to a pressure increase are observed when both anode and cathode are humidified. Yan et al. [5] extended the study of the effects of operating conditions to PEMFC with interdigitated flow field. Nafion-based PEM fuel cell performance analysis with various reactant humidification levels, which varied from no external humidification to a fully saturated anode and cathode, was carried out by Williams et al. [6]. Klika et al. [7] have developed a thermodynamically-consistent model based on polynomial functions to study the behavior of water sorption in Nafion membranes. A three-dimensional multiphase numerical model was developed by Fan et al. [8] to study the PEMFC performance at low external humidification. It was found that the dependency on external humidification of a PEMFC can be cut down at high current density, due to the water produced in the cathode catalyst layer that is sufficient to be employed to humidify both the cathode and anode polymer electrolyte.

In addition to operating conditions, there are other parameters affecting the fuel cell performance. Bayrakçeken et al. [9] found that membrane thickness, hot-pressing conditions of the gas diffusion layer (GDL), and the Teflon to carbon ratio in the GDL, are also significant parameters to provide good PEMFC performance. The study showed that thinner membrane thickness and higher Teflon:carbon ratio in the GDLs give better performances. Jiang et al. [10] implemented an effective "elementary effect" (EE) method based on Monte Carlo random experiments to analyze 22 uncertain parameters involved in their two-phase 1D analytical PEMFC model. Among all of the parameters, membrane thickness and volume fractions were found to be the most important factors influencing the cell performance. The effect of catalyst layer microstructure was recently investigated numerically by Carcadea et al. [11]. A CFD model was used to study the behavior of a PEMFC as a function of Pt loading, Pt particle radius, ionomer volume fraction, and carbon support, and to establish the optimum range of these parameters. It was observed that increasing the ionomer volume fraction in the catalyst layer (CL) leads to better performance due to the fact that the ionomer acts as a network for the mass and charge transport. Moreover, higher Pt loading and a lower particle radius are recommended to achieve better PEMFC performance. Lee et al. [12] investigated the performance improvement of a PEMFC as a function of gas diffusion layer porosity and impregnation of the Nafion solution.

The previously mentioned studies confirm the significance of various parameters on the operation of the PEMFC. It is, therefore, crucial to select the optimum values in order to achieve a high-performance fuel cell. Efforts have been made by researchers toward the optimization of critical parameters influencing the PEMFC operation using different approaches. Salva et al. [13] developed a one-dimensional analytical model and used it to obtain the operating conditions under which a single PEMFC provides the maximum power output for different current intensities. The optimization was carried out for every value of current intensity by solving the parametric table consisting of all possible combinations obtained from modifying the stoichiometry in the cathode and anode, relative humidity in the anode and cathode, and the operating temperature, while keeping the pressure constant. Wu et al. [14] employed a multi-resolution approach and the radial basis function (RBF) surrogate model for simulation and optimization of operating conditions for hydrogen polymer electrolyte fuel cells. Zervas et al. [15] performed a phosphoric acid fuel cell (PAFC) system optimization study based on meta-models that were derived by applying the linear regression and the RBF neural network methodology on the results produced by a CFD model. The optimization of different operating and design parameters on PEMFC using the Taguchi method was performed by Karthikeyan et al. [16], Solehati et al. [17], and Sasmito et al. [18]. Grujicic and Chittajallu [19] utilized a single-phase two-dimensional electrochemical model, coupled with a nonlinear constrained optimization algorithm, which was solved using sequential quadratic programming (SQP) to obtain the operational and geometric parameters for achieving the maximum electric current in a PEMFC. The parameters investigated include air inlet pressures and cathode thickness, cathode length for each shoulder segment of flow channel, and a fraction of cathode length associated with the flow channel. Similarly, Na and Gou [20] used SQP to optimize the fuel cell system efficiency and cost. Guo et al. [21] proposed an optimization algorithm that combines the teaching–learning based optimization (TLBO) with a differential evolution (DE) algorithm, known as the TLBO-DE method, to promote the efficiency of PEMFC. Behrou et al. [22] demonstrated the use of density-based topology optimization for the practical design of flow fields for PEMFCs, with goals to maximize both the output power and homogeneity of the current density distribution, as well as permit reduced costs and higher durability. The response surface methodology (RSM) has been employed by Kanani et al. [23] and Xuan et al. [24] to maximize the performance of a PEMFC system. Recently, a comprehensive evaluation of different optimization scenarios for a PEMFC is provided by Sohani et al. [25].

Optimization of controlling parameters at the fuel cell system level, like membrane thickness, size of cathode catalyst particle, and protonic conductivity coefficient of the membrane, however, have been very limited. This can be attributed to the fact that they cannot be changed during the cell utilization [4], which makes it tedious and uneconomical to perform these studies experimentally. Moreover, the complex structure of MEA, comprising of polymer electrolyte membrane sandwiched by agglomerate catalyst layers at both the anode and cathode, adds complexity for the design of experiments due to a multiscale nature of the system. In addition, none of the research studies had focused on the membrane electrode assembly coupled with agglomerate catalyst layer parameters. This paper aims to develop a numerical model to simulate a polymer electrolyte membrane fuel cell with detailed multiscale MEA with an agglomerate catalyst layer model, and determine the optimum values of the previously mentioned parameters that provide maximum current density for various voltage values in the ideal range of operation. The study focuses on sensitivity analysis of design parameters of the membrane electrode assembly, including membrane thickness, membrane equivalent weight, the radius of the cathode catalyst particle, catalyst ionomer resistance, cathode catalyst porosity, a membrane protonic conductivity coefficient, a cathode catalyst hydrophobic angle, ionomer tortuosity, and a cathode catalyst volume fraction. The parameters which have significant impact on the current density magnitude are considered. Once the model is validated, it is used to carry out parameter optimization for better cell performance. In the optimization, the response surface methodology is employed for meta-modelling. RSM is a collection of statistical and mathematical methods for optimizing and predicting responses with limited experimental data at various input factors, as well as performing sensitivity analysis [23]. It is extensively used in the industrial world, particularly in situations where the output is swayed by several input parameters. This method has been widely used in different fields and applications such as metals removal [26], chemical extraction [27,28], and the chemical and environmental engineering field [29–31]. As compared to other methods, RSM being a collection of mathematical and statistical techniques, gives a better understanding into the role of different parameters at play and generates a continuous model for visualizing the effect of parameters on the entire range as opposed to the average value of the response [32]. This study pioneers the sensitivity analysis of parameters of MEA coupled with agglomerate catalyst layers using the design of the experiment RSM method, along with the validated three-dimensional numerical model. The total numbers of simulations for combinations of 10 parameters would have been computationally very expensive and could not have been done using traditional parametric studies. The RSM method reduces the number of simulations significantly.

## 2. Methodology

### 2.1. Mathematical Formulation

The schematic figure of a typical PEMFC and its functional layers is illustrated in Figure 1. The system consists of a proton exchange membrane (m), sandwiched between two catalyst layers (cl), two gas diffusion layers (gdl), two porous metal foam flowfields (ff), and two terminal plates (tp). The main assumptions/approximations adopted in the model are:

- Thermal equilibrium: Local thermal equilibrium between all the phases.
- Membrane: The membrane model takes into account the water flux due to electro-osmatic drag and diffusion.
- Catalyst layers: A cathode particle/agglomerate model is implemented to account for the mass transfer inside the cathode catalyst layer. It is assumed that the particle is spherical and covered by a thin layer of ionomer and water film [33–36]. The Butler-Volmer equation is employed to calculate the volumetric current transfer or exchange current density.

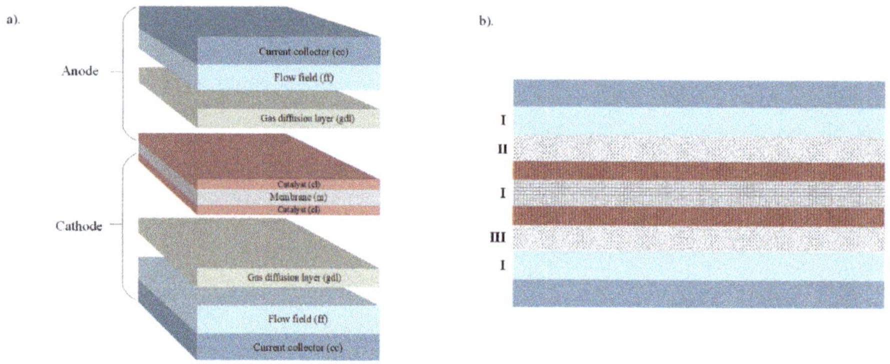


**Figure 1.** Schematic view of a single PEMFC: (**a**) components of PEMFC, (**b**) computational domains with boundaries: I—insulation/wall, II—anode inlet, III—cathode inlet, and IV—outlets.

The mathematical model is comprised of governing equations for the conservation of mass, momentum, species, energy, charge, and water transport in the membrane. The physical parameters, geometry, and operating conditions for two different PEMFC experimental data sets that are used later for validation purposes can be found in Tables 1 and 2.

**Table 1.** Geometrical and operating parameters for two PEMFC experimental data sets.

| | Case a Segmented Cell | Case Single-Gas Diffusion Layer [37] |
|---|---|---|
| Physical parameters | | |
| $\kappa_{cl}, \kappa_{gdl}$ | $7.3 \times 10^{-13}$ m$^2$ | $6.1 \times 10^{-11}$ m$^2$ |
| $\epsilon_{ff}$ | 0.9 | 0.635 |
| $\epsilon_{gdl}$ | 0.4 | 0.77 |
| $\sigma_{s,gdl}$ | 500 S m$^{-1}$ | 491 S m$^{-1}$ |
| $\sigma_{s,cl}$ | 500 S m$^{-1}$ | 491 S m$^{-1}$ |
| $r_{agg}$ | $10^{-7}$ m | $5 \times 10^{-6}$ m (adapted) |
| $j_{c,0}^{ref}$ | $10^3$ A m$^{-3}$ (adapted) | $3.5 \times 10^4$ A m$^{-3}$ (adapted) |
| $\alpha_c$ | 2 (adapted) | 0.65 |
| $\beta_m$ | 0.9 (adapted) | 0.2 (adapted) |

**Table 1.** *Cont.*

| | Case a Segmented Cell | Case Single-Gas Diffusion Layer [37] |
|---|---|---|
| Geometry | | |
| $h_{cc}$ | $5 \times 10^{-4}$ m | $6 \times 10^{-4}$ m |
| $h_{ff}$ | $5 \times 10^{-4}$ m | $6 \times 10^{-4}$ m |
| $h_{gdl}$ | $3 \times 10^{-4}$ m | $1.1 \times 10^{-4}$ m |
| $h_{cl}$ | $10^{-5}$ m | $2 \times 10^{-5}$ m |
| $h_m$ | $3 \times 10^{-5}$ m | $5.1 \times 10^{-5}$ m |
| L | 0.09 m | 0.015 m |
| Operating conditions | | |
| $\eta_{a,c}$ | 95%, 95% | 100%, 100% |
| $T_{a,c}^{in}$ | 333 K, 333 K | 338 K, 328 K |
| $p^{ref}$ | 101,325 Pa | 1.5 bars |
| $\xi_{a,c}$ | 3.35, 2.3 | – |
| $U_{a,c}^{out}$ | – | 0.03, 0.16 m $s^{-1}$ |
| $E_{cell}$ | 0.15–0.85 V | 0–0.95 V |

**Table 2.** Additional parameters for all cases.

| Parameter | Value |
|---|---|
| $c_F$ | 0.55 [38] |
| $c_{H_2}^{ref}$ | 40.88 mol $m^{-3}$ [39] |
| $c_{O_2}^{ref}$ | $\frac{1}{H_{O_{2,pol}}}$ mol $m^{-3}$ [40] |
| $D_{H_2}^0, D_{H_2O}^0, D_{O_2}^0$ | $(11.03, 7.35, 3.23) \times 10^{-5}$ $m^2$ $s^{-1}$ [39] |
| $D_{O_2,m}^0$ | $3.1 \times 10^{-7} e^{(-\frac{1768}{T})}$ $m^2$ $s^{-1}$ [40] |
| $E_a$ | 73,269 J $mol^{-1}$ [41] |
| $E_{rev}^0$ | 1.23 V [39] |
| F | 96,487 A s $mol^{-1}$ |
| $H_{O_2,pol}, H_{O_2,liq}$ | $1.33 \exp(-666/T)$, $5.08 \exp(-498/T)$ atm $m^3$ $mol^{-1}$ [40] |
| $j_{a,0}^{ref}$ | $10^9$ A $m^{-3}$ [39] |
| $k_{cc}, k_{cl}, k_{ff}, k_{gdl}, k_m$ | 16.3, 1.5, 13.3, 1.5, 0.1 W $m^{-1}K^{-1}$ [42–45] |
| $k_{H_2}, k_{H_2O}, k_{N_2}, k_{O_2}$ | $(20.285, 2.16, 2.82, 2.89) \times 10^{-2}$ W $m^{-1}K^{-1}$ [46] |
| $M_{H_2}, M_{H_2O}, M_{N_2}, M_{O_2}$ | $(2, 18, 28, 32) \times 10^{-3}$ kg $mol^{-1}$ |
| $M_m$ | 1.1 kg $mol^{-1}$ [39] |
| $m_{pol}$ | $10^{-2}$ kg $m^{-2}$ [37] |
| R | 8.314 J $mol^{-1}K^{-1}$ |
| $\alpha_a$ | 1 K [39] |
| $\gamma_{gd}, \gamma_{ld}$ | 0.5 |
| $\vartheta$ | 1 |
| $k_{ff}, k_m$ | $10^{-8}$, $10^{-18}$ $m^2$ [47] |
| $\mu$ | $1.9 \times 10^{-5}$ kg $m^{-1}s^{-1}$ [45] |
| $\mu_{H_2}, \mu_{H_2O}, \mu_{N_2}, \mu_{O_2}$ | $(9.656, 10.98, 19.39, 22.62) \times 10^{-6}$ kg $m^{-1}s^{-1}$ [46] |
| $\rho_C, \rho_m, \rho_{Pt}$ | $(1.8, 2, 21.45) \times 10^3$ kg $m^{-3}$ [39,40] |
| $\sigma_{s,cc}, \sigma_{s,ff}$ | $(1.37, 0.1) \times 10^6$ S $m^{-1}$ [42,45] |
| $c_1, c_2, c_3, c_4$ | $-2.1794$, $2.953 \times 10^{-2}$, $-9.1837 \times 10^{-5}$, $1.4544 \times 10^{-7}$; –, $K^{-1}$, $K^{-2}$, $K^{-3}$ [47] |
| $T_0, T_1, T_2$ | 273.15, 353.15, 298.15 K [39] |

#### 2.1.1. Governing Equations

Conservation of Mass [34]:

$$\frac{\partial \rho}{\partial t} + \nabla \cdot \left(\rho \vec{v}\right) = S_m \tag{1}$$

where $S_m$ is the mass source added from the continuous phase to the dispersed second phase and any other user-defined sources.

Conservation of Momentum [34]:

$$\frac{\partial}{\partial t}\left(\rho \vec{v}\right) + \nabla \cdot \left(\rho \vec{v} \vec{v}\right) = -\nabla p + \nabla \cdot \left(\bar{\bar{\tau}}\right) + \rho \vec{g} + S_{mom} \quad (2)$$

where $p$ denotes pressure, $\bar{\bar{\tau}}$ is the stress tensor, $\rho \vec{g}$ denotes the gravitational body forces, and $S_{mom}$ is the momentum source term for porous media, which includes the gas diffusion layer, catalyst layer, and membrane. The stress tensor $\bar{\bar{\tau}}$ is given by the equation below [34].

$$\bar{\bar{\tau}} = \mu\left[\left(\nabla \vec{v} + \nabla \vec{v}^{T}\right) - \frac{2}{3}\nabla \cdot \vec{v} I\right] \quad (3)$$

where $\mu$ is the molecular viscosity and $I$ is the unit tensor. The second term in the right-hand side of the equation represents the effects of volume dilation.

Species Transport [34]:

$$\frac{\partial}{\partial t}\left(\rho Y_i\right) + \nabla \cdot \left(\rho \vec{v} Y_i\right) = -\nabla \cdot D_{i,eff} \nabla Y_i + S_m \quad (4)$$

where $Y_i$ denotes the local mass fraction of species $i$ and $D_{i,eff}$ is the effective diffusivity of the species. Note that the total species mixture should conserve the total mass, and, thus, the source terms in the conservation of mass should be equal to the source terms in the conservation of species [48].

Electric Potential [34]:

$$\nabla \cdot (\sigma \nabla \psi) + S = 0 \quad (5)$$

where $\psi$ is the electric potential, $\sigma$ is the electric conductivity in a solid zone or ionic conductivity in a fluid zone, and $S$ is a source term.

Conservation of Energy [34]:

$$\frac{\partial}{\partial t}\left(\rho E\right) + \nabla \cdot \left(\rho C_p \vec{v} T\right) = \nabla \cdot \left(k_{eff} \nabla T\right) + S_h \quad (6)$$

where $C_p$ is the specific heat capacity and $k_{eff}$ is the effective thermal conductivity. The first term on the right-hand side of the equation represents the energy transfer due to conduction.

Volumetric source terms ($S_m$) for $H_2$ and $O_2$ and the dissolved water content $\lambda$ in the triple-phase boundaries (catalyst layers) due to electrochemical reactions are shown below [34].

$$S_{H_2} = -\frac{M_{wH_2}}{2F} R_{an} < 0 \quad (7)$$

$$S_{O_2} = -\frac{M_{wO_2}}{2F} R_{cat} < 0 \quad (8)$$

$$S_{\lambda} = -\frac{M_{wH_2O}}{2F} R_{cat} > 0 \quad (9)$$

where $M_{wH_2}$, $M_{wH_2O}$, and $M_{wH_2}$ are the molecular mass of hydrogen, oxygen, and water, respectively, and $F$ is the Faraday constant.

Mass transfer and water transport occurring in the PEMFC model is considered to be in two different phases, which are discussed below.

1. Dissolved phase

The dissolved phase exists in the catalyst layers (ionomers) and in the membrane. The generation and transport of the dissolved water is described by the equation below [49].

$$\frac{\partial}{\partial t}\left(\epsilon_i M_{w,H_2O} \frac{\rho_i}{EW} \lambda\right) + \nabla \cdot \left(\vec{i}_m \frac{\eta_d}{F} M_w\right) = \nabla \cdot \left(M_w D^i{}_w \nabla \lambda\right) + S_\lambda + S_{gd} + S_{ld} \quad (10)$$

where $\epsilon_i$ denotes the porous media porosity, $\lambda$ denotes the dissolved water content, $\eta$ is the osmotic drag coefficient, and $\vec{\imath}_m$ is the ionic current density, calculated as $\vec{\imath}_m = -\beta_m \sigma_{mem} \nabla \phi_{mem}$. In the right hand-side of the equation, $D^i{}_w$ represents the diffusion coefficient of the water content, $S_\lambda$ denotes the water generation rate due to the cathode side reaction in the catalyst layer, $S_{gd}$ is the rate of mass change between gas and dissolved phases, and $S_{ld}$ is the rate of mass change between the liquid and the dissolved phases. The mass change rates are shown in the equations below [50].

$$S_{gd} = \left(1 - s^{\theta}\right) \gamma_{gd} M_{w,H_2O} \frac{\rho_i}{EW}\left(\lambda_{eq} - \lambda\right) \tag{11}$$

$$S_{ld} = \left(s^{\theta}\right) \gamma_{ld} M_{w,H_2O} \frac{\rho_i}{EW}\left(\lambda_{eq} - \lambda\right) \tag{12}$$

where $\rho_i$ is the dry ionomer or membrane density, $EW$ is the equivalent weight of the membrane, $s$ denotes the liquid saturation, $\lambda_{eq}$ denotes the equilibrium water content, $\theta$ is the exponential liquid coverage, and $\gamma_{gd}$, $\gamma_{ld}$ are the gas and liquid mass exchange rate constants and are user-specified parameters. The equilibrium water content $\lambda_{eq}$ can be calculated using the equation below [50].

$$\begin{aligned} \lambda_{eq} = {} & 0.3 + 6a(1 - tanh(a - 0.5)) + 0.69(\lambda_{a=1} - 3.52)a^{0.5}\left(1 + \tanh\left(\tfrac{a-0.89}{0.23}\right)\right) \\ & + s.(\lambda_{s=1} - \lambda_{a=1}) \end{aligned} \tag{13}$$

where $a$ is the water activity, defined as:

$$a = \frac{p_{wv}}{p_{sat}} \tag{14}$$

where $p_{wv}$ is the water vapour partial pressure and $p_{sat}$ is the saturation pressure. Both $\lambda_{s=1}$ and $\lambda_{a=1}$ in Equation (13) are user-specified parameters.

2. Liquid Phase

Liquid is present in all the porous electrodes and gas channels. The driving force of the liquid water transport is the liquid pressure gradient $\nabla p_l$, as shown in the liquid water transport equation below [50].

$$\frac{\partial}{\partial t}(\epsilon_i \rho_l\, s) = \nabla \cdot \left(\frac{\rho_l K K_r}{\mu_l} \nabla p_l\right) + S_{gl} - S_{ld} \tag{15}$$

In Equation (15), $\rho_l$ is the liquid water density, $\mu_l$ is the liquid dynamic viscosity, $K$ is the absolute permeability, $K_r$ is the relative permeability, $p_l$ is liquid pressure, and $S_{gl}$ is the rate of mass change between the gas and liquid phases. Replacing the liquid pressure with the sum of capillary pressure $p_c$ and gas pressure $p$, Equation (15) can be written as the equation below.

$$\frac{\partial}{\partial t}(\epsilon_i \rho_l\, s) = \nabla \cdot \left(\frac{\rho_l K K_r}{\mu_l} \nabla (p_c + p)\right) + S_{gl} - S_{ld} \tag{16}$$

The mass transfer rate between the gas and liquid phases can be calculated based on the unidirectional diffusion theory [50,51].

$$S_{gl} \begin{cases} \gamma_{gl}\, \epsilon\, s D_{gl} \frac{M_{w,H_2O}}{RT} p ln\left(\frac{p - p_{sat}}{p - p_{wv}}\right), & p_{wv} \leq p_{sat} \\ \gamma_{gl}\, \epsilon\, (1 - s) D_{gl} \frac{M_{w,H_2O}}{RT} p ln\left(\frac{p - p_{sat}}{p - p_{wv}}\right), & p_{wv} > p_{sat} \end{cases} \tag{17}$$

where $\epsilon$ is porosity, $\gamma_{gl}$ is the geometric factor of the droplet size, and $D_{gl}$ $\left[m^2/s\right]$, in the function of temperature [K] and pressure [Pa], takes the following form.

$$D_{gl} \begin{cases} 0.365 \cdot 10^{-4} \left(\frac{T}{343}\right)^{2.334} \cdot \left(\frac{10^5}{p}\right), & cathode \\ 1.79 \cdot 10^{-4} \left(\frac{T}{343}\right)^{2.334} \cdot \left(\frac{10^5}{p}\right), & anode \end{cases} \tag{18}$$

Equation (16) is solved in all the regions from the anode GDL-channel interface to the cathode GDL-channel interface. At both interface boundaries, the liquid water flux is considered to leave the gas diffusion layers into the gas channel only. The flux is assumed to be driven by the capillary pressure [50].

$$f_{liq} = max[\Theta\epsilon\, sp_c, 0] \tag{19}$$

where $\Theta$ is the coefficient of liquid water removal.

Liquid saturation in the channels is calculated from the Leverett function.

$$p_c \begin{cases} \sigma\, cos\theta_c \sqrt{\frac{\epsilon}{K}}\, J\,(1-s), & \theta_c < 90^{\circ} \\ \sigma\, cos\theta_c \sqrt{\frac{\epsilon}{K}}\, J\, s, & \theta_c > 90^{\circ} \end{cases} \tag{20}$$

where

$$J(x) = 1.417x - 2.12x^2 + 1.263\,x^3 \tag{21}$$

Liquid water transport in the gas channels is determined to predict the pressure drop increase using the following correlation.

$$\frac{\partial}{\partial t}(\rho_l\, s) + \nabla\cdot\left(p_l \vec{v}_l s\right) = \nabla\cdot\left(D_{liq}\, \nabla s\right) \tag{22}$$

where $D_{liq}$ is the liquid water diffusion coefficient in the gas channel and the liquid velocity $\vec{v}_l$ is assumed to be a fraction of the gas velocity $\vec{v}_g$, i.e., $\vec{v}_l = \chi\vec{v}_g$.

Since the flow-channels (ff) in our model are porous in nature, user-defined function (UDF) is used to add the corresponding source term to X, Y, and Z momentum for the inertial losses.

$$S_i = -\left(\frac{\mu}{\alpha}\, v_i + C_2\, \frac{1}{2}\, \rho\, |v|\, v_i\right) \tag{23}$$

where $S_i$ denotes the source for the *ith* (x, y, or z) momentum equation, $|v|$ denotes the magnitude of the velocity, $\alpha$ is the permeability, and $C_2$ is the inertial resistance factor.

In laminar flows through porous media, the pressure drop is typically proportional to the velocity and the constant $C_2$ can be considered zero. Ignoring the convective acceleration and diffusion, the porous media model is reduced to Darcy's law.

$$S_i = -\frac{\mu}{\alpha}\, v_i \tag{24}$$

The volumetric heat sources in various zones can be found in Table 3, where the variables $i_s$ and $i_m$ represent the magnitudes of the solid phase and membrane phase current density, respectively, and $L$ (<0) is the latent heat due to water condensation.

**Table 3.** Energy source terms for the governing equations of conservation energy [50].

| Zone | Additional Source Terms |
|---|---|
| GDL+MPL | $\frac{i_s^2}{\sigma_{sol}} - S_{gl}\cdot L$ |
| Anode catalyst layer | $R_{an}\left(\eta_{an} - \frac{T\Delta S_{an}}{2F}\right) + \frac{i_s^2}{\sigma_{sol}} + \frac{i_m^2}{\sigma_{mem}} - \left(S_{dl} + S_{gl}\right)\cdot L$ |
| Cathode catalyst layer | $R_{cat}\left(-\eta_{cat} - \frac{T\Delta S_{cat}}{2F}\right) + \frac{i_s^2}{\sigma_{sol}} + \frac{i_m^2}{\sigma_{mem}} - \left(S_{dl} + S_{gl}\right)\cdot L$ |
| Membrane (solid) | $\frac{i_m^2}{\sigma_{mem}}$ |
| Current collector (solid) | $\frac{i_s^2}{\sigma_{sol}}$ |
| Gas channels | - |

### 2.1.2. Electrochemistry and Cathode Particle/Agglomerate Model

The driving force of these reactions is the surface overpotential, which is the difference between the phase potential of the solid and of the electrolyte or membrane. The phenomenon is accounted for in two equations: one for the electron transport in the catalyst layer, solid grids of porous media, and the current collector, and the other for the protonic conduction or transport of $H^+$ at the catalyst and the membrane [52,53].

$$\nabla\cdot(\sigma_{sol}\ \nabla\phi_{sol}) + R_{sol} = 0 \tag{25}$$

$$\nabla\cdot(\beta_m\sigma_{mem}\ \nabla\phi_{mem}) + R_{mem} = 0 \tag{26}$$

where $\sigma$ denotes the electrical conductivity in ohm-m$^{-1}$, $\phi$ denotes the electric potential in volts, and $R$ denotes the volumetric transfer current in A·m$^{-3}$, which is also known as exchange current density, expressed as:

$$R_{an} = (\ \zeta_{an}\ j_{an}\ (T))\left(\frac{[A]}{[A]_{ref}}\right)^{\gamma_{an}}\left(e^{\alpha_{an}^{an}F\eta_{an}/RT} - e^{-\alpha_{cat}^{an}F\eta_{an}/RT}\right) \tag{27}$$

$$R_{cat} = (\ \zeta_{cat}\ j_{cat}\ (T))\left(\frac{[C]}{[C]_{ref}}\right)^{\gamma_{cat}}\left(-e^{\alpha_{an}^{cat}F\eta_{cat}/RT} + e^{-\alpha_{cat}^{cat}F\eta_{cat}/RT}\right) \tag{28}$$

In the above equations, $j\ (T)$ is the reference exchange current density per active surface area [A·m$^{-2}$], $\zeta$ is the specific active surface area [m$^{-1}$], [ ] and $[\ ]_{ref}$ are the species local concentration and its reference value [kmol·m$^{-3}$], $\gamma$ is the concentration dependence, $\alpha_{an}^{an}$ and $\alpha_{cat}^{an}$ are anode and cathode dimensionless transfer coefficients of the anode electrode, respectively, $\alpha_{an}^{cat}$ and $\alpha_{cat}^{cat}$ are the anode and cathode dimensionless transfer coefficients of cathode electrode, $\eta_{an}$ is the surface overpotential, F is the Faraday constant (9.65 × $10^7$ C·kmol$^{-1}$), R is the universal gas constant, and $T$ is the temperature.

The reference exchange current densities $j_{an}(T)$ and $j_{cat}(T)$ are dependent on a local temperature, described as follows [52,53].

$$j_{an}(T) = j_{an}^{ref}\ e^{-E_{an}/RT(1-T/T_{an}^{ref})} \tag{29}$$

$$j_{cat}(T) = j_{cat}^{ref}\ e^{-E_{cat}/RT(1-T/T_{cat}^{ref})} \tag{30}$$

where $E$ and $T_{ref}$ are user-specified activation energy and reference temperature, respectively, and $j_{an}^{ref}$ and $j_{cat}^{ref}$ are the associated exchange current densities at the specified reference temperature.

The driving force for the kinetics is the local surface overpotential $\eta$, also known as the activation loss. It is defined as the difference between the solid and membrane potentials, $\phi_{sol}$ and $\phi_m$.

$$\eta_{an} = \phi_{sol} - \phi_m - U_{an}^0 \tag{31}$$

$$\eta_{cat} = \phi_{sol} - \phi_m - U_{cat}^0 \tag{32}$$

The half-cell potentials at the cathode and anode, $U_{an}^0$ and $U_{cat}^0$, can be calculated using the Nernst equation [50].

$$U_{an}^0 = E_{an}^0 - \frac{\Delta S_{an}}{2F}\left(T - T^0\right) - \frac{RT}{2F}ln\frac{p_{H_2}}{p^0} \tag{33}$$

$$U_{cat}^0 = E_{cat}^0 - \frac{\Delta S_{cat}}{2F}\left(T - T^0\right) - \frac{RT}{2F}ln\frac{p_{H_2O}}{p_{sat}\sqrt{p_{O_2}/p^0}} \tag{34}$$

where $E^0$ denotes the reversible potential, $\Delta S$ denotes the reaction entropy, $p_{sat}$ denotes the saturation pressure of water, $T^0$ and $p^0$ are standard temperature and pressure, and $p_{H_2}$, $p_{O_2}$, and $p_{H_2O}$ are the partial pressures of hydrogen, oxygen, and water vapor, respectively.

In computing the cathode transfer current using Equation (28), the mass transport resistance of the microstructure is not taken into account in the equation [50]. The resistance consists of two parts, which includes the resistance due to ionomer film $R_{ion}$ and the resistance due to liquid water film surrounding the particles $R_{liq}$. The volumetric transfer current in the cathode layer is represented by the formula below.

$$R_{cat} = 4F\frac{c_{O_2}}{c_{O_2}/j_{O_2}^{ideal} + R_{ion} + R_{liq}} \quad (35)$$

where $c_{O_2}$ is the oxygen concentration at the wall, $R_{ion}$ is a user-specified value, and $R_{liq}$ can be calculated as the equation below.

$$R_{liq} = \frac{\zeta_{cat} r_p^2}{K_w D_w} \cdot \frac{\sqrt[3]{1 + \frac{s\epsilon}{1-\epsilon}}}{3(1-\epsilon)} \quad (36)$$

where $\zeta_{cat}$ is the specific active surface area for the cathode catalyst [$m^{-1}$], $s$ is the liquid saturation, $\epsilon$ is the porosity, $r_p$ is the particle diameter, and $K_w D_w$ is the product of oxygen solubility and diffusivity in liquid water.

The parameter $j_{O_2}^{ideal}$ in Equation (35) is defined as the equation below.

$$j_{O_2}^{ideal} = \frac{R_{cat}^0}{4F} \quad (37)$$

In this case, $R_{cat}^0$ is the ideal current transfer, computed using Equation (28) without considering the resistances.

### 2.1.3. Constitutive Relations

The density of the gas is given by the equation below.

$$\rho = \frac{p\,M}{R\,T}$$

where the mixture molecular weight, expressed in terms of molar fraction of individual species $x_i$, is given by the equation below.

$$M = M_{O_2}x_{O_2} + M_{H_2}x_{H_2} + M_{H_2O}x_{H_2O} + M_{N_2}x_{N_2}$$

The molar fractions are related to the mass fractions shown below.

$$\omega_i = \frac{M_i x_i}{M}$$

Molar concentration of species $i$ is defined as:

$$c_i = \frac{P}{RT} \times x_i$$

and can be calculated as:

$$c_i = x_i\left(c_{O_2} + c_{O_2} + c_{H_2O} + c_{N_2}\right)$$

The relative humidity percentage $\eta$ is a function of water saturation pressure $p_{H_2O}^{sat}$.

$$\eta = \frac{p x_{H_2O}}{p_{H_2O}^{sat}} \times 100$$

$$p_{H_2O}^{sat} = 101325 \times 10^{c_1 + c_2(T-T_0) + c_3(T-T_o)^2 + c_4(T-T_0)^3}$$

Assuming the ratio $\frac{x_{O_2}}{x_{N_2}} = \frac{21}{79}$, the oxygen molar fraction at the inlet can be determined from the equation below.

$$x_{O_2}^{in} = 1 - \frac{x_{H_2O}^{in}}{1 + 79/21}$$

while the mole fraction on the anode side is defined below.

$$x_{H_2,a} = 1 - x_{H_2O,a}$$

The inlet velocities at the cathode and anode side are given by the following equations, respectively.

$$v_a^{in} = \xi_a^{in} \frac{i_{avg} A_{mem}}{2FA_{inlet}} \times \frac{1}{c_{H_2}^{in}}$$

$$v_c^{in} = \xi_c^{in} \frac{i_{avg} A_{mem}}{4FA_{inlet}} \times \frac{1}{c_{O_2}^{in}}$$

The average current density $i_{avg}$ is given by the equation below.

$$i_{avg} = \frac{1}{L} \int_0^L i \cdot e_y \, dx$$

where $L$ is the length of the fuel cell.

#### 2.1.4. Boundary Conditions

The boundaries of the system as illustrated in Figure 1b are as follows.

I. At the side walls:

$$u^{(g)} = 0, \quad \frac{\partial \omega_i^{(g)}}{\partial x} = \frac{\partial \phi^{(s)}}{\partial x} = \frac{\partial \phi^{(m)}}{\partial x} = \frac{\partial s}{\partial x} = \frac{\partial T}{\partial x} = 0$$

II. At the anode inlet:

$$\dot{m}_a^{(g)} = \dot{m}_a^{in}, \ \omega_{H_2}^{(g)} = \omega_{H_2,a}^{in}, \ \omega_{H_2O}^{(g)} = \omega_{H_2O,a}^{in}, \ T = T^{in}, s = 0$$

III. At the cathode inlet:

$$\dot{m}_c^{(g)} = \dot{m}_c^{in}, \ \omega_{H_2}^{(g)} = \omega_{H_2,c}^{in}, \ \omega_{H_2O}^{(g)} = \omega_{H_2O,c}^{in}, \ T = T^{in}, s = 0$$

IV. At the outlets:

$$p^{(g)} = p^{ref}, \quad \frac{\partial \omega_i^{(g)}}{\partial x} = \frac{\partial \phi^{(s)}}{\partial x} = \frac{\partial s}{\partial x} = \frac{\partial T}{\partial x} = 0$$

V. At the anode wall terminal:

$$\phi^{(s)} = 0$$

VI. At the cathode wall terminal:

$$-\sigma^{(s)} \frac{\partial \phi^{(s)}}{\partial n} = i^{set}$$

The governing equations together with the constitutive relations and appropriate boundary conditions are then solved numerically.

## 2.2. Numerical Method

The developed mathematical model is implemented and customized in the commercial computational software ANSYS Fluent and its PEMFC module together with user-defined functions. The latter allows for changes in constitutive relations, parameters, and—to some extent—the governing equations. Fluent solves all the equations throughout the domain, so variables in the layers that should not be solved are set to zero.

The computational domains (Figure 1b) are created in the commercial software ANSYS Design Modeller and ANSYS Meshing. The whole domain is defined porous, except for the current collectors (cc) and the wall terminals. The whole domain is partitioned into smaller domains for running it in a parallel mode. The computational model is partitioned using the Cartesian Z-direction method to prevent any floating-point exceptions or errors. With convergence criteria of $10^{-6}$ for the residuals of all the conservation equations, iterations are performed, after the mesh independence test to ensure an accurate solution.

## 2.3. Response Surface Methodology

In general concern of a process or system involving a response $y$ that depends on many other controllable input variables $\xi_1$, $\xi_2$, $\xi_3$, ..., $\xi_k$ , its relationship with $y$ is given by the equation below [54].

$$y = f(\xi_1, \xi_2, \xi_3, \ldots \xi_k) + \epsilon$$

where $\epsilon$ represents other sources for variability unaccounted for in $f$. Treating $\epsilon$ as a statistical error, we assume it to have a normal distribution with a mean zero and variance $\sigma^2$.

## 2.4. Kriging-Based Response Surface Methodology

Design and analysis of computer experiments (DACE) is also called the kriging method of response surface generation. It involves training points in estimating the unknown parameters $\alpha$ and predicting new response points. It interpolates the model at all the training points. This method is used to generate response surfaces for voltage values of 0.6 and 0.8 V. The response can be expressed by the equation below [54].

$$\hat{y}(x) = f(x) + z(x)$$

where $f(x)$ is a low-order polynomial that interpolates the design points. Typically, a constant value is found in order to predict for modelling complex input-output relations. Hence, the output can be viewed as:

$$\hat{y}(x) = \beta + z(x)$$

$z(x)$ is a Gaussian stochastic function representing the realization of a random process with zero mean, variance $\sigma^2$, and its covariance is given by the equation below.

$$Cov(Z) = \sigma^2 R\left(x^i, x^j\right)$$

where $R\left(x^i, x^j\right)$ is the correlation matrix, defined by the equation below.

$$R\left(x^i, x^j\right) = exp\left[-d\left(x^i, x^j\right)\right]$$

$$d\left(x^i, x^j\right) = \sum_{l=1}^{k} \theta_l\left(\left|x_l^i - x_l^j\right|^p\right)$$

where $i, j$ denote the two training points, $l$ refers to the design parameter, $\theta$ is the positive weight factor related to each design parameter, and $k$ denotes the number of design parameters. The mean parameter $\beta$ is evaluated by the equation below.

$$\beta = \left[A^T R^{-1} A\right]^{-1} A^T R^{-1} y$$

where $A$ is an $n \times n$ matrix of training set points depending on the choice of the function $f(x)$. The parameters $\theta$ and $p$ ensure best fit to the training data. They are evaluated by using a maximizing likelihood estimation (MLE) [55].

$$-\frac{1}{2}\left[n \ln(2\pi) + n\, ln\sigma^2 + \ln|R| + \frac{1}{2\sigma^2}(y - A\,\beta)^T R^{-1}(y - A\beta)\right]$$

where the maximum likelihood $\sigma^2$ is expressed by the equation below.

$$\sigma^2 = \frac{1}{n}(y - A\beta)^T R^{-1}(y - A\beta)$$

The response at a new point $\overline{x}$, $\hat{y}(\overline{x})$ is directly evaluated by applying the equation below.

$$\hat{y}(\overline{x}) = \beta + r^T(\overline{x}) R^{-1}(y - A\beta)$$

where $r(\overline{x})$ is a correlation vector between $\overline{x}$ and all the training points.

## 3. Results and Discussion

### 3.1. Validation

Due to the complexity of the model being solved, model validation with experimental data is imperative to prevent misleading conclusions in predicting the behavior of the fuel cell system. In this work, we aimed to validate our model with two experimental single cells. The first experimental cell was equipped with a Gore Primea 5510 membrane, which is a microscopically reinforced composite membrane. The expressions for various phenomenological membrane models are generally based on the Nafion membrane and so need to be adapted to account for the Gore membrane used in the experiment. Two parameters have been, therefore, adapted to validate the model using both the polarization curve and its iR-corrected counterpart.

The iR-corrected potential is given by the equation below.

$$E_{IR} = E_{rev} - \eta_a - \eta_c$$

where $\eta_a$ ($> 0$) and $\eta_c$ ($< 0$) are the corresponding overpotentials of the anode and the cathode catalyst layers, respectively. In this case, we focused on parameter adaption of the cathode reference exchange current density, $j_{cat}^{ref}$, and the cathode transfer coefficient, $\alpha_{cat}$, and retained the anode counterparts from the work of Wang and co-workers [39,56].

Two points from the experimentally determined iR-corrected polarization curve were chosen: one at a low current density and the other at a higher current density. Initially, the membrane protonic conductivity coefficient, $\beta_m$, was set to one. Once a good agreement for the two points was obtained, we predicted the complete iR-corrected polarization curve. A good agreement for the whole range was achieved, as can be seen in Figure 2. Subsequently, $\beta_m$ was varied to finish the validation for the full polarization curve, where the cell voltage is defined below.

$$E_{cell} = E_{rev} - \eta_a - \eta_c - \sum i^{(s)}\left(\frac{1}{\sigma_{eff}}\right) - \sum i^{(m)}\left(\frac{1}{\sigma_{eff}}\right)$$

In this case, the last two terms on the right-hand side of the equation account for the various ohmic losses in the solid and membrane functional layers, respectively.

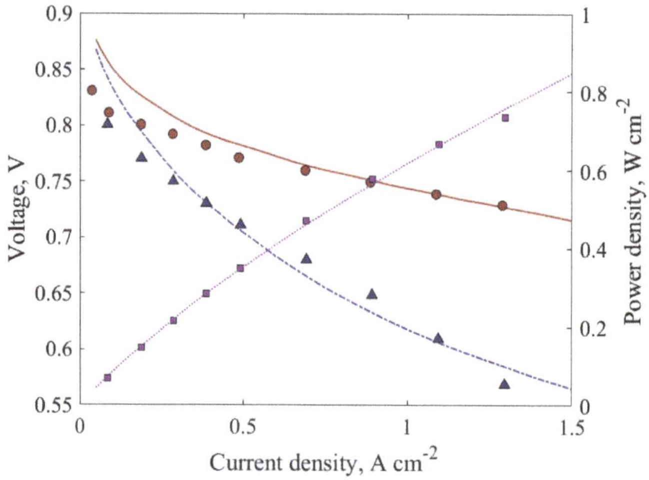


**Figure 2.** Polarization curves: [▲] is the experimental potential, [●] is the iR-corrected experimental potential, [■] is the experimental power density, [—·] is the predicted potential, [—] is the predicted iR-corrected potential, and [···] is the predicted power density.

Furthermore, the predicted local current density distribution along the top of the anode terminal was compared with the experimental counterpart. Figure 3 illustrates this local current density distribution for each measured average current density in Figure 2. It can be observed that better prediction is achieved at lower currents, while the model is less accurate at higher current densities, with the most deviation observed near the inlet and outlet. This could be due to the inlet boundary condition in the simulation that does not represent correctly the position of the inlet manifolds location as in the experiments.

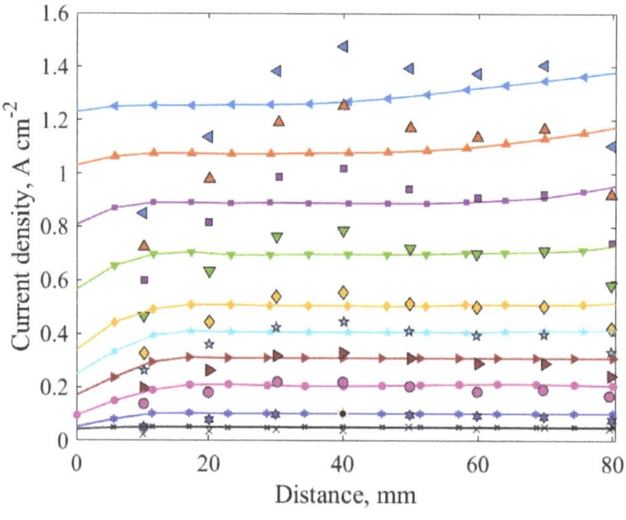


**Figure 3.** Local current density distribution. Predicted (lines) and experimental (points) for current densities: 1.3 Acm$^{-2}$ (◀), 1.1 A cm$^{-2}$ (▲), 0.9 A cm$^{-2}$ (■), 0.7 A cm$^{-2}$ (▼), 0.5 A cm$^{-2}$ (◆), 0.4 A cm$^{-2}$ (★), 0.3 A cm$^{-2}$ (▶), 0.2 A cm$^{-2}$ (●), 0.1 A cm$^{-2}$ (✱), and 0.05 A cm$^{-2}$ (✖).

The model is also validated against the second set of experimental polarization curve, especially at a limiting current density to justify the MEA model with an agglomerate catalyst layer. The model was validated with experimental fuel cells with a single-layered gas diffusion studied by Han et al. [37], as depicted in Figure 4. It shows that the model has good agreement with the curve and is able to predict the limiting current density due to mass transport limitations and the presence of two-phase liquid water.

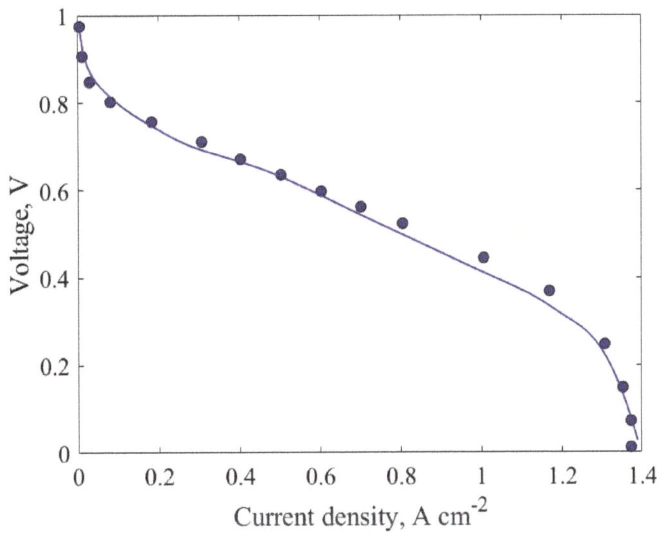


**Figure 4.** Experimental (points) and predicted (line) polarization curve for case b.

### 3.2. Response Surface Generation and Optimization

To determine the dominant parameters affecting the PEMFC performance and optimize them, response surface generation and local sensitivity analysis were carried out at medium (0.6 V) and high (0.8 V) voltages. The response surfaces were created using the design of experiments (DOE) method of central composite design (CCD) and the kriging method of response surface generation. The CCD was employed to capture the non-linear interactions that cannot otherwise be described by linear functions. Hence, experimental designs for quadratic response surfaces, like three-level factorial, Box-Behnken, central composite, and Doehlert designs [57], should be used instead. The list of design variables considered for the DOE and their upper and lower bounds are tabulated in Table 4. The base case parameters correspond to case (a) in Table 1, which has been validated for both global and local current densities (Figures 2 and 3). The inputs used for the base-case correspond to a current density of 1 A/cm$^2$, i.e., anode stoichiometry of 3.35 ($v_a^{in}$ = 0.173 m/s) and cathode stoichiometry of 2.3 ($v_c^{in}$ = 1.052 m/s), which are calculated from constitutive relations, as explained in the previous section. The response surface generated was then carefully evaluated. If the response surface was not within the desired limits of accuracy, determined by the "goodness of fit," it was modified by adding refinement points to the kriging method. Additional CFD simulations were run to generate the response surface data to improve the confidence of the response surface. This iterative method was done until a good fit and reliable response surface was achieved and is illustrated in Figure 5.

**Table 4.** Design variable values for the DOE method.

| Design Variable | Base Case | Lower Bound | Upper Bound |
|---|---|---|---|
| Proton Exchange Membrane thickness | 0.03 mm | 0.005 mm | 0.05 mm |
| Equivalent weight of membrane | 1100 kg $kmol^{-1}$ | 700 kg $kmol^{-1}$ | 1500 kg $kmol^{-1}$ |
| Radius of cathode catalyst particle | $10^{-7}$ m | $10^{-8}$ m | $10^{-7}$ m |
| Cathode Catalyst Ionomer resistance | 25 s $m^{-1}$ | 10 s $m^{-1}$ | 100 s $m^{-1}$ |
| Porosity of the cathode catalyst layer | 0.4 | 0.2 | 0.7 |
| Protonic conduction coefficient of the membrane | 0.9 | 0.5 | 1.5 |
| Hydrophobic angle of the cathode catalyst layer | 95° | 90° | 180° |
| Ionomer tortuosity of the cathode catalyst layer | 1 | 0.7 | 1.5 |
| Ionomer volume fraction of the cathode catalyst layer | 1 | 0.5 | 1.0 |

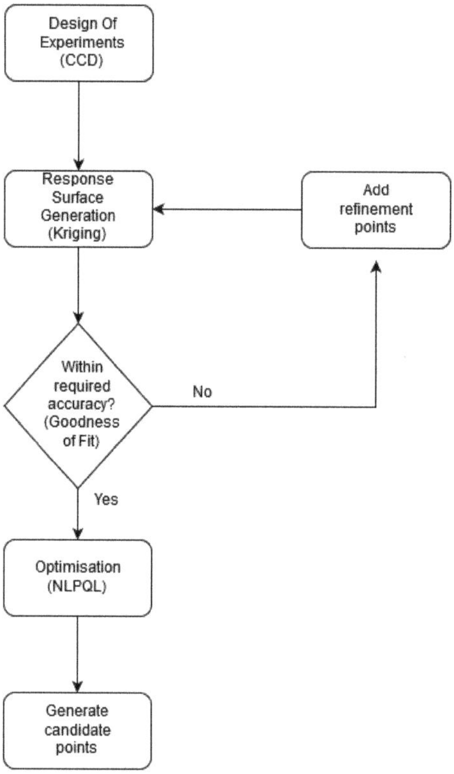


**Figure 5.** Flow diagram of response surface methodology.

Goodness of fit was used to determine the reliability of the response surface predicted, i.e., how close the predicted values are to the observed values. The predicted values from the response surface were compared against the values observed from design points for both 0.6 and 0.8 V. The results demonstrate good agreement for both voltage values, which is shown in Figure 6.

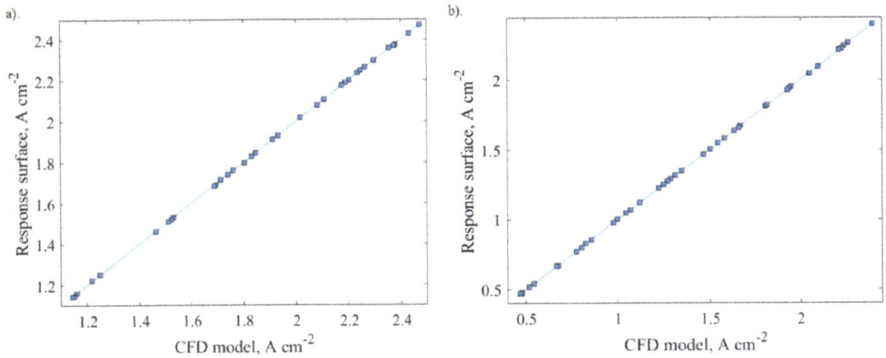


**Figure 6.** Goodness of fit for response surface generated at: (**a**) 0.8 V and (**b**) 0.6 V.

## 3.3. Local Sensitivity

In determining which variables influence the output current density the most, the following relation was used.

$$local\ sensitivity = \frac{Output_{max} - Output_{min}}{Output_{average}}$$

The local sensitivity analysis was carried out using outputs obtained from the DOE. Figures 7a and 8a illustrate the relative impacts of the different input parameters on the local sensitivity for the 0.6 and 0.8 V, respectively. The corresponding sensitivity curves (Figures 7b and 8b) show the output variation with changes in one input parameter, while keeping the other parameters constant. The results show that, for both voltage values, parameters that have the most impact are the membrane protonic conductivity coefficient and the membrane thickness. The cathode catalyst ionomer volume fraction, cathode catalyst porosity, cathode catalyst ionomer tortuosity, volume fraction, cathode catalyst hydrophobic angle, and the radius of cathode agglomerate particles have a minor impact on the output that can be considered negligible when compared to these two parameters.

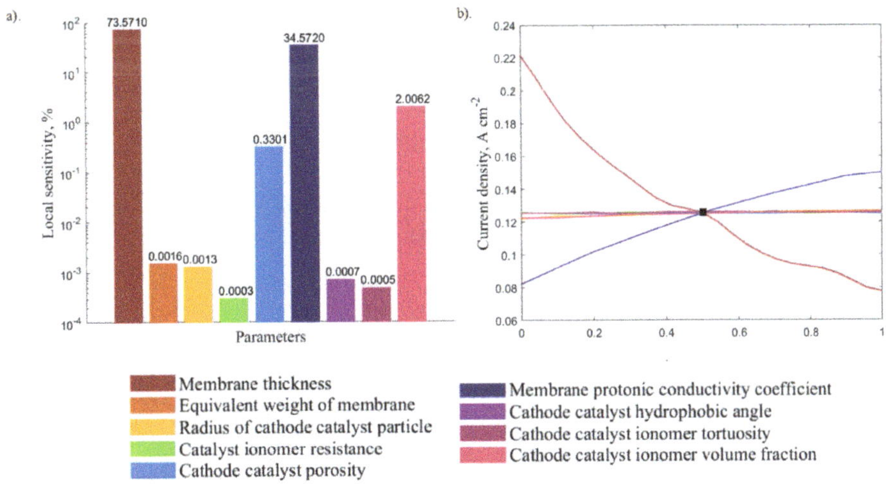


**Figure 7.** Local sensitivity analysis at 0.6 V presented in: (**a**) bar plot and (**b**) sensitivity curve.

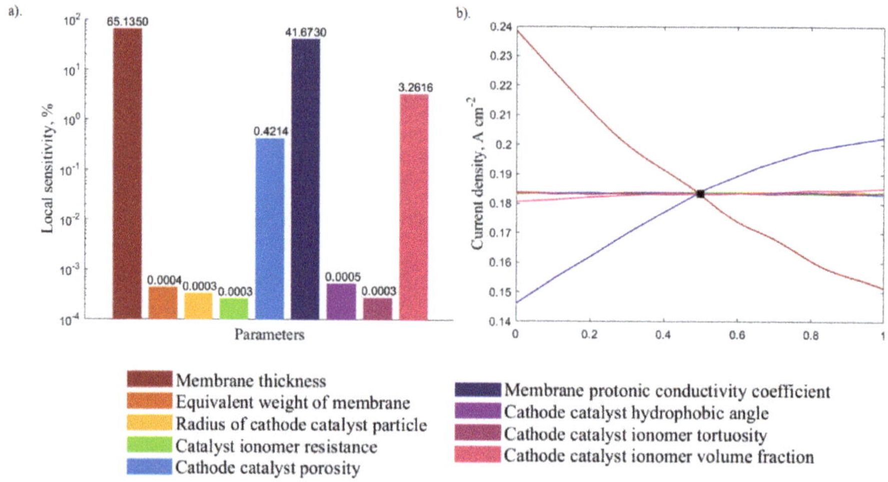


**Figure 8.** Local sensitivity analysis at 0.8 V presented in: (**a**) the bar plot and the (**b**) sensitivity curve.

Based on these results, the membrane thickness and the membrane protonic conductivity coefficient are chosen as the varying parameters to study the interrelated response, as well as to perform the optimization. The three-dimensional response charts illustrating how the two variables affect the current output are depicted in Figures 9 and 10 for both medium and high voltages, respectively. These response surfaces are used for the system optimization.

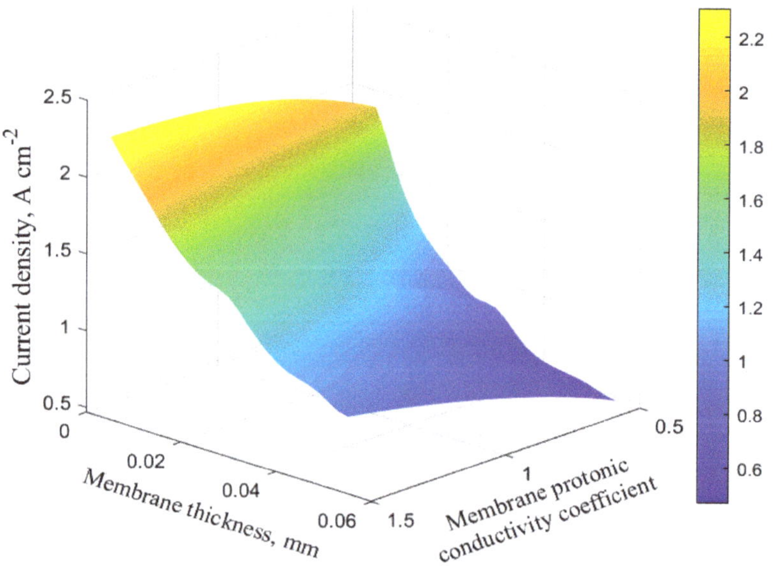


**Figure 9.** Three-dimensional response surface plot at 0.6 V.

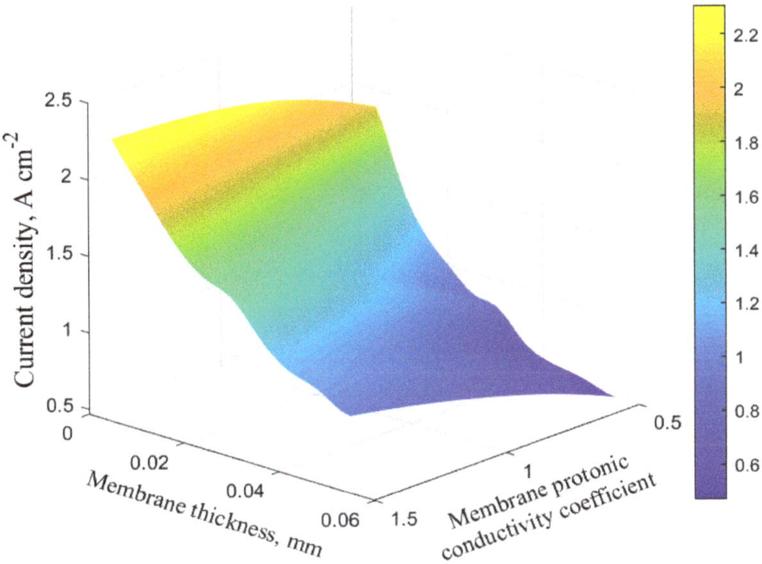


**Figure 10.** Three-dimensional response surface plot at 0.8 V.

### 3.4. Optimization

Optimization was carried out using the non-linear programming through a quadratic Lagrangian (NLPQL) approach based on the response surfaces generated previously. NLPQL is a gradient-based algorithm that provides a refined local optimization result. It supports a single constraint on the output parameter and is limited to continuous parameters. The NLPQL approximates derivatives by a central difference scheme and finds candidate points by iterations. This approach was used to determine the optimum values of the variables considered.

Table 5 shows the optimization result that provides maximum current output for three candidate points at 0.8 V. It can be observed that the response surface prediction agrees well with the CFD model, with a maximum error of approximately 0.17%, occurred at candidate point 1 (case i). The current density output increased from 0.18 A cm$^{-2}$ at the base to an average of 0.2472 A cm$^{-2}$, which indicates almost a 40% increase. This increase is attributed to the change in the protonic conductivity coefficient from 0.9 to 1.45 and attributed to the decrease in membrane thickness from 0.03 mm to the lowest bound value of 0.005 mm.

Similarly, a good agreement between the response surface and the computational fluid dynamics (CFD) model is obtained for the medium-voltage case, as shown in Table 6. At 0.6 V, the error is slightly higher, with a maximum error of 0.57%, which occurred at candidate point 1. As in the case at a high voltage value, the protonic conduction coefficient increased from 0.9 to 1.5 m and the membrane thickness reduced to the lowest bound value. However, the current density output for the 0.6 V almost doubled, from 1.23 A cm$^{-2}$ at the base case to 2.4 A cm$^{-2}$, which gives approximately a 96% increase. This large improvement demonstrates the significance of optimization at medium voltage levels, as compared to high voltage values.

**Table 5.** Optimization result for three candidate points at 0.8 V.

| Parameter | Range of Values | Case i | | Case ii | | Case iii | |
|---|---|---|---|---|---|---|---|
| Membrane thickness | 0.005–0.05 mm | 0.005 mm | | 0.005 mm | | 0.005 mm | |
| Equivalent weight of membrane | 700–1500 kg $kmol^{-1}$ | 700 kg $kmol^{-1}$ | | 746.65 kg $kmol^{-1}$ | | 824.12 kg $kmol^{-1}$ | |
| Radius of the cathode catalyst particle | $10^{-8}$–$10^{-7}$ m | $10^{-7}$ m | | $9.4613 \times 10^{-8}$ m | | $86,287 \times 10^{-8}$ m | |
| Cathode catalyst ionomer resistance | 10–100 s $m^{-1}$ | 10 s $m^{-1}$ | | 10 s $m^{-1}$ | | 10 s $m^{-1}$ | |
| Porosity of cathode catalyst layer | 0.2–0.7 | 0.7 | | 0.7 | | 0.7 | |
| Protonic conduction Coefficient of the membrane | 0.5–1.5 | 1.4537 | | 1.443 | | 1.4533 | |
| Hydrophobic angle of the cathode catalyst layer | 90°–180° | 180° | | 180° | | 180° | |
| Ionomer tortuosity | 0.7–1.5 | 0.7 | | 1.0354 | | 1.07 | |
| Ionomer volume fraction | 0.5–1.0 | 1 | | 1 | | 1 | |
| Current density magnitude | | Response Surface | Numerical Simulation | Response Surface | Numerical Simulation | Response Surface | Numerical Simulation |
| | | **2477 A $m^{-2}$** | 2472.8 A $m^{-2}$ | **2475.2 A $m^{-2}$** | 2471.9 A $m^{-2}$ | **2474.6 A $m^{-2}$** | 2472.7 A $m^{-2}$ |
| Error % | | 0.1698% | | 0.1335% | | 0.0768% | |
| % Increase from base case output (1780.6 A $m^{-2}$) | | 38.87% | | 39% | | 38.87% | |

**Table 6.** Optimization result for three candidate points at 0.6 V.

| Parameter | Range of Values | Case i | | Case ii | | Case iii | |
|---|---|---|---|---|---|---|---|
| Membrane thickness | 0.005–0.05 mm | 0.005 mm | | 0.005 mm | | 0.005 mm | |
| Equivalent weight of membrane | 700–1500 kg $kmol^{-1}$ | 1500 kg $kmol^{-1}$ | | 1420.1 kg $kmol^{-1}$ | | 1301.1 kg $kmol^{-1}$ | |
| Radius of the cathode catalyst particle | $10^{-8}$–$10^{-7}$ m | $10^{-8}$ m | | $10^{-8}$ m | | $10^{-8}$ m | |
| Cathode catalyst ionomer resistance | 10–100 s $m^{-1}$ | 100 s $m^{-1}$ | | 100 s $m^{-1}$ | | 100 s $m^{-1}$ | |
| Porosity of the cathode catalyst layer | 0.2–0.7 | 0.2 | | 0.2 | | 0.2 | |
| Protonic conduction coefficient of the membrane | 0.5–1.5 | 1.5 | | 1.5 | | 1.5 | |
| Hydrophobic angle of the cathode catalyst layer | 90°–180° | 90° | | 90° | | 90° | |
| Ionomer tortuosity | 0.7–1.5 | 1.5 | | 1.5 | | 1.5 | |
| Ionomer volume fraction | 0.5–1.0 | 1 | | 1 | | 1 | |
| Current density magnitude | | Response Surface | Numerical Simulation | Response Surface | Numerical Simulation | Response Surface | Numerical Simulation |
| | | **24166 A $m^{-2}$** | 24029 A $m^{-2}$ | **24154 A $m^{-2}$** | 24029 A $m^{-2}$ | **24140 A $m^{-2}$** | 24029 |
| Error % | | 0.57% | | 0.52% | | 0.46% | |
| % Increase from the base case output (12273 A $m^{-2}$) | | 95.78% | | 95.78% | | 95.78% | |

It can be observed that the extent of optimization required increased greatly when the voltage is reduced from a high to a medium level. This is due to the fact that an increase in current density requires better conductivity of the $H^+$ ions through the membrane to maintain system net neutrality. In addition, the reduction in membrane thickness decreases the resistance offered by the membrane for the hydrogen ions to move from the anode to the cathode side. The sensitivity analysis also shows that the influence of membrane thickness on the output is greater than that of the protonic conductivity coefficient and rises slightly when the voltage is reduced from 0.8 to 0.6 V. Furthermore, the different values of parameters, other than the two mentioned, in Tables 5 and 6, confirms the insignificance of these variables to the current output and, hence, varying them would not affect the output largely.

In short, it can be deduced that, in designing high performance PEMFC, one needs to aim for thinner MEA with higher membrane protonic conductivity, which can be achieved by using carbon-reinforced membrane or water absorbent materials including polytetrafluoroethylene (PTFE), polyvinylidene fluoride (PVDF), and fluorinated ethylene propylene (FEP) or silica gels.

## 4. Conclusions

A numerical study of the two-phase PEMFC with a detailed multiscale agglomerate catalyst layer model was developed and validated against two sets of experimental data for iR-corrected and full polarization curves, including at limiting current densities due to mass transport limitation and the local current density distributions. The model is then extended and coupled with response surface methodology to optimize the design of the membrane electrode assembly (MEA), i.e., membrane thickness, equivalent weight of membrane, radius of agglomerate catalyst particle, cathode catalyst ionomer resistance, porosity of the catalyst layer, a membrane protonic conductivity coefficient, a hydrophobic angle of the catalyst layer, ionomer tortuosity, and a catalyst layer, at high and medium voltages. From sensitivity analysis, it was found that the membrane thickness and membrane protonic conductivity coefficient yield the most significant factor. Reducing the membrane thickness by 40% and increasing protonic conductivity by 50% gives rise to a current density of up to 40% at a higher voltage and up to 100% at a medium voltage. This finding could help fuel cell engineers and designers to carefully manufacture MEA with optimum parameters for a high-performance fuel cell system.

Future work will focus on a combined optimization of design and operating parameters simultaneously for better MEA design, thermal, water, and gas management. A more advanced optimization algorithm including artificial intelligence and machine learning will be considered as well.

**Author Contributions:** Conceptualization, A.P.S. and M.A. Methodology, A.P.S. and R.K.S.S.V. Software, R.K.S.S.V. and B.A.C. Validation, A.P.S., R.K.S.S.V. and B.A.C. Formal analysis, R.K.S.S.V., B.A.C., A.P.S. and M.A. Investigation, R.K.S.S.V., B.A.C. and A.P.S. Resources, A.P.S., M.A., L.J. and L.C. Data curation, A.P.S. and M.A. Writing—original draft preparation, R.K.S.S.V. and B.A.C. Writing—review and editing, A.P.S., M.A., L.J. and L.C. Visualization, R.K.S.S.V. and B.A.C. Supervision, A.P.S. and M.A. Project administration, L.J. and L.C. Funding acquisition, A.P.S., L.J. and L.C.

**Funding:** This research received no external funding.

**Acknowledgments:** R.K.S.S.V. and B.A.C. gratefully acknowledge support from MITACS Globalink Research Award. A.P.S., L.J., and L.C. acknowledge the support from the SDUST Open Grant. We acknowledge anonymous reviewers for the valuable feedback and suggestions.

**Conflicts of Interest:** The authors declare no conflict of interest.

## Nomenclature

| | |
|---|---|
| $A_{cl}$ | catalyst area, $m^2$ |
| $c_i^{(g)}$ | molar concentration of species $i$, mol $m^{-3}$ |
| $C_{i,ref}^{(g)}$ | reference molar concentration of species $i$, mol $m^{-3}$ |
| $c_p^{(g)}$ | specific heat capacity, J $kg^{-1}$ $K^{-1}$ |
| $c_1, c_2, c_3, c_4$ | constants for saturation pressure of water, -, $K^{-1}$, $K^{-2}$, $K^{-3}$ |
| $D^{(c)}$ | capillary diffusion, $m^2$ $s^{-1}$ |

| | |
|---|---|
| $D_i$ | diffusivity of species $i$, $m^2\ s^{-1}$ |
| $E_{cell}$ | cell voltage, V |
| $E_a$ | activation energy, J $mol^{-1}$ |
| $E_{rev}$ | reversible cell potential, V |
| $F$ | Faraday constant, C $mol^{-1}$ |
| $h_j$ | height of layer $j$, m |
| $H^{(l)}_{O_2}, H^{(p)}_{O_2}$ | Henry's constant for air-water and air-polymer interfaces, Pa $m^3\ mol^{-1}$ |
| $\eta_{a,c}$ | relative humidity, % |
| $i$ | current density, A $m^{-2}$ |
| $j^{ref}_{a,c}$ | anode and cathode volumetric reference exchange current density, A $m^{-3}$ |
| $J$ | volumetric reference current density, A $m^{-3}$ |
| $L$ | length of channel, m |
| $m_{H_2O}$ | interphase mass transport, kg $m^{-3}\ s^{-1}$ |
| $M^{(g)}$ | mean molecular mass of the gas phase, kg $mol^{-1}$ |
| $M_i$ | molecular mass of species, kg $mol^{-1}$ |
| $M^{(m)}$ | equivalent weight of the dry membrane, kg $mol^{-1}$ |
| $n_d$ | electroosmotic drag coefficient |
| $p^{sat}_{H_2O}$ | saturation pressure of water, Pa |
| $R$ | gas constant, J $mol^{-1}\ K^{-1}$ |
| $s$ | liquid saturation |
| $S$ | source term |
| $T_0, T_1, T_2$ | constants, K |
| $T$ | temperature, K |
| $V$ | volume, $m^3$ |
| $\omega^{(g)}_i$ | mass fraction of species $i$ |
| **Greek symbols** | |
| $\alpha$ | transfer coefficient |
| $\beta_m$ | membrane modification coefficient |
| $\gamma$ | volume fraction |
| $\delta$ | thickness of film, m |
| $\epsilon$ | porosity |
| $\eta$ | overpotential, V |
| $\theta$ | wetting angle |
| $\kappa$ | permeability, $m^2$ |
| $\lambda$ | membrane water content |
| $\mu$ | dynamic viscosity, kg $m^{-1}\ s^{-1}$ |
| $\xi$ | stoichiometry |
| $\rho$ | density, kg $m^{-3}$ |
| $\tau$ | surface tension, Pa |
| $\sigma$ | total stress tensor, Pa |
| $\phi$ | potential, V |
| **Superscripts** | |
| sat | saturation |
| (g) | gas phase |
| (m) | membrane |
| (l) | liquid phase |
| (s) | solid |
| in | inlet |
| ref | reference |
| (c) | capillary |
| **Subscripts** | |
| $\alpha, \beta$ | index for species |
| a | anode |
| c | cathode |

| | |
|---|---|
| cc | current collector |
| cl | catalyst layer |
| ff | flow field |
| gdl | gas diffusion layer |
| $H_2$ | hydrogen |
| $H_2O$ | water |
| i | species i |
| j | functional layer j |
| m | membrane |
| $N_2$ | nitrogen |
| $O_2$ | oxygen |
| pot | potential |
| ref | reference |
| o | standard condition |

## References


1. Wang, L.; Husar, A.; Zhou, T.; Liu, H. A parametric study of PEM fuel cell performances. *Int. J. Hydrogen Energy* **2003**, *28*, 1263–1272. [CrossRef]
2. Ferng, Y.M.; Tzang, Y.C.; Pei, B.S.; Sun, C.C.; Su, A. Analytical and experimental investigations of a proton exchange membrane fuel cell. *Int. J. Hydrogen Energy* **2004**, *29*, 381–391. [CrossRef]
3. Amirinejad, M.; Rowshanzamir, S.; Eikani, M.H. Effects of operating parameters on performance of a proton exchange membrane fuel cell. *J. Power Sources* **2006**, *161*, 872–875. [CrossRef]
4. Santarelli, M.G.; Torchio, M.F. Experimental analysis of the effects of the operating variables on the performance of a single PEMFC. *Energy Convers. Manag.* **2007**, *48*, 40–51. [CrossRef]
5. Yan, W.-M.; Chen, C.-Y.; Mei, S.-C.; Soong, C.-Y.; Chen, F. Effects of operating conditions on cell performance of PEM fuel cells with conventional or interdigitated flow field. *J. Power Sources* **2006**, *162*, 1157–1164. [CrossRef]
6. Williams, M.V.; Kunz, H.R.; Fenton, J.M. Operation of Nafion®-based PEM fuel cells with no external humidification: Influence of operating conditions and gas diffusion layers. *J. Power Sources* **2004**, *135*, 122–134. [CrossRef]
7. Klika, V.; Kubant, J.; Pavelka, M.; Benziger, J.B. Non-equilibrium thermodynamic model of water sorption in Nafion membranes. *J. Membr. Sci.* **2017**, *540*, 35–49. [CrossRef]
8. Fan, L.; Zhang, G.; Jiao, K. Characteristics of PEMFC operating at high current density with low external humidification. *Energy Convers. Manag.* **2017**, *150*, 763–774. [CrossRef]
9. Bayrakçeken, A.; Erkan, S.; Türker, L.; Eroğlu, İ. Effects of membrane electrode assembly components on proton exchange membrane fuel cell performance. *Int. J. Hydrogen Energy* **2008**, *33*, 165–170. [CrossRef]
10. Jiang, Y.; Yang, Z.; Jiao, K.; Du, Q. Sensitivity analysis of uncertain parameters based on an improved proton exchange membrane fuel cell analytical model. *Energy Convers. Manag.* **2018**, *164*, 639–654. [CrossRef]
11. Carcadea, E.; Varlam, M.; Marinoiu, A.; Raceanu, M.; Ismail, M.S.; Ingham, D.B. Influence of catalyst structure on PEM fuel cell performance—A numerical investigation. *Int. J. Hydrogen Energy* **2019**, *44*, 12829–12841. [CrossRef]
12. Lee, H.-K.; Park, J.-H.; Kim, D.-Y.; Lee, T.-H. A study on the characteristics of the diffusion layer thickness and porosity of the PEMFC. *J. Power Sources* **2004**, *131*, 200–206. [CrossRef]
13. Salva, J.A.; Iranzo, A.; Rosa, F.; Tapia, E.; Lopez, E.; Isorna, F. Optimization of a PEM fuel cell operating conditions: Obtaining the maximum performance polarization curve. *Int. J. Hydrogen Energy* **2016**, *41*, 19713–19723. [CrossRef]
14. Wu, J.; Liu, Q.; Fang, H. Toward the optimization of operating conditions for hydrogen polymer electrolyte fuel cells. *J. Power Sources* **2006**, *156*, 388–399. [CrossRef]
15. Zervas, P.L.; Tatsis, A.; Sarimveis, H.; Markatos, N.C.G. Development of a novel computational tool for optimizing the operation of fuel cells systems: Application for phosphoric acid fuel cells. *J. Power Sources* **2008**, *185*, 345–355. [CrossRef]

16. Karthikeyan, P.; Muthukumar, M.; Vignesh Shanmugam, S.; Pravin Kumar, P.; Murali, S.; Senthil Kumar, A.P. Optimization of Operating and Design Parameters on Proton Exchange Membrane Fuel Cell by using Taguchi method. *Procedia Eng.* **2013**, *64*, 409–418. [CrossRef]
17. Solehati, N.; Bae, J.; Sasmito, A.P. Optimization of operating parameters for liquid-cooled PEM fuel cell stacks using Taguchi method. *J. Ind. Eng. Chem.* **2012**, *18*, 1039–1050. [CrossRef]
18. Sasmito, A.P.; Kurnia, J.C.; Shamim, T.; Mujumdar, A.S. Optimization of an open-cathode polymer electrolyte fuel cells stack utilizing Taguchi method. *Appl. Energy* **2017**, *185*, 1225–1232. [CrossRef]
19. Grujicic, M.; Chittajallu, K.M. Design and optimization of polymer electrolyte membrane (PEM) fuel cells. *Appl. Surf. Sci.* **2004**, *227*, 56–72. [CrossRef]
20. Na, W.; Gou, B. The efficient and economic design of PEM fuel cell systems by multi-objective optimization. *J. Power Sources* **2007**, *166*, 411–418. [CrossRef]
21. Guo, C.; Lu, J.; Tian, Z.; Guo, W.; Darvishan, A. Optimization of critical parameters of PEM fuel cell using TLBO-DE based on Elman neural network. *Energy Convers. Manag.* **2019**, *183*, 149–158. [CrossRef]
22. Behrou, R.; Pizzolato, A.; Forner-Cuenca, A. Topology optimization as a powerful tool to design advanced PEMFCs flow fields. *Int. J. Heat Mass Transf.* **2019**, *135*, 72–92. [CrossRef]
23. Kanani, H.; Shams, M.; Hasheminasab, M.; Bozorgnezhad, A. Model development and optimization of operating conditions to maximize PEMFC performance by response surface methodology. *Energy Convers. Manag.* **2015**, *93*, 9–22. [CrossRef]
24. Xuan, D.; Li, Z.; Cheng, T.; Shen, Y. Optimization of PEM Fuel Cell systems with RSM. In Proceedings of the Power and Energy Engineering Conference, Chengdu, China, 28–31 March 2010; pp. 341–344.
25. Sohani, A.; Naderi, S.; Torabi, F. Comprehensive comparative evaluation of different possible optimization scenarios for a polymer electrolyte membrane fuel cell. *Energy Convers. Manag.* **2019**, *191*, 247–260. [CrossRef]
26. Gugushe, A.S.; Nqombolo, A.; Nomngongo, P.N. Application of Response Surface Methodology and Desirability Function in the Optimization of Adsorptive Remediation of Arsenic from Acid Mine Drainage Using Magnetic Nanocomposite: Equilibrium Studies and Application to Real Samples. *Molecules* **2019**, *24*, 1792. [CrossRef]
27. Cui, H.; Lu, T.; Wang, M.; Zou, X.; Zhang, Y.; Yang, X.; Dong, Y.; Zhou, H. Flavonoids from Morus alba L. Leaves: Optimization of Extraction by Response Surface Methodology and Comprehensive Evaluation of Their Antioxidant, Antimicrobial, and Inhibition of α-Amylase Activities through Analytical Hierarchy Process. *Molecules* **2019**, *24*, 2398. [CrossRef]
28. Cui, F.-J.; Qian, L.-S.; Sun, W.-J.; Zhang, J.-S.; Yang, Y.; Li, N.; Zhuang, H.-N.; Wu, D. Ultrasound-Assisted Extraction of Polysaccharides from Volvariella volvacea: Process Optimization and Structural Characterization. *Molecules* **2018**, *23*, 1706. [CrossRef]
29. Silva, G.F.; Camargo, F.L.; Ferreira, A.L.O. Application of response surface methodology for optimization of biodiesel production by transesterification of soybean oil with ethanol. *Fuel Proc. Technol.* **2011**, *92*, 407–413. [CrossRef]
30. Auta, M.; Hameed, B.H. Optimized waste tea activated carbon for adsorption of Methylene Blue and Acid Blue 29 dyes using response surface methodology. *Chem. Eng. J.* **2011**, *175*, 233–243. [CrossRef]
31. Singh, K.P.; Gupta, S.; Singh, A.K.; Sinha, S. Optimizing adsorption of crystal violet dye from water by magnetic nanocomposite using response surface modeling approach. *J. Hazard. Mater.* **2011**, *186*, 1462–1473. [CrossRef]
32. Subramonian, S.; Milkey, K.R.; Samsudin, A.R.; Dubey, A.K.; Kidd, P. Comparison between Taguchi Method and Response Surface Methodology (RSM) in Modelling CO2 Laser Machining. *Jordan J. Mech. Ind. Eng.* **2014**, *8*, 35–42.
33. Ly, H.; Birgersson, E.; Vynnycky, M.; Sasmito, A.P. Validated Reduction and Accelerated Numerical Computation of a Model for the Proton Exchange Membrane Fuel Cell. *J. Electrochem. Soc.* **2009**, *156*, B1156–B1168. [CrossRef]
34. Sasmito, A.P.; Birgersson, E.; Mujumdar, A.S. Numerical Investigation of Liquid Water Cooling for a Proton Exchange Membrane Fuel Cell Stack. *Heat Transf. Eng.* **2011**, *32*, 151–167. [CrossRef]
35. Harvey, D.; Pharoah, J.G.; Karan, K. A comparison of different approaches to modelling the PEMFC catalyst layer. *J. Power Sources* **2008**, *179*, 209–219. [CrossRef]
36. Secanell, M.; Karan, K.; Suleman, A.; Djilali, N. Multi-variable optimization of PEMFC cathodes using an agglomerate model. *Electrochim. Acta* **2007**, *52*, 6318–6337. [CrossRef]

37. Han, M.; Xu, J.H.; Chan, S.H.; Jiang, S.P. Characterization of gas diffusion layers for PEMFC. *Electrochim. Acta* **2008**, *53*, 5361–5367. [CrossRef]
38. Nield, D.A.; Bejan, A. *Convection in Porous Media*, 4th ed.; Springer: New York, NY, USA, 2013; ISBN 978-1-4899-9822-4.
39. Ju, H.; Wang, C.-Y.; Cleghorn, S.; Beuscher, U. Nonisothermal Modeling of Polymer Electrolyte Fuel Cells I. Experimental Validation. *J. Electrochem. Soc.* **2005**, *152*, A1645–A1653. [CrossRef]
40. Rao, R.M.; Bhattacharyya, D.; Rengaswamy, R.; Choudhury, S.R. A two-dimensional steady state model including the effect of liquid water for a PEM fuel cell cathode. *J. Power Sources* **2007**, *173*, 375–393. [CrossRef]
41. Parthasarathy, A.; Srinivasan, S.; Appleby, A.J.; Martin, C.R. Temperature Dependence of the Electrode Kinetics of Oxygen Reduction at the Platinum/Nafion® Interface—A Microelectrode Investigation. *J. Electrochem. Soc.* **1992**, *139*, 2530–2537. [CrossRef]
42. Padhy, B.R.; Reddy, R.G. Performance of DMFC with SS 316 bipolar/end plates. *J. Power Sources* **2006**, *153*, 125–129. [CrossRef]
43. Birgersson, E.; Nordlund, J.; Ekström, H.; Vynnycky, M.; Lindbergh, G. Reduced Two-Dimensional One-Phase Model for Analysis of the Anode of a DMFC. *J. Electrochem. Soc.* **2003**, *150*, A1368–A1376. [CrossRef]
44. Karawacki, E.; Suleiman, B.M. Dynamic plane source technique for simultaneous determination of specific heat, thermal conductivity and thermal diffusivity of metallic samples. *Meas. Sci. Technol.* **1991**, *2*, 744–750. [CrossRef]
45. Noponen, M.; Birgersson, E.; Ihonen, J.; Vynnycky, M.; Lundblad, A.; Lindbergh, G. A Two-Phase Non-Isothermal PEFC Model: Theory and Validation. *Fuel Cells* **2004**, *4*, 365–377. [CrossRef]
46. Rankin, D.W.H. *CRC Handbook of Chemistry and Physics*, 89th ed.; David, R.L., Ed.; Crystallography Reviews; CRC Press: Boca Raton, FL, USA, 2009; Volume 15, pp. 223–224.
47. Liu, Z.; Mao, Z.; Wang, C.; Zhuge, W.; Zhang, Y. Numerical simulation of a mini PEMFC stack. *J. Power Sources* **2006**, *160*, 1111–1121. [CrossRef]
48. Klika, V. A Guide through Available Mixture Theories for Applications. *Crit. Rev. Solid State Mater. Sci.* **2014**, *39*, 154–174. [CrossRef]
49. Wu, H.; Li, X.; Berg, P. On the modeling of water transport in polymer electrolyte membrane fuel cells. *Electrochim. Acta* **2009**, *54*, 6913–6927. [CrossRef]
50. Scholz, H.; Stolten, D.; Jupke, A. *Modellierung und Untersuchung von Flutungsphänomenen in Niedertemperatur-PEM-Brennstoffzellen*; AutoUni-Schriftenreihe; Logos-Verl: Berlin, Germany, 2015; ISBN 978-3-8325-4122-4.
51. Barrett, J.C.; Clement, C.F. Growth rates for liquid drops. *J. Aerosol Sci.* **1988**, *19*, 223–242. [CrossRef]
52. Kulikovsky, A.A.; Divisek, J.; Kornyshev, A.A. Modeling the Cathode Compartment of Polymer Electrolyte Fuel Cells: Dead and Active Reaction Zones. *J. Electrochem. Soc.* **1999**, *146*, 3981–3991. [CrossRef]
53. Um, S.; Wang, C.-Y.; Chen, K.S. Computational Fluid Dynamics Modeling of Proton Exchange Membrane Fuel Cells. *J. Electrochem. Soc.* **2000**, *147*, 4485–4493. [CrossRef]
54. Myers, R.H.; Montgomery, D.C.; Anderson-Cook, C.M. *Response Surface Methodology: Process and Product Optimization Using Designed Experiments*; John Wiley & Sons: Hoboken, NJ, USA, 2016; ISBN 978-1-118-91603-2.
55. Keane, A.; Nair, P. *Computational Approaches for Aerospace Design: The Pursuit of Excellence*; John Wiley & Sons: Hoboken, NJ, USA, 2005; ISBN 978-0-470-85547-8.
56. Wang, Y.; Wang, C.-Y. A Nonisothermal, Two-Phase Model for Polymer Electrolyte Fuel Cells. *J. Electrochem. Soc.* **2006**, *153*, A1193–A1200. [CrossRef]
57. Bezerra, M.A.; Santelli, R.E.; Oliveira, E.P.; Villar, L.S.; Escaleira, L.A. Response surface methodology (RSM) as a tool for optimization in analytical chemistry. *Talanta* **2008**, *76*, 965–977. [CrossRef]


**Sample Availability:** Not Available.

molecules

MDPI

*Article*

# Development of a Current Collector with a Graphene Thin Film for a Proton Exchange Membrane Fuel Cell Module

**Yean-Der Kuan [1,*], Ting-Ru Ke [1], Jyun-Long Lyu [1], Min-Feng Sung [2] and Jing-Shan Do [3]**

1 Department of Refrigeration, Air-Conditioning and Energy Engineering, National Chin-Yi University of Technology, Taichung City 41170, Taiwan; muke1002@gmail.com (T.-R.K.); hyt5406@gmail.com (J.-L.L.)
2 Kenda Rubber Ind. Co., Ltd., Yuan-Lin 51064, Taiwan; song221@gmail.com
3 Department of Chemical and Materials Engineering, National Chin-Yi University of Technology, Taichung City 41170, Taiwan; jsdo@ncut.edu.tw
* Correspondence: ydkuan@ncut.edu.tw; Tel.: +886-4-23924505 (ext. 8256)

Academic Editors: Jean St-Pierre and Shangfeng Du
Received: 10 October 2019; Accepted: 19 February 2020; Published: 20 February 2020

**Abstract:** This paper constructs planar-type graphene thin film current collectors for proton exchange membrane fuel cells (PEMFCs). The present planar-type current collector adopts FR-4 as the substrate and coats a copper thin film using thermal evaporation for the electric-conduction layer. A graphene thin film is then coated onto the current collector to prevent corrosion due to electrochemical reactions. Three different coating techniques are conducted and compared: Spin coating, RF magnetron sputtering, and screen printing. The corrosion rates and surface resistances are tested and compared for the different coating techniques. Single cell PEMFCs with the developed current collectors are assembled and tested. A PEMFC module with two cells is also designed and constructed. The cell performances are measured to investigate the device feasibility.

**Keywords:** proton exchange membrane fuel cell; graphene thin film; current collector; module

## 1. Introduction

The proton exchange membrane fuel cell (PEFMC) adopts hydrogen as its fuel, and the fuel energy is converted into electrical energy through an electrochemical reaction. In portable applications, bipolar plates or current collectors account for the major weight percentage among PEMFC components [1]. In portable applications, the PEMFC power requirement is merely a few watts. The PEMFC Balance of Plant (BOP) consumption should be as low as possible; therefore, the PEMFC planar-type module could be more suitable than the conventional, vertical PEMFC stack, as it could omit the air fan or air-pump at the cathode side via the self-air breathing design, significantly reducing the device thickness.

In order to shrink the fuel cell size for portable applications, Lee et al. first proposed the micro-electro-mechanical system (MEMS) technique to construct a micro fuel cell [2]. After that, MEMS techniques were intensively investigated for micro fuel cell fabrication. MEMS techniques were applied in planar-type current collector fabrications, such as the metal lift-off process [3], metal powers deposition onto the wafer surface [4], coating gold-titanium and gold-nickel onto a stainless steel thin plate [5], coating Au onto 316L stainless steel mesh via electro-deposition [6], coating TiN, TiAlN mono-layers, and TiN/TiAlN bi-layers onto 316L stainless steel plates via the physical vapor deposition (PVD) process [7]. The application of MEMS techniques to fabricate micro channels for micro fuel cells was also widely studied, including micro channel fabrication onto a silicon substrate [8], constructing micro channels and metallization onto a polymethyl methacrylate (PMMA) substrate [9], or adopting the electroforming process to make micro channels on a thin copper substrate [10].

Although good electrical conductivity is important for micro fuel cell current collectors, corrosion resistance is also very important, as poor corrosion resistance degrades micro fuel cell durability. Related research has been conducted on current collector electrical conductivity, such as coating gold or TiN/TiAlN layers onto stainless steel expanded meshes [11], depositing a NieP (10e12%P) thin layer onto aluminum thin sheets [12], and coating Ag Nano wires with polydimethylsiloxane (PDMS), as current collector flow channels [13,14].

The printed-circuit board (PCB) technique was also applied to construct the planar-type portable fuel cell, which was first proposed by O'Hare et al. [15] and Schmitz et al. [16]. Some related studies were conducted using flexible PCB as a current collector for air-breathing planar fuel cell stacks [17], process-induced fuel cell damage due to the PCB assembly process [18], and a disc type current collector with copper clad aluminum surface coated with gold using the PCB process [19].

To avoid distortion due to the mismatch of significantly different material properties between the metal current collector and FR-4 substrate in the PCB fuel cell, a lightweight current collector was proposed by Sung et al. [20] and Kuan et al. [21], coated with a copper thin film for electric conduction and a nickel thin film for corrosion resistance, on a FR-4 plate as the substrate, via the thermal evaporation technique. The authors' research team subsequently coated a graphene thin film instead of a nickel thin film as the corrosion-resistance layer. The graphene suspension was adopted and coated onto to the current collector electric-conduction layer via the spinning coating and vacuum oven processes.

A single cell PEMFC was also assembled and tested [22]. Ning et al. [23] proposed a light and flexible air breathing PEMFC using a carbon nanotube (CNT) membrane with holes as the current collectors, to form flexible composite electrodes. They also discovered that thicker carbon nanotube membranes would show higher cell performance especially for large size flexible power sources. In addition, adjusting the directions of the current collectors could also significantly increase the cell performance due to the proper electron transfer pathways that might effectively reduce electric resistance [24]. Mallick et al. [25] made a critical review of current collectors for passive direct methanol fuel cell (DMFC). They indicated that the open ratio of current collector is a major parameter of the DMFCs, the reduction of weight is valuable to increase the gravimetric energy density, and the metal mesh and porous metal are potential substitutes to the perforated plate current collectors in the passive DMFC.

In addition, the precious metal coating of the current collector is essential for improving corrosion resistance; however, it would increase the total cost of the cell. Optimizing the coating thickness is essential to maintain the balance between the corrosion resistance and cost. Surfaces and interfaces present a significant portion of the workable area and a network of energetically mismatched or metastable molecules, which could be exploited to either control surface reactions, engineer bulk stability, or reveal new fundamental details of those now well understood processes or systems [26]. Lee et al. [27] adopted chemical vapor deposition (CVD) to produce graphene coated bipolar plates. The experimental results show that the ultra-thin graphene layer on the copper bipolar plate could act as a very thin passivation layer that could minimize the surface oxidation on the copper plate without performance degradation of the polymer electrolyte fuel cells.

The main objective of this paper is to develop lightweight planar-type current collectors for a PEMFC module with two cells. The lightweight current collector also adopts FR-4 as the substrate material; the copper thin film is coated via the thermal evaporation process that functions as the electric-conduction layer. A graphene thin film is further coated onto the copper thin film as the corrosion-resistance layer. The proposed planar-type current collectors were fabricated with three different graphene thin film coating processes. The completed current collectors were assembled into a two-cell PEMFC module. Related performance and stability experiments were also performed.

## 2. The Lightweight Planar-Type Current Collector Construction

The present lightweight current collector for the two-cell PEMFC module consists of three layers. The first layer is the FR-4 substrate. The second layer is the electric-conduction layer–copper thin film fabricated by adopting the thermal evaporation process to coat copper particles onto the FR-4 surface. The third layer is the corrosion-resistance layer–graphene thin film coating via three different processes discussed and compared in this paper.

A geometric drawing of the two-cell current collector is shown in Figure 1. The current collector outline is 150 × 80 mm and the reaction area of each membrane electrolyte assembly (MEA) is 50 × 50 mm. Holes of two diameters, 3 and 1.2 mm, are arrayed in the reaction area. Two 25 × 15 mm cuboids for in/out are connected among the electric cells.

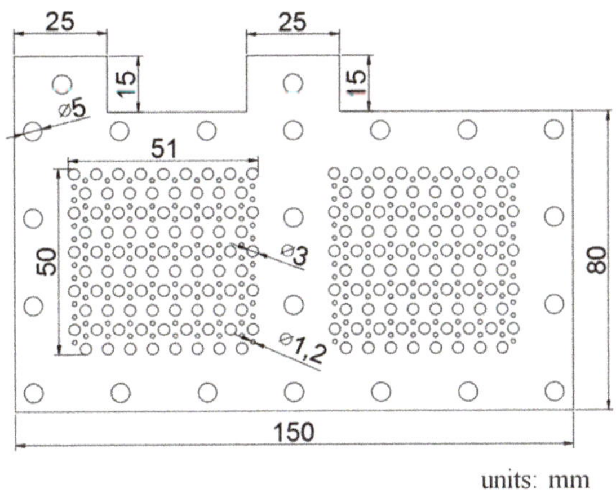


**Figure 1.** Geometric schematic drawing of the current collectors.

The fabrication process for the electrical conduction layer is shown in Figure 2. The machined FR-4 substrate has a heat-resistant tape pasted onto the back side with copper ingots placed into the evaporator vacuum chamber. Once the chamber is vacuumed to 5E-5 torr pressure, the thermal evaporation process begins, and the copper ingots are evaporated and then coated onto the FR-4 substrate to a 50 Åm thickness. The evaporation rate is controlled at 1 Åm/s in the 0–2 kÅm evaporation thickness range and increased to 3 kÅm in the 2–50 kÅm evaporation thickness range. When the copper thin film is completely coated as the electrical conduction layer, the graphene thin film coating as the corrosion-resistance layer is then added. Three different types of graphene thin films are investigated, graphene ink, graphene suspension, and graphene dispersion.

The fabrication process for the graphene thin film using graphene ink is shown in Figure 3. A screen is placed to cover the FR-4 current collector substrate with copper thin film. The graphene ink is then coated onto the surface through screen printing. Then, the current collector with the graphene thin film is placed into the vacuum oven, the vacuum degree is 76 cm Hg, and the oven temperature is kept at 100 °C for 30 min to evaporate the water contained in the graphene thin film. Additionally, the vacuum temperature is reduced to room temperature and the completed current collector is removed from the oven. The graphene ink adopted in this paper is a commercial graphene ink produced by Enerage Inc. The appearance is a black paste, the adhesion is larger than 4B (for PET film), the pencil hardness is larger than HB, the sheet resistance is less than 1968.5 Ω/sq/mm, and the viscosity is 35,000 ± 10,000 cP [28].

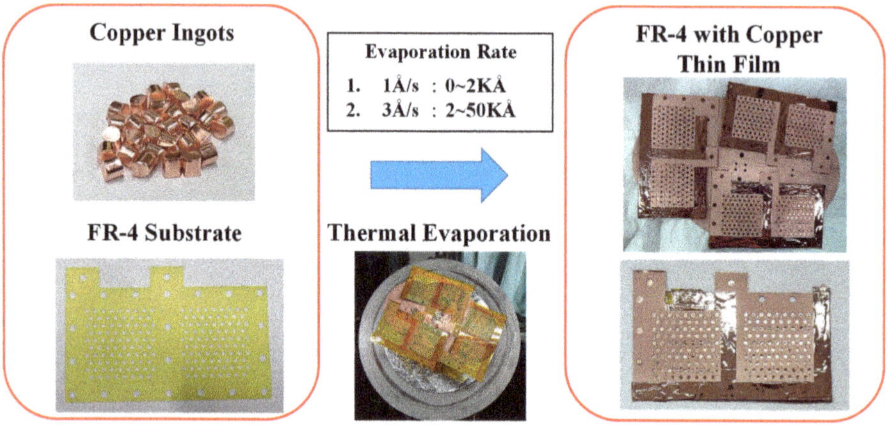


**Figure 2.** Fabrication process for the electric-conduction layer.

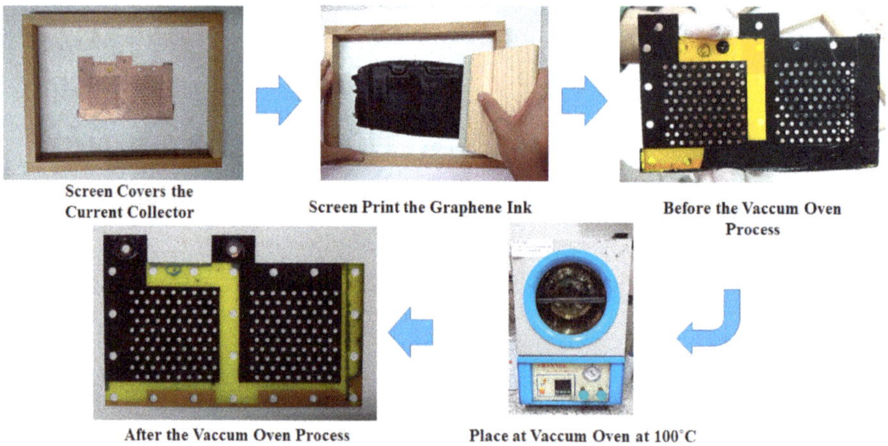


**Figure 3.** Fabrication process for a graphene thin film using graphene ink.

The graphene thin film fabrication process uses a graphene suspension, as shown in Figure 4. The current collector FR-4 substrate has a precoated copper thin film on the surface produced by a spin coater rotating disc at 1100 rpm for 60 s. A 9 mL graphene suspension is slowly dripped onto the FR-4 copper thin film substrate. Then, the current collector with the graphene thin film is placed into a vacuum oven. The vacuum degree is 76 cm Hg and the oven temperature is kept at 100 °C for 30 min to evaporate the water contained in the graphene thin film. The vacuum temperature is then reduced to room temperature with the complete current collector removed from the oven. The graphene suspension adopted in this paper is a commercial graphene suspension by Enerage Inc., the appearance is a gray black liquid, and the solvent is water. The solid content is 5 wt% and the additive is less than 2 wt%. The graphene suspension viscosity is 2500 ± 500 cP. The PH value is approximately 8.0. The average lateral size is less than 15 μm, and the fineness is 5 μm [29].

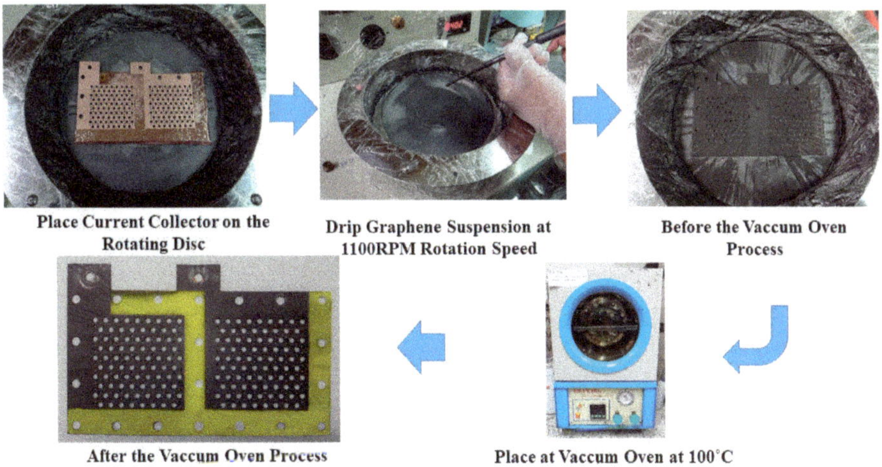


**Figure 4.** Fabrication process for a graphene thin film using graphene suspension.

The graphene thin film fabrication process using graphene dispersion is shown in Figure 5. The current collector FR-4 substrate with a precoated copper thin film on the surface is placed on a spin coater rotating disc at 1000 rpm for 60 s. The 6 mL graphene dispersion is slowly dripped onto the FR-4 substrate copper thin film. Then, the current collector with the graphene thin film is placed into a vacuum oven at 100 °C for 30 min to evaporate the dispersant contained in the graphene thin film. The vacuum temperature is reduced to room temperature and the complete current collector is removed from the oven. The graphene dispersion adopted in this paper is a commercial graphene suspension produced by GI Business. The liquid is black and adapts EAC (ethyl acetate) as the dispersant. The graphene sheets are 0.1–20 μm in diameter, with a less than 3 nm average thickness, and a concentration larger than 5000 ppm [30]. As the graphene dispersion is very dilute, the graphene film spin coating is ultra-thin; therefore, although the surface color does not change distinctly, the ultra-thin film could be a very thin passivation layer on the copper thin film, which would minimize the surface oxidation [27].

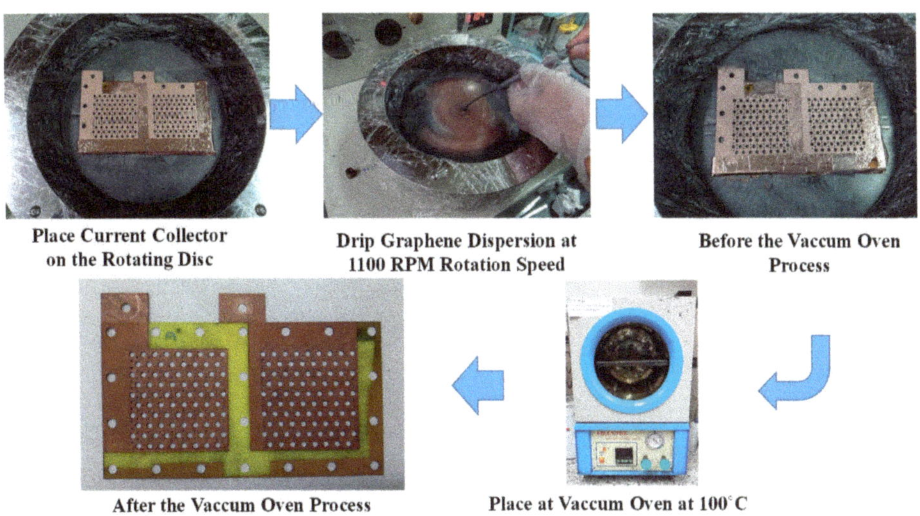


**Figure 5.** Fabrication process for a graphene thin film using graphene dispersion.

## 3. Results and Discussion

The measurements on the weight and thickness of the current collectors are shown in Table 1. The thickness and weight of the FR-4 substrate of the current collector are 0.5 mm and 9.015 g, respectively. After coating the copper thin film, the thickness and weight are 0.51 mm and 9.134 g, respectively. The increment of the thickness after further coating of the graphene is very small. For both the graphene suspension and graphene coated with thin film, the total thickness is not changed via the Vernier caliper measurements. Only the graphene ink coating could be measured at the 0.01 mm thickness increment. After coating, the current collector with the graphene ink thin film shows the highest and most significant weight increment, followed by the current collector with the graphene suspension thin film; the current collector with the graphene-dispersion thin film has the lowest weight increment. Therefore, it can be inferred that the graphene ink thin film has the largest thickness, the graphene suspension has a medium thickness, and the graphene-dispersion thin film has the smallest thickness.

**Table 1.** Weight and thickness of the current collectors.

| Item | FR-4 Substrate | Copper Thin Film Coated | Graphene Ink Thin Film Coated | Graphene Suspension Thin Film Coated | Graphene-Dispersion Thin Film Coated |
|---|---|---|---|---|---|
| Weight (g) | 9.105 | 9.134 | 9.55 | 9.176 | 9.168 |
| Thickness (mm) | 0.5 | 0.51 | 0.52 | 0.51 | 0.51 |

The four probe resistance measurements were conducted to ensure electrical conductibility uniformity for the complete current collectors. An illustration of the five current collector monitoring points for surface resistance measurements is shown in Figure 6.

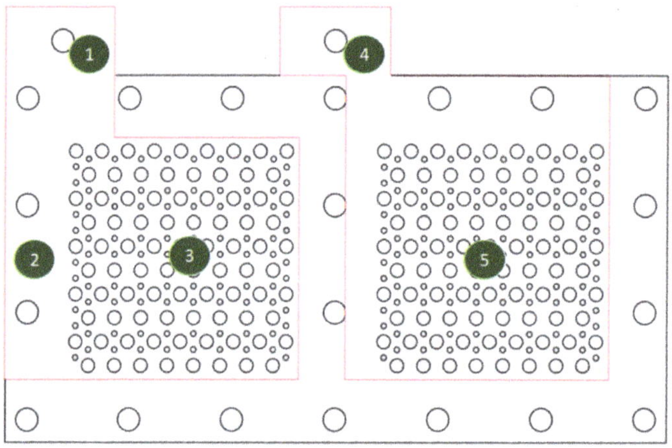


**Figure 6.** The illustration of five monitoring points for the surface resistance.

The current collector surface resistance values with a graphene thin film using graphene ink, graphene suspension, or graphene dispersion, for the complete current collectors, are shown in Tables 2–4, respectively. The results show that all surface resistances are low, and the values for the current collector with a graphene thin film using graphene ink are in the range between 11.19 and 15.40 mΩ/sq. The values for the current collector with a graphene thin film using graphene suspension are in the range between 11.55 and 19.12 mΩ/sq. The values for the current collector with a graphene thin film using graphene dispersion are in the range between 11.33 and 17.58 mΩ/sq, which also shows that the coating is quite uniform in all three processes. In addition, the surface resistance values before and after coating the current collector with a copper thin film and graphene thin film are quite close in all three processes. This implies that the corrosion-resistance layer would not reduce the electrical conduction capability.

**Table 2.** Surface resistances of a current collector with a graphene thin film using graphene ink.

| Layers \ Monitoring Point | 1 | 2 | 3 | 4 | 5 | Standard Deviation (STD DEV) | Unit |
|---|---|---|---|---|---|---|---|
| Copper Thin Film Layer | 15.09 | 15.40 | 14.50 | 11.33 | 14.04 | 1.62 | mΩ/sq |
| Graphene Thin Film Layer | 15.27 | 15.31 | 15.40 | 11.19 | 14.23 | 1.79 | mΩ/sq |

**Table 3.** Surface resistances of a current collector with a graphene thin film using graphene suspension.

| Layers \ Monitoring Point | 1 | 2 | 3 | 4 | 5 | STD DEV | Unit |
|---|---|---|---|---|---|---|---|
| Copper Thin Film Layer | 15.01 | 18.33 | 18.21 | 11.24 | 16.08 | 2.90 | mΩ/sq |
| Graphene Thin Film Layer | 15.49 | 19.12 | 18.67 | 11.55 | 16.04 | 3.03 | mΩ/sq |

**Table 4.** Surface resistances of a current collector with a graphene thin film using graphene dispersion.

| Layers \ Monitoring Point | 1 | 2 | 3 | 4 | 5 | STD DEV | Unit |
|---|---|---|---|---|---|---|---|
| Copper Thin Film Layer | 15.45 | 15.86 | 17.81 | 10.78 | 16.04 | 2.62 | mΩ/sq |
| Graphene Thin Film Layer | 15.81 | 16.63 | 17.58 | 11.33 | 15.45 | 2.40 | mΩ/sq |

Preventing corrosion is also very important for the current collectors. Therefore, the Tafel extrapolation method is conducted to measure the corrosion characteristics. An aqueous methanol solution and de-ionized water are first prepared and mixed as an acidic aqueous methanol solution, the mix ratio of methanol to de-ionized water is 1:9 and the PH value is 6.2. A sample of the fabricated current collector would then be placed in the solution. The Ag/AgCl (3.5M KCl) is adopted as the reference electrode with a platinum counter electrode. Impedance measurements were conducted adopting the BioLogic SP-150AC instrument. The scan rate is 0.166 mV/s and the voltage is −0.025–0.025 V. The results of the Tafel curves are shown in Figure 7, the corrosion voltage (Ecorr) and corrosion current (Icorr) results are shown in Table 5. The higher corrosion resistance would show a higher Ecorr value and the higher corrosion rate would show a higher Icorr value. The results show that adopting the graphene dispersion to fabricate the current collector presents the highest corrosion resistance, adopting the graphene ink second, and adopting the graphene suspension presents the lowest corrosion resistance.

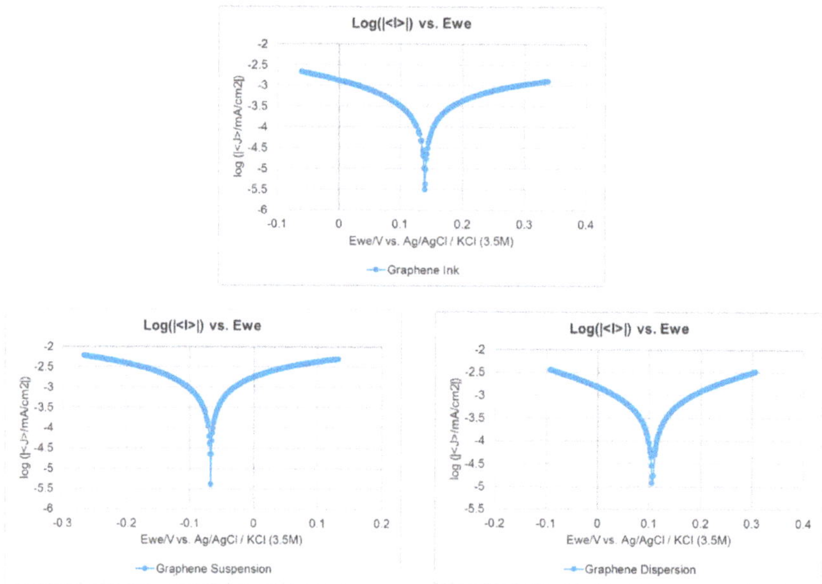


**Figure 7.** The Tafel curves of the current collectors with different graphene films.

**Table 5.** Corrosion characterization of the current collectors.

| | |
|---|---|
| **Current Collector with a Corrosion-Resistance Layer Using Graphene Ink** | |
| Ecorr (mV) | 139.5 |
| Icorr (uA/cm$^2$) | 1.19 |
| Corrosion Rate (mmpy) | 11.7 e$^{-6}$ |
| **Current COLLECTOR with a Corrosion-Resistance Layer Using Graphene Suspension** | |
| Ecorr (mV) | −68.4 |
| Icorr (uA/cm$^2$) | 1.46 |
| Corrosion Rate (mmpy) | 14.4 e$^{-6}$ |
| **Current Collector with a Corrosion-Resistance Layer Using Graphene Dispersion** | |
| Ecorr (mV) | 106.5 |
| Icorr (uA/cm$^2$) | 0.85 |
| Corrosion Rate (mmpy) | 8.3 e$^{-6}$ |

Scanning electron microscopy (SEM) is adopted to investigate the micrographs of the cross-sections of the current collectors coated with different graphene thin films, and this is shown in Figure 8a–c. The graphene ink thin film is the most uneven at the surface and there are some obvious voids inside, the graphene suspension thin film is smoother at the surface with much fewer voids inside, and the graphene-dispersion thin film is the thinnest and smoothest at the surface.

After completed current collector fabrication, the proposed current collectors were assembled into a two-cell PEMFC module to measure cell performance. PEMFC module with both forced convection air-breathing cathodes and self-air-breathing cathodes were studied. The MEA adopted in this paper is a three-layer catalyst coated type with a DuPont Nafion HP membrane and carbon paper diffusion layers. The reaction area was 5 × 5 cm, the catalyst load was Pt 0.1 mg cm$^{-2}$ at the anode, and Pt 0.4 mg cm$^{-2}$ at the cathode. Silicon gaskets with a 0.15 mm thickness were used for sealing. The cell compression ratio was 34.78%.

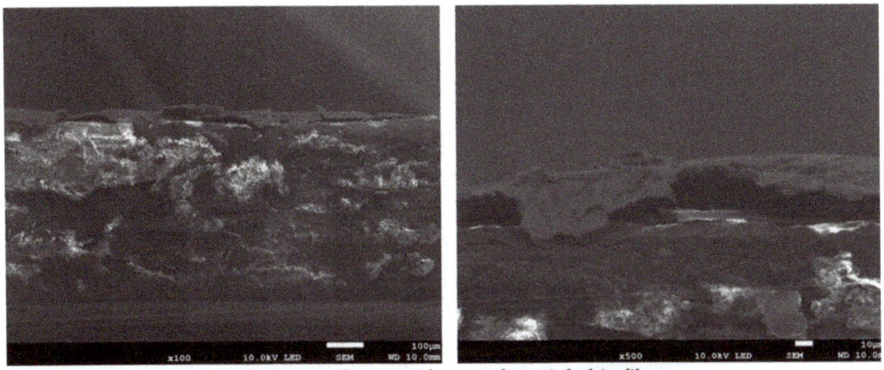

(**a**) Current collector with a graphene ink thin film.

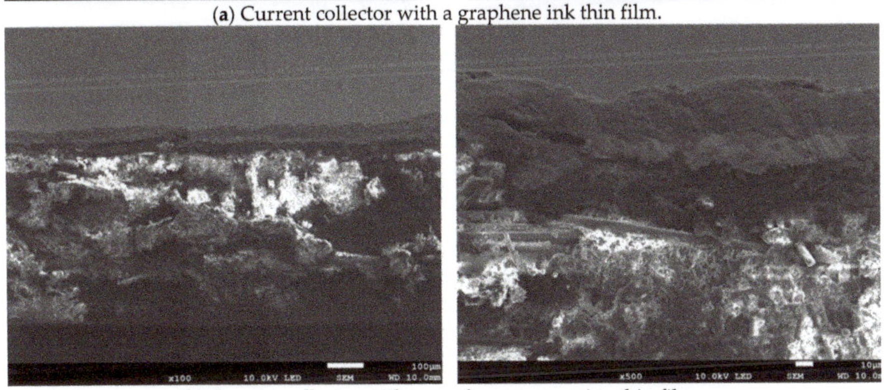

(**b**) Current collector with a graphene suspension thin film.

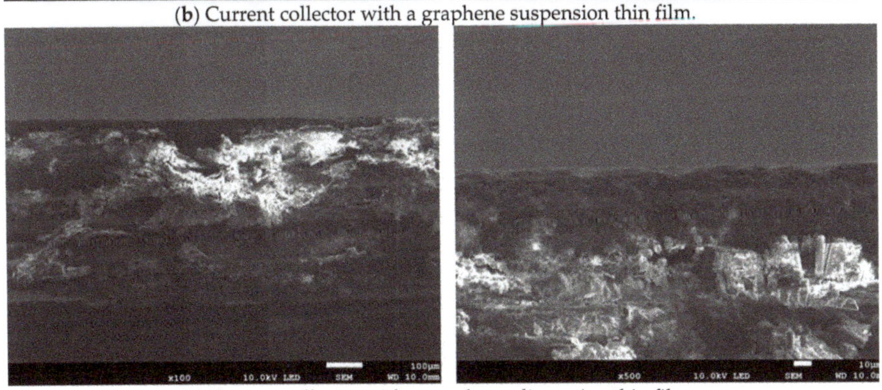

(**c**) Current collector with a graphene-dispersion thin film.

**Figure 8.** The scanning electron microscopy (SEM) images of the cross sections of the current collectors with different graphene thin films. (**a**) Graphene ink; (**b**) Graphene suspension; (**c**) Graphene dispersion.

An exploded drawing of the two-cell PEMFC module with a forced convection air-breathing cathode is shown in Figure 9. From cathode to anode, the components are cathode flow board, gasket, cathode current collector, gasket, MEA, gasket, anode current collector, gasket, and anode flow board, respectively. A picture of the assembled PEMFC module with a forced convection air-breathing cathode is shown in Figure 10. An exploded drawing of the two-cell PEMFC module with a self-air-breathing cathode is shown in Figure 11. From cathode to anode, the components are cathode end plate, gasket, cathode current collector, gasket, MEA, gasket, anode current collector, gasket, and anode flow board, respectively. A picture of the assembled PEMFC module with self-air-breathing cathode is shown in Figure 12.

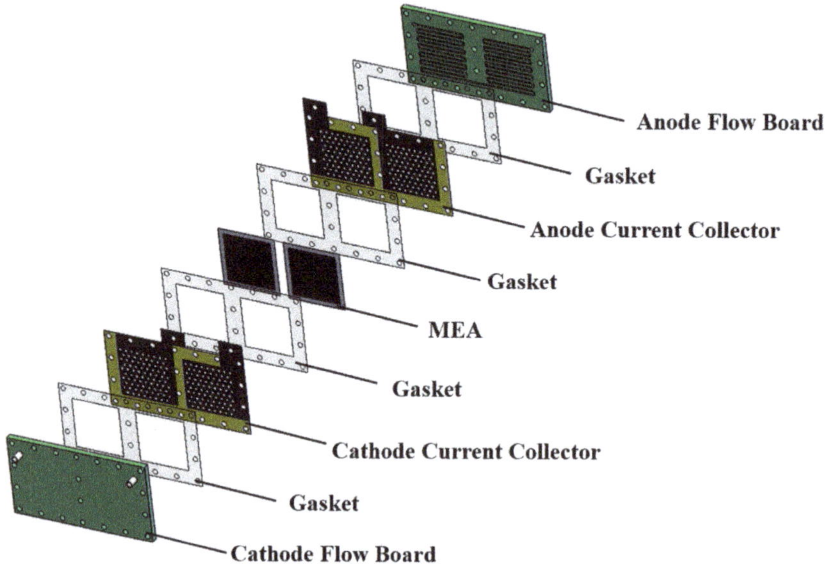


**Figure 9.** The exploded view of the Proton Exchange Membrane Fuel Cell (PEMFC) with forced convection air-breathing cathode. Membrane electrolyte assembly (MEA).

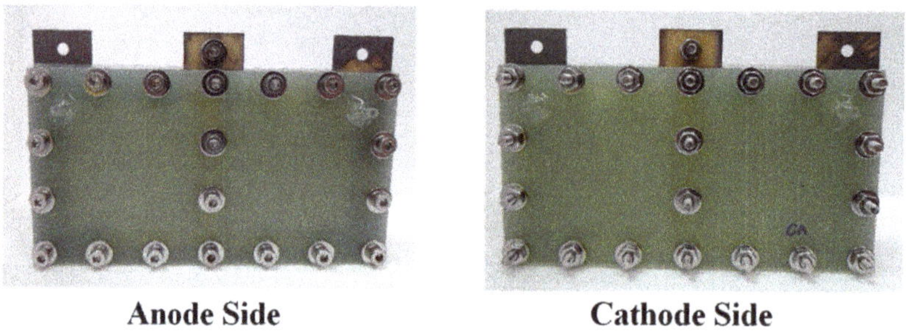


**Figure 10.** The anode side and cathode side of the PEMFC module with a forced convection air-breathing cathode.

The PEMFC module performances with forced convection air-breathing and three different types of graphene coatings were conducted at different anode hydrogen and cathode air flowrates. The anode hydrogen flowrate and cathode air flow rate ratios were kept at 1:2 in the experiment. Both the anode hydrogen flow and cathode airflow were humidified and kept at 50 °C. The anode hydrogen/cathode air flow rates of 25/50, 50/100, 100/200, 200/400, 300/600, and 400/800 sccm were investigated.

A performance comparison of the PEMFC module with a forced air-breathing cathode and the corrosion-resistance layer using graphene ink coating is shown in Figure 13. The results show that the PEMFC module performance increased with the increase in the low flow rate and reached the highest performance at the anode/cathode flow rates of 100/200 sccm. At low flow rates, increasing the flow rate is helpful to decrease the low mass transfer effect that would increase the cell performance. Whereas at high flow rates, increasing the flow rate reduced the cell performance, due to sub-saturated streams, which dehydrates the membrane, which is caused by the cell temperature inside exceeding the dew point.

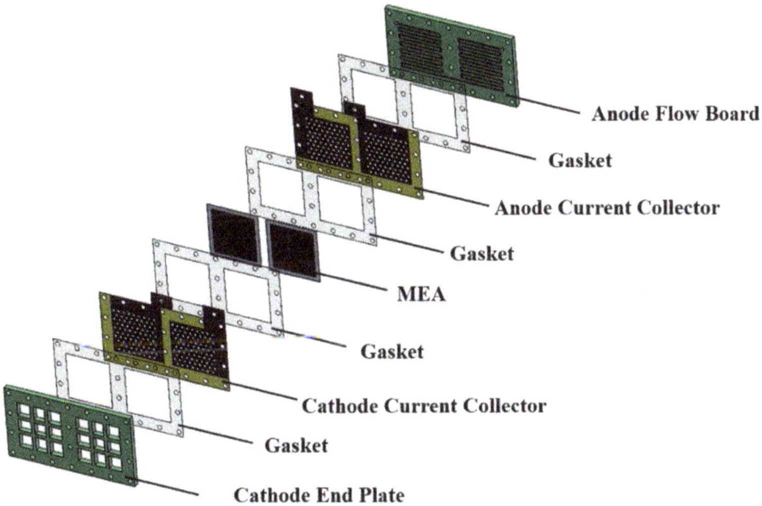


**Figure 11.** The exploded view of the PEMFC module with a self-air-breathing cathode.

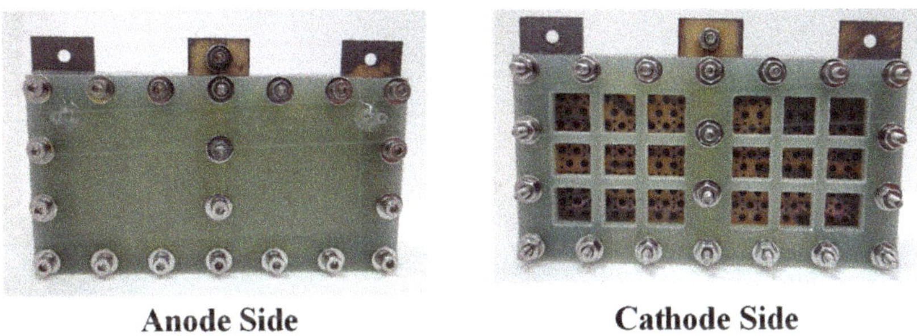


**Figure 12.** The anode side and cathode side of PEMFC module with a self-air breathing cathode.

In the high-current range of higher than 75 mA $cm^{-2}$ at the low flow rate (25/50 sccm), the further reduction of the operating voltage leads to a sharp drop of the current. This is due to the insufficient reactant, and the fuel cell becomes highly unstable, the cell performance is sharply dropped and is without a smooth transition through the ohmic loss to the concentration loss. Therefore, the low flow rate condition is not suitable for the cell operation at the high-current range for the current collector with the graphene ink thin film.

The PEMFC module performance is decreased when the flow rates are increased. The PEMFC module performance comparison with a forced air-breathing cathode and a corrosion-resistance layer using a graphene-suspension coating is shown in Figure 14. The results also show that the PEMFC module performance is increased at the low flowrate and reaches the highest performance at the anode/cathode flow rates of 100/200 sccm. The PEMFC module performance is decreased when the flow rates are further increased. At low flow rates, increasing the flow rates is helpful to decrease the low mass transfer effect that increases the cell performance. The results show a similar trend as the current collectors with graphene ink coatings and the related explanation refers to the Figure 13 explanation.

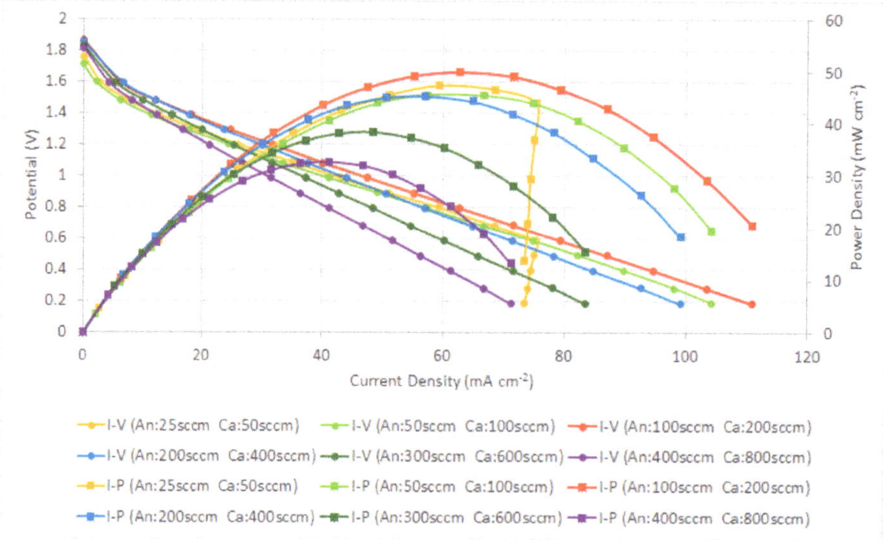


**Figure 13.** PEMFC module performance comparison with a forced air-breathing cathode and the corrosion-resistance layer using a graphene ink coating.

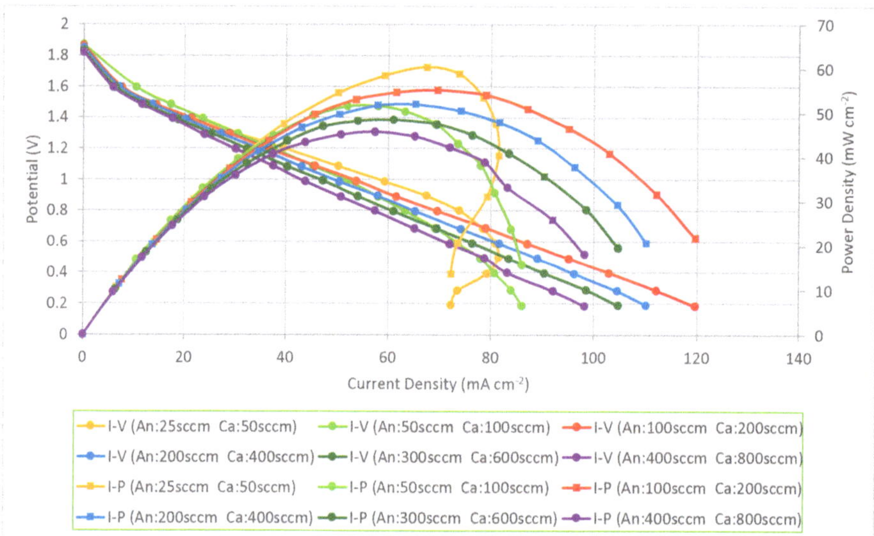


**Figure 14.** PEMFC module performance comparison with a forced air-breathing cathode and the corrosion-resistance layer using a graphene-suspension coating.

The performance comparison of the PEMFC module with a forced air-breathing cathode and the corrosion-resistance layer using a graphene-dispersion coating is shown in Figure 15. Similar to the previous two cases, the results still show that the PEMFC module performance is increased when increasing at the low flow rate and reaches the highest performance at the anode/cathode flow rates of 100/200 sccm. The performance comparison of the PEMFC module with a forced air-breathing cathode and a corrosion-resistance layer, at the anode/cathode flow rates is 100/200 sccm. At low flow rates, increasing the flow rates is helpful to decrease the low mass transfer effect that can increase the cell performance. Whereas at high flow rates, increasing the flow rate would reduce the cell performance due to sub-saturated streams, which dehydrates the membrane, and is caused by the cell temperature inside exceeding the dew point. In the high-current range of higher than 75 mA cm$^{-2}$ at the low flow rate (25/50 sccm), the further reduction of the operating voltage will lead to a sharp drop of the current. The results also show a similar trend to the current collectors with a graphene ink coating and the related explanation refers to the Figure 13 explanation.

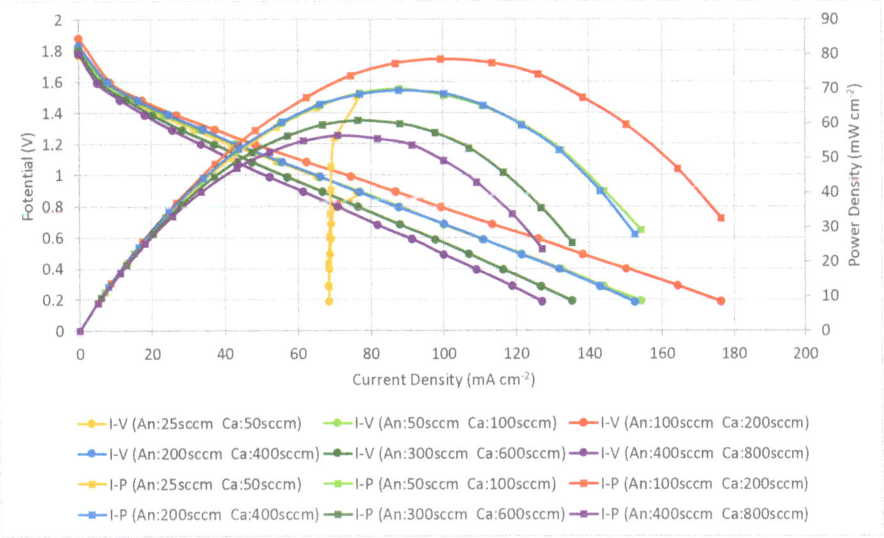


**Figure 15.** PEMFC module performance comparison with a forced air-breathing cathode and the corrosion-resistance layer using a graphene-dispersion coating.

The results show that the current collectors with three different types of graphene thin film show similar phenomena and trends. In addition, the current collector using the graphene-dispersion coating presents the highest performance, using graphene suspension is second, and using graphene ink presents the lowest PEMFC module performance.

The PEMFC module performances with self-breathing and three different types of graphene coatings were further conducted at the anode hydrogen flow rate of 100 sccm. The anode hydrogen flow was humidified, kept at 50 °C and a 100 sccm flow rate. The cathode side was open for self-air-breathing and the ambient temperature was 25 °C. The performance comparison of the PEMFC module with self-air-breathing cathode and the corrosion-resistance layer is shown in Figure 16. Similar to the force convection air-breathing cathode, the results also show that the current collector using graphene-dispersion coating presents the highest performance, using graphene suspension second, and using graphene ink presents the lowest PEMFC module performance. Cell performance of the self-air breathing cathode is lower than the forced air-breathing. This is due to the lower air stoichiometry, a lower temperature caused by the lower room temperature would reduce the kinetic activation, reduce the mass transfer, and increase the ohmic loss.

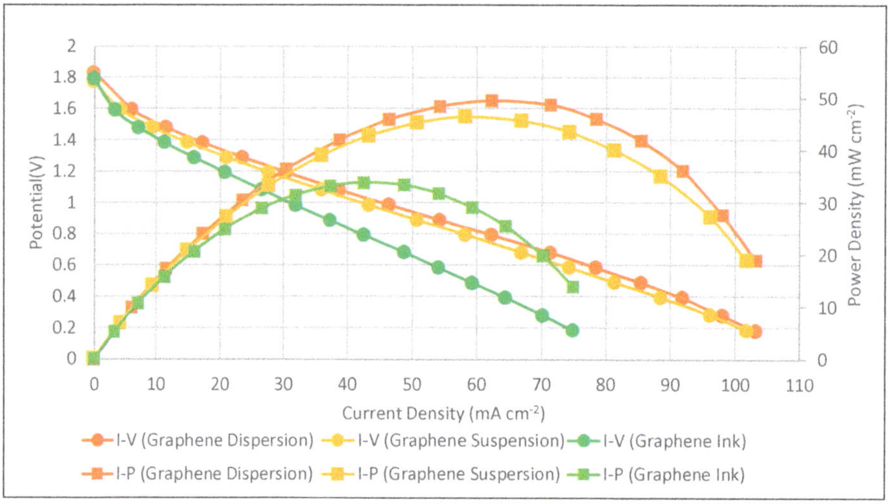


**Figure 16.** PEMFC module performance comparison with a self-air-breathing cathode and different types of the corrosion-resistance layers (anode fuel rate: 100 sccm).

Although the surface resistances are roughly the same for all cells via different graphene thin film coatings in Tables 2–4, there are still differences in the polarization resistances at the ohmic regime. This implies the surface resistance, which reflects the in-plane resistance, might not play the most important role. In the design concept, the graphene thin film is a corrosion-resistance layer, which is coated on the copper thin film layer, which functions as the electric-conduction layer. Therefore, the thinner the graphene layer, the better, and the interface between the graphene thin film and copper layer will also likely be better. If the interface is not good, it might degrade the electric conductivity.

According to the measurements in Table 1 and the SEM images in Figure 8, the thickness and weight of the graphene ink thin film is significantly larger than the other two thin films, with some obvious voids inside the graphene ink thin film, and the surface of the graphene ink thin film is much more uneven than the other two thin films. The graphene-dispersion thin film has the lowest weight and the most flatness. The graphene-dispersion thin film has a little bit higher weight and the surface is uneven. The above could be why the current collector with the graphene ink thin film shows the significantly lowest performance, the performances of the other two are close to each other; however, the current collector with the graphene-dispersion thin film has the best performance overall.

After investigating the PEMFC module performance with the developed current collectors for both forced air-breathing and self-air-breathing cathodes, long-term stability tests of the PEMFC modules, with self-air-breathing cathodes, using three different graphene coatings as corrosion-resistance layers, were further conducted. Experiments were performed at a 100 sccm anode hydrogen flow rate and the hydrogen was humidified and kept at an inlet fuel temperature of 50 °C. The cathode side is open for self-air-breathing and the ambient temperature is 25 °C. The PEMFC module stability tests with self-air-breathing cathodes and different types of corrosion-resistance layers under 1 V loads, are shown in Figure 17.

The results show that both the PEMFC modules with current collectors using graphene-dispersion and graphene-suspension coatings are quite stable. The PEMFC module performance with the collectors using the graphene-dispersion coating is slightly higher than that using the graphene suspension. Although the PEMFC module with current collectors using the graphene ink coating also showed good stability, the PEMFC module performance is distinctly lower than those using the graphene dispersion and graphene suspension. The results also successfully show the feasibility of the developed current collectors in portable applications. However, the tests only represent approximately a 10 h operation period, which is much less than the anticipated life specification. Therefore, additional longer duration tests, particularly under realistic cycling conditions that accelerate degradation, should be made in the future for realistic commercialized applications [31].

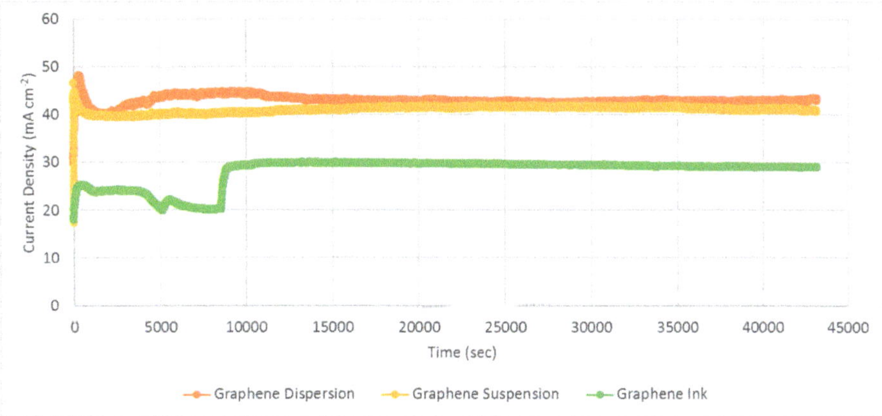


**Figure 17.** PEMFC module stability tests with a self-air-breathing cathode and different types of corrosion-resistance layers under 1 V load (anode fuel rate: 100 sccm).

## 4. Conclusions

This paper proposes lightweight current collectors and constructed a PEMFC module with two cells. The current collectors consist of a FR-4 substrate, a copper thin film as the electric-conduction layer, and a graphene thin film as the corrosion-resistance layer. The copper thin film is accomplished by a thermal evaporation process. Three different graphene coating processes to fabricate the graphene thin film were investigated. The results show that adopting a graphene-dispersion coating has the best corrosion-resistance capability, the graphene-suspension coating is second, and graphene ink is the lowest. Further study of the fabricated current collectors assembled into a two-cell PEMFC module show that the current collectors with a graphene-dispersion coating have the highest PEMFC module performance, the graphene-suspension coating is second, and graphene ink also has the lowest PEMFC module performance.

The graphene-dispersion thin film shows the lowest coating weight and superior surface flatness, and the graphene ink thin film shows a significantly higher coating weight, voids inside, and an uneven surface in the testing results. Long-time stability studies were also conducted, and all three graphene coating processes showed good PEMFC module stability after long-term operations. During the long-time stability tests, the PEMFC module performances for the current collectors that used a graphene-dispersion coating are a little bit higher than for the current collectors that used a graphene-suspension coating. These performances were both clearly higher than those of current collectors that used a graphene ink coating. The feasibility studies also showed the potential of the developed current collectors in portable applications.

**Author Contributions:** Conceptualization, Y.-D.K. and M.-F.S.; Methodology, Y.-D.K., T.-R.K. and J.-L.L.; Software, T.-R.K. and J.-L.L.; Validation, Y.-D.K.; Investigation, Y.-D.K.; Resources, Y.-D.K., M.-F.S. and J.-S.D.; Data Curation, T.-R.K. and J.-L.L.; Writing-Original Draft Preparation, Y.-D.K.; Writing-Review & Editing, Y.-D.K., M.-F.S. and J.-S.D.; Visualization, T.-R.K. and J.-L.L.; Supervision, Y.-D.K. and J.S.D.; Project Administration, Y.-D.K. and J.-S.D.; Funding Acquisition, Y.-D.K. All authors have read and agreed to the published version of the manuscript.

**Funding:** This work was supported by the Ministry of Science and Technology of Taiwan, grant number MOST 106-2221-E-167-023 and MOST 107-2221-E-167-017.

**Conflicts of Interest:** The authors declare no conflict of interest.

## References


1. O'Hayre, R.; Cha, S.W.; Colella, W.; Prinz, F.B. *Fuel Cell Fundamentals*, 3rd ed.; John Wiley & Sons: New York, NY, USA, 2016.
2. Lee, S.J.; Chang-Chien, A.; Cha, S.W.; O'Hayre, R.; Park, Y.I.; Saito, Y.; Prinz, F.B. Design and fabrication of a micro fuel cell array with flip-flop interconnection. *J. Power Sources* **2002**, *110*, 410–418. [CrossRef]
3. Cha, H.Y.; Choi, H.G.; Nam, J.D.; Lee, Y.; Cho, S.M.; Lee, E.S.; Lee, J.K.; Chung, C.H. Fabrication of all-polymer micro-DMFCs using UV-sensitive photoresist. *Electrochim. Acta* **2004**, *242*, 795–799. [CrossRef]
4. Lu, G.Q.; Wang, C.Y.; Yen, T.J.; Zhang, X. Development and characterization of a silicon-based micro direct methanol fuel cell. *Electrochim. Acta* **2004**, *49*, 821–828. [CrossRef]
5. Yun, Y.H. Deposition of gold-titanium and gold-nickel coatings on electro polished 316L stainless steel bipolar plates for proton exchange membrane fuel cells. *J. Hydrogen Energy* **2010**, *35*, 1713–1718. [CrossRef]
6. Feng, L.; Cai, W.; Li, C.; Zhang, J.; Liu, C.; Xing, W. Fabrication and performance evaluation for a novel small planar passive direct methanol fuel cell stack. *Fuel* **2012**, *94*, 401–408. [CrossRef]
7. Mani, S.P.; Srinivasan, A.; Rajendran, N. Effect of nitrides on the corrosion behavior of 316L SS bipolar plates for Proton Exchange Membrane Fuel Cell (PEMFC). *Int. J. Hydrogen Energy* **2015**, *40*, 3359–3369. [CrossRef]
8. Chen, C.H.; Chen, T.Y.; Cheng, C.W.; Peng, R.G. An experimental study on micro proton exchange membrane fuel cell. *J. Fuel Cell Sci.* **2011**, *9*, 031001-1–031001-7. [CrossRef]
9. Alanis-Navarro, J.A.; Reyes-Betanzo, C.; Moreira, J.; Sebastian, P.J. Fabrication and characterization of a micro-fuel cell made of metalized PMMA. *J. Power Sources* **2013**, *242*, 1–6. [CrossRef]
10. Hsieh, S.S.; Huang, C.F. Design; fabrication and performance test of a planar array module-type micro fuel cell stack. *J. Energy Convers. Manag.* **2013**, *76*, 971–979. [CrossRef]
11. Wang, A.; Yuan, W.; Huang, S.; Tang, Y.; Chen, Y. Structural effects of expanded metal mesh used as a flow field for a passive direct methanol fuel cell. *Appl. Energy* **2017**, *208*, 184–194. [CrossRef]
12. Renau, J.; Barroso, J.; Lozano, A.; Nueno, A.; Sanchez, F.; Martin, J.; Barreras, F. Design and manufacture of a high-temperature PEMFC and its cooling system to power a lightweight UAV for a high altitude mission. *Int. J. Hydrogen Energy* **2015**, *41*, 19702–19712. [CrossRef]
13. Chang, I.; Park, T.; Lee, J.; Lee, H.B.; Ji, S.; Lee, M.H.; Ko, S.H.; Cha, S.W. Performance enhancement in bendable fuel cell using highly conductive Ag nanowires. *Int. J. Hydrogen Energy* **2014**, *39*, 7422–7427. [CrossRef]
14. Park, T.; Chang, I.; Lee, H.B.; Ko, S.H.; Cha, S.W. Performance variation of bendable polymer electrolyte fuel cell based on Ag nanowire current collector under mixed bending and twisting load. *Int. J. Hydrogen Energy* **2017**, *42*, 1884–1890. [CrossRef]
15. O'Hayre, R.; Braithwaite, D.; Hermann, W.; Lee, S.J.; Fabian, T.; Cha, S.W.; Saito, Y.; Prinz, F.B. Development of portable fuel cell arrays with printed-circuit technology. *J. Power Sources* **2003**, *124*, 459–472. [CrossRef]
16. Schmitz, A.; Tranitz, M.; Wagner, S.; Hahn, R.; Hebling, C. Planar self-breathing fuel cells. *J. Power Sources* **2003**, *118*, 162–171. [CrossRef]
17. Kim, S.H.; Cha, H.Y.; Miesse, C.M.; Jang, J.H.; Oh, Y.S.; Cha, S.W. Air-breathing miniature planar stack using the flexible printed circuit board as a current collector. *Int. J. Hydrogen Energy* **2009**, *34*, 459–466. [CrossRef]
18. Kuan, Y.D.; Chang, C.H. Experimental investigation on the process-induced damage of a DMFC assembled by the printed circuit board technique. *J. Fuel Cell Sci. Technol.* **2009**, *6*, 011016-1–011016-9. [CrossRef]
19. Yuan, W.; Zhang, X.; Zhang, S.; Hu, J.; Li, Z.; Tang, Y. Lightweight current collector based on printed-circuit-board technology and its structural effects on the passive air-breathing direct methanol fuel cell. *J. Renew. Energy* **2015**, *81*, 664–670. [CrossRef]
20. Sung, M.F.; Kuan, Y.D.; Chen, B.X.; Lee, S.M. Design and fabrication of light weight current collectors for direct methanol fuel cells using the micro-electro mechanical system technique. *J. Power Sources* **2011**, *196*, 5897–5902. [CrossRef]
21. Kuan, Y.D.; Lee, S.M.; Sung, M.F. Development of a direct methanol fuel cell with lightweight disc type current collectors. *J. Energy* **2014**, *7*, 3136–3147. [CrossRef]
22. Kuan, Y.D.; Lyu, J.L.; Ke, T.R.; Sung, M.F.; Do, J.S. Planar current collector design and fabrication for proton exchange membrane fuel cell. *Int. J. Hydrogen Energy* **2019**, *44*, 10071–10081. [CrossRef]
23. Ning, F.; He, X.; Shen, Y.; Jin, H.; Li, Q.; Li, D.; Li, S.; Zhan, Y.; Du, Y.; Jiang, J.; et al. Flexible and lightweight fuel cell with high specific power density. *ACS Nano* **2017**, *11*, 5982–5991. [CrossRef] [PubMed]

24. Ning, F.; Shen, Y.; Bai, C.; Wei, J.; Lu, G.; Gui, Y.; Zhou, Z. Critical importance of current collector property to the performance of flexible electrochemical power sources. *Chin. Chem. Lett.* **2019**, *30*, 1282–1288. [CrossRef]
25. Mallick, R.K.; Thombre, S.B.; Shrivastava, N.K. A critical review of the current collector for passive direct methanol fuel cells. *J. Power Sources* **2015**, *285*, 510–529. [CrossRef]
26. Chang, B.; Martin, A.; Gregory, P.; Kundu, S.; Du, C.; Orondo, C.; Thuo, M. Functional materials through surfaces and Interfaces. *Mar. Adv.* **2018**, *3*, 2221–2233. [CrossRef]
27. Lee, Y.H.; Noh, S.; Lee, J.-H.; Chun, S.-H.; Cha, S.W.; Chang, I. Durable graphene-coated bipolar plates for polymer electrolyte fuel cells. *Int. J. Hydrogen Energy* **2017**, *42*, 27350–27353. [CrossRef]
28. Graphage Graphene. Available online: http://www.graphene.com.tw/I-MS18.php (accessed on 11 December 2018).
29. Graphage Graphene. Available online: http://www.graphene.com.tw/S-WB30.php (accessed on 11 December 2018).
30. Date Provided by the Golden Innovation Business; New Taipei, Taiwan. Available online: http://www.gibusiness.com (accessed on 18 December 2018).
31. Zhao, J.; Li, X. A review of polymer electrolyte membrane fuel cell durability for vehicular applications: Degradation modes and experimental techniques. *Energy Convers. Manag.* **2019**, *199*, 112022. [CrossRef]


**Sample Availability:** Samples of the compounds are not available from the authors.

molecules

Article

# Tolerance and Recovery of Ultralow-Loaded Platinum Anode Electrodes upon Carbon Monoxide and Hydrogen Sulfide Exposure

**Sebastian Prass [1,*], Kaspar Andreas Friedrich [2,3] and Nada Zamel [1]**

1 Fraunhofer Institute for Solar Energy Systems ISE, Heidenhofstr. 2, 79110 Freiburg, Germany; nada.zamel@ise.fraunhofer.de
2 German Aerospace Center, Institute of Engineering Thermodynamics, Pfaffenwaldring 38-40, 70569 Stuttgart, Germany; andreas.friedrich@dlr.de
3 University of Stuttgart, Institute of Building Energetics, Thermal Engineering and Energy Storage (IGTE), Pfaffenwaldring 31, 70569 Stuttgart, Germany
* Correspondence: sebastian.prass@ise.fraunhofer.de; Tel.: +49-761-4588-5310; Fax: +49-761-4588-9320

Academic Editors: Jean St-Pierre and Shangfeng Du
Received: 22 August 2019; Accepted: 26 September 2019; Published: 27 September 2019

**Abstract:** The effects of carbon monoxide (CO) and hydrogen sulfide ($H_2S$) in concentrations close to their respective limits in the Hydrogen Quality Standard ISO 14687-2:2012 on the performance of proton exchange membrane fuel cells (PEMFCs) with ultralow-loaded platinum anode catalyst layers (CLs) were investigated. The anodic loadings were 50, 25, and 15 μg/cm$^2$, which represent the current state-of-the-art, target, and stretch target, respectively, for future automotive PEMFCs. Additionally, the effect of shut-down and start-up (SD/SU) processes on recovery from sulfur poisoning was investigated. CO at an ISO concentration of 0.2 ppm caused severe voltage losses of ~40–50% for ultralow-loaded anode CLs. When $H_2S$ was in the fuel, these anode CLs exhibited both a nonlinear decrease in tolerance toward sulfur and an improved self-recovery during shut-down and start-up (SD/SU) processes. This observation was hypothesized to have resulted from the decrease in the ratio between CL thickness and geometric cell area, as interfacial effects of water in the pores increasingly impacted the performance of ultrathin CLs. The results indicate that during the next discussions on the Hydrogen Quality Standard, a reduction in the CO limit could be a reasonable alternative considering future PEMFC anodic loadings, while the $H_2S$ limit might not require modification.

**Keywords:** fuel impurities; ISO concentration; ultralow-loaded anode catalyst layer; platinum electrode; shut-down and start-up process

## 1. Introduction

Proton exchange membrane fuel cells (PEMFCs) are a promising clean energy alternative for applications in the transport sector, as they combine high-power density and efficiency with the significant advantage of fast system refueling times. Hydrogen ($H_2$) as a fuel might, however, contain low concentrations of impurities stemming from production and infrastructure. Impurities such as carbon monoxide (CO) and hydrogen sulfide ($H_2S$) can deteriorate the performance and lifetime of PEMFCs. Naturally, the severity of an impurity is not only affected by its concentration (or rather, dose), but also by the catalyst type, operational parameters, cross-effects, and active or passive mitigation strategies [1,2]. For example, air-bleeding is an effective strategy to provide oxygen ($O_2$) for the oxidation of adsorbed contaminant species in the anode electrode [3], while catalyst alloys containing platinum (Pt) and other platinum group metals (PGMs) can provide higher tolerances versus certain contaminants [4–7]. Although they are very effective, such mitigation strategies partially come with implications about performance or durability. For example, a fraction of the $O_2$ introduced by air-bleeding readily reacts with $H_2$ in the anode compartment and thereby lowers the fuel efficiency while simultaneously accelerating membrane degradation through additional peroxide and radical formation [8]. Moreover, alloy catalysts containing PGMs or metals other than Pt usually offer a lower stability, as the alloying components exhibit higher leaching rates. What typically remains is a catalyst particle with a Pt-enriched surface [9], while the leaching cations eventually have impacts on the protonic conductivity or even integrity of the ionomer in the electrode or membrane [10].

Apart from active or passive PEMFC system internal contamination mitigation techniques, adjusting the allowed impurity limits in the Hydrogen Quality Standard ISO 14687-2:2012 poses an additional layer in accommodating enhanced PEMFC requirements versus fuel contaminants. If electrode design or system internal strategies are exhausted, the allowed impurity level for the respective contaminant could be lowered at reasonable levels based upon tangible experimental PEMFC data. Although this option eventually leads to higher $H_2$ production costs, it helps to avoid higher PEMFC system costs per vehicle or implications coming from internal tolerance improvement strategies.

Some of the major cost drivers in mass-produced PEMFC vehicles are the catalyst layers (CLs) attached to the membrane. The choice of CL materials, the electrode design, and production are primary levers in reducing PEMFC costs while simultaneously increasing the lifetime. Although substantial reductions in PGM catalyst loading per cell area have already been achieved, further reductions are required as a consequence of increasing PGM prizes with higher FC vehicle market penetration. The stipulated reductions range from 50% to 75% compared to the approximate state-of-the-art, resulting in PGM targets for 2020 of about 125 and 62.5 $\mu g/cm^2$ depending on the contemplated scenario [11]. In both cases, the loading of the anode electrode is expected to account for 20% (i.e., 25 and 12.5 $\mu g/cm^2$ of PGMs, respectively): this is called ultralow loading in the present study hereafter.

Generally, lower anodic catalyst loadings are less tolerant toward catalyst contaminants, as both fuel and contaminants compete for fewer active sites in the electrode. For pure Pt electrodes, the voltage drop was found to increase by 25% when the Pt-loading decreased from 400 to 50 $\mu g/cm^2$ if 1 ppm CO was introduced [12,13]. A similar trend was observed for $H_2S$, where the tolerance of the electrode was found to decreased proportionally with the reduction in the anode loading [14]. It is expected that this trend would continue for ultralow loadings (<50 $\mu g_{PGM}/cm^2$), but so far there has been no study in the literature that has investigated the tolerance of such ultralow anodic loadings. Additionally, processes such as the shut-down and start-up (SD/SU) of FC vehicles are expected to affect the poisoning phenomenon of the electrodes. During downtime, reactants can diffuse from the anode to the cathode, and conversely, mixed potentials arise at the electrodes and poisoned catalysts eventually recover. However, there are limited experimental data available in the literature on recovery due to SD/SU processes, which is especially of interest in the case of recovery from sulfur contamination. Cyclic voltammetry (CV)-like methods triggering oxidative processes at ~0.9–1.1 V count as a recovery strategy for sulfur-contaminated electrodes [6,15,16], but this strategy also induces carbon corrosion and therefore destruction of the electrode itself.

The study presented here therefore seeks to add to the studies by Hashimasa et al. [12,14] by investigating the tolerance of ultralow-loaded anodic platinum catalyst layers. Two different types of contaminants were selected: CO, as its poisoning effect is fully reversible, while in contrast, $H_2S$ typically poisons the catalyst irreversibly during regular fuel cell operation. Additionally, recovery from sulfur poisoning through simple shut-down and start-up (SD/SU) processes was examined in more detail for ultralow anodic catalyst loadings.

## 2. Materials and Methods

### *2.1. Test Station and Contaminant Introduction*

Single-cell tests were carried out in an in-house-built test station with an integrated potentiostat (Zahner Zennium Pro) and an electric load (Kikusui PLZ664WA) with fluidics (shown schematically in Figure 1).

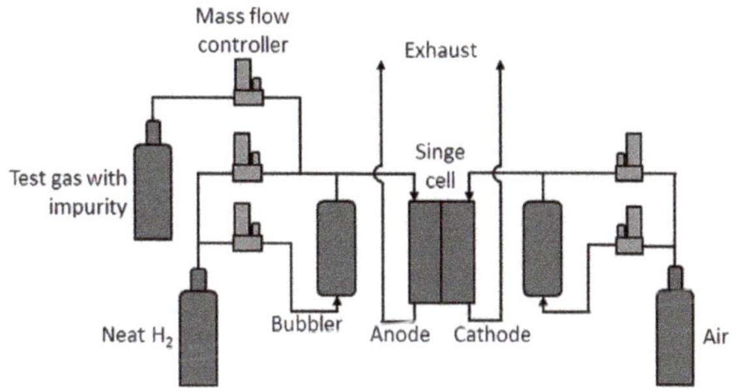


**Figure 1.** Single-cell test station scheme.

In principle, the test station was comparable to the one used by Hashimasa et al. [12], but with a different humidification system for the anode, a different position of the test gas feed inlet (here, the test gases were not fed through the humidifier), and no gas analysis system. In the present study, a differential cell (Baltic qCF type with automotive linear-channel flow field) with an active area of 20.25 cm$^2$ was employed, which allowed for the minimization of in-plane effects such as gradients in partial gas pressures, relative humidity, and temperature and therefore enhanced focus on the contamination effect at a given concentration. Although the effects of very low concentrations of impurities eventually become less visible in such a cell [17], a rather uniform coverage of the contaminant on the catalyst throughout the active area was expected.

Low concentrations of impurities were achieved by mixing precontaminated test gases with neat $H_2$. Therefore, carbon monoxide (CO, 10 ppm in $H_2$ 5.0) and hydrogen sulfide ($H_2S$, 0.5 ppm in $N_2$ 5.0) were mixed via mass flow controllers with house-supply high-purity hydrogen (all gases provided by Linde AG) in the required fractions.

### *2.2. Materials*

The variations in the anode-loading on the catalyst-coated membranes (CCMs, provided by Greenerity GmbH) were achieved through different thicknesses of the anode catalyst layers (CLs), while the cathode loading was kept constant at 400 µg/cm$^2$. The catalyst material for both electrodes, the anode and the cathode, was pure Pt on carbon. The membrane electrode assembly (MEA) specifications are shown in Table 1.

**Table 1.** Membrane electrode assembly (MEA) specifications.

| **Active Cell Area** | | **20.25 cm$^2$** |
|---|---|---|
| Catalyst | Anode | Pt/C |
| | Cathode | Pt/C |
| Electrode loading | Anode | 50/25/15 µg/cm$^2$ (named A, B, and C hereafter) |
| | Cathode | 400 µg/cm$^2$ |
| Membrane thickness | | ~15 µm |
| Gas diffusion layer | | Freudenberg H23C9 |

## 2.3. *Testing Procedure and Conditions*

For every test with a different type of contaminant gas, a fresh MEA sample was assembled into the test cell. To measure the effect of the impurities, the test cell was operated with a constant load to detect the voltage drop associated with the contaminant species and concentration. In the following figures, the cell voltage drop is defined as the relative change based on the initial cell voltage. The effect of CO was tested at three different concentrations, namely 0.1, 0.2, and 0.4 ppm (50%, 100%, and 200% of the impurity limit noted in the $H_2$ Quality Standard). Before and after the actual contamination, the fuel cell was operated with neat $H_2$ to establish a baseline voltage and to detect eventual irreversible degradation of the electrodes. The effect of $H_2S$ was tested at two concentrations, which were 4 and 20 ppb (100% and 500% of the limit in the Quality Standard), with neat $H_2$ operation only at the start of the contaminant test. The conditions during the contaminant tests are shown in Table 2.

**Table 2.** Operating conditions during contamination.

| | |
|---|---|
| Cell Temperature | 80 °C |
| Outlet pressure anode/cathode | 1.2/1.2 bara |
| Relative humidity anode/cathode | 90%/75% |
| Current density | 1.0 A/cm$^2$ |
| Stoichiometry anode/cathode | 12/14 |

The MEAs were characterized, including cyclic voltammetry (CV) on the anode and cathode side at the beginning and end of life (BOL and EOL), as were the polarization curves at the BOL, to compare the performance between the MEA types before starting the contaminant test. The gas pressure during contamination was selected in reference to the studies by Hashimasa et al. [12,14], while the pressure during the polarization curves was chosen according to in-house standardized testing protocols. The conditions during the polarization curves are shown in Table 3.

**Table 3.** Polarization curve conditions.

| | |
|---|---|
| Cell Temperature | 80 °C |
| Outlet pressure anode/cathode | 2/2 bara |
| Relative humidity anode/cathode | 95%/75% |
| Gas flow anode/cathode | 3/7 l/min |

CV measurements were performed to determine the electrochemically active surface area (ECSA) of the CLs before and after contamination and recovery procedures, specifically from $H_2S$ poisoning. The CVs were performed on both the anode and cathode electrodes under the conditions summarized in Table 4. To conduct an anode CV, the test cell was purged with nitrogen in order to exchange the gas supply and the electric connectors of the anode and cathode compartment and then reconditioned with fully humidified $H_2$ and $N_2$ for 12 min prior to the CV. Following the anode CV, the cell was purged again and reconnected in a regular anode/cathode configuration for subsequent tests.

**Table 4.** Cyclic voltammetry (CV) conditions.

| | |
|---|---|
| Cell Temperature | 80 °C |
| Outlet pressure anode/cathode | 1.05/1.05 bara |
| Relative humidity anode/cathode | 95%/95% |
| Gas flow anode/cathode | 1.0/0 l/min (1.0 l/min $N_2$ for 12 min prior to CV on cathode) |
| Scan range | 50–700 mV |
| Sweep rate | 100 mV/s |

An upper CV boundary of 700 mV was selected to avoid the oxidation of adsorbed foreign species, especially during the $H_2S$ recovery tests, and to solely focus on the recovery from SD/SU processes. Moreover, the $N_2$ flow was stopped during the actual CV to avoid disproportionally high $H_2$ evolution currents during the anodic sweep, which were observed especially for the lowest anodic loading. Figure 2 shows exemplary BOL CVs of the three different anode electrodes and one cathode electrode for comparison.

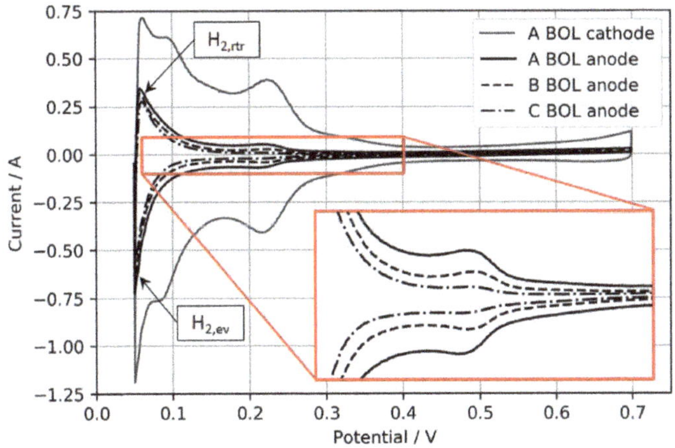


**Figure 2.** Anode CVs of MEA types A, B, and C with an MEA type A cathode CV for reference. The inset expands the $H_2$ adsorption/desorption regions of the anode catalyst layers (CLs) for visual comparison.

Normally, the ECSA is determined through integration of the charge transfer between voltage boundaries, starting from ~0.08 to 0.1 V to the minima or maxima of the respective double-layer charging current, which typically is somewhere between 0.3 and 0.6 V [18]. However, in this study, these boundaries were considered less suitable for CVs on ultralow-loaded anode CLs. High currents associated with $H_2$ evolution during the cathodic sweep ($H_{2,ev}$) and the coherent reverse-transport of eventually evolved $H_2$ during the anodic sweep ($H_{2,rtr}$) would account for relatively large errors in the ECSA. Hence, the voltage boundaries for the determination of the ECSA were chosen as 0.15 to 0.3 V, as shown in Figure 3.

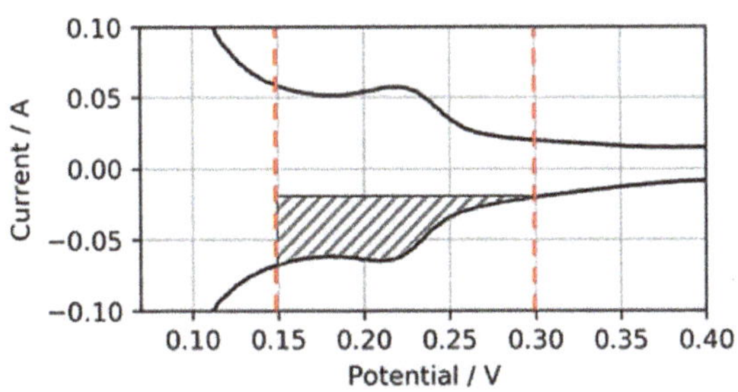


**Figure 3.** Electrochemically active surface area (ECSA) determination from reduced $H_2$ adsorption area.

Using this narrowed voltage range, the anode ECSA was determined from the anodic sweeps associated with the adsorption of $H_2$ on the catalyst surfaces. Although this procedure cuts the measured ECSA compared to integration between regular voltage ranges, it was found that it would increase the accuracy of the ECSA determination and its changes in the case of ultralow-loaded anodes (as tested in the present study).

## 3. Results and Discussion

### *3.1. Performance and Stability of Ultralow-Loaded Anodic CLs*

Before the actual contamination tests, the BOL performance and voltage stability of the MEA samples with ultralow-loaded anodes were established. Figure 4 shows the BOL polarization curves of the three different MEAs when neat $H_2$ was supplied to the test cells.

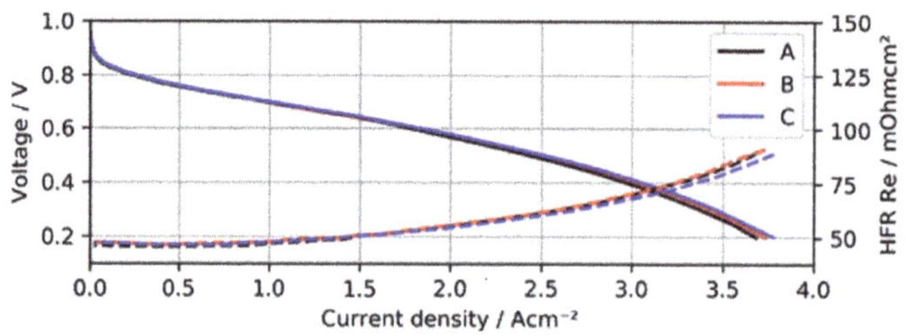


**Figure 4.** Polarization curves of MEA types A, B, and C using neat $H_2$, with high-frequency resistance (HFR) as dashed lines.

As can be seen in the figure, the polarization curves of the different MEAs overlap quite well, indicating that overpotentials arising due to a lack of active catalyst sites for the hydrogen oxidation reaction (HOR) were not significant for ultralow anodic loadings. In fact, MEA type C (15 μg/cm$^2$) even showed a slightly better performance at current densities above 2.5 A/cm$^2$, (~15 mV at 3 A/cm$^2$), which might have been a result of minimal differences in humidification characteristics of this specific sample and the lower measured high-frequency resistance (HFR).

In addition to the BOL performance, the cell voltage stability of the three MEA types over a testing time of 100 h of continuous operation at a constant load with neat $H_2$ was established, which is shown in Figure 5.

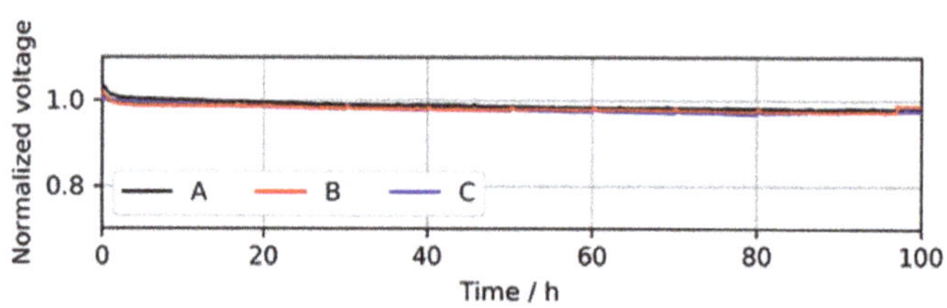


**Figure 5.** Voltage decay over 100 h of continuous operation in neat $H_2$ at 1.0 A/cm$^2$, showing similar voltage stabilities of the MEA types. The voltages were normalized to the initial cell voltage at time = 0 h.

During these stability tests, no significant difference between the voltage drops of the MEA types was observed. A slight voltage drop during the first ~2 h was visible for all three MEA types and was associated with the consumption of reactants, which saturated in the electrode before the current was increased.

Overall, the comparability of the different MEA types at the BOL under operation with neat $H_2$ was considered satisfactory and was accepted for subsequent tests with contaminants. Before each contamination test, the cell was operated for 20 h with neat $H_2$ to establish a baseline voltage. In the case of CO, the first concentration of contaminant was introduced and increased at time steps of 20 h, before we finally shut off the impurity for an additional 20 h of operation with neat $H_2$. In the case of $H_2S$, after operation with neat $H_2$, a single concentration of $H_2S$ was introduced until the cell voltage broke down, and subsequently SD/SU recovery tests were conducted. For all tests, the anode bubbler required a refill with fresh deionized (DI) water every 10 h. This DI water contained dissolved $O_2$, which was driven out as soon as it was heated in the bubbler and was consequently available for the recovery of poisoned Pt sites, which is visible as voltage peaks in the following figures.

### 3.2. *Effect of CO on Ultralow-Loaded Anode CLs*

Essentially, CO adsorbs on Pt and thereby competes with the actual HOR for active sites on the catalyst surfaces, as shown in Equations (1)–(3):

$$2Pt + H_2 \leftrightarrow 2(Pt - H), \tag{1}$$

$$Pt + CO \leftrightarrow Pt - CO, \tag{2}$$

$$2Pt + CO \leftrightarrow (Pt)_2 = CO. \tag{3}$$

Depending on the coverage of CO, each molecule blocks one or two active Pt sites via linear or bridge bonds (Equations (2) and (3), respectively) [5,19]. At lower coverages, a higher fraction of bridge bonds is expected, while at higher coverages, an adlayer with CO linear bonds dominates [20]. However, the adlayer CO structure depends on particle sizes, adsorption potentials, facet orientations, and temperature in a complex way because dipole–dipole interactions are important [21]. The effect of different CO concentrations on the voltage decay rates of the three ultralow-loaded anodic CLs is shown in Figure 6.

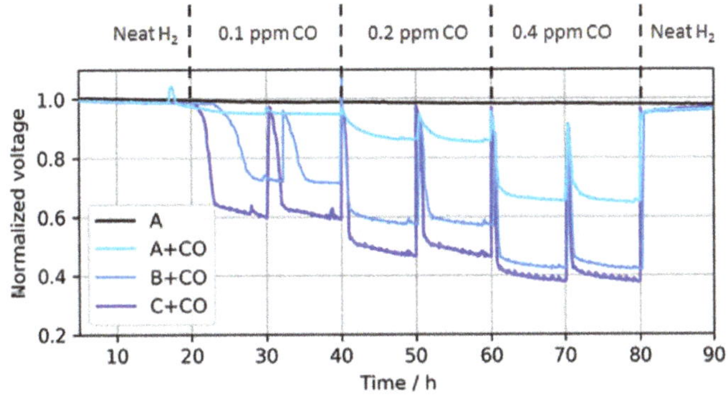


**Figure 6.** Voltage drops induced by different CO concentrations in MEA types A, B, and C at a constant load of 1.0 A/cm$^2$. The voltage peaks (every 10 h) were caused by anode bubbler refills and coherent recovery of Pt sites with $O_2$ dissolved in DI water. Again, the voltages were normalized to the initial cell voltage at time = 0 h, while the results are shown starting from $t$ = 5 h.

As expected, the effect of CO in the fuel generally increased for lower anodic loadings, including both a faster and more severe voltage drop. The leveling of the potentials, i.e., the initial decline toward a plateau, depended on the contaminant concentration and the CL thickness [22,23]. For thinner CLs, the reaction front increasingly corresponded with the actual CL thickness, and therefore the local potential was more uniform while contaminants competed throughout the layer with hydrogen for adsorption sites, which resulted in a lower tolerance for thinner (and lower-loaded) CLs. At the ISO concentration (0.2 ppm), the voltage loss due to CO poisoning accounted for ~8%, 41%, and 51% when the anodic loading decreased from 50 to 25 and 15 µg/cm$^2$, respectively. Slight potential oscillations of the ultralow-loaded anode MEA types (type B and especially C) at high CO concentrations between normalized voltage ratios of 0.4 and 0.6 were also visible. At these potentials, overpotentials induced by CO poisoning forced the anode potential to shift frequently toward the cathode potential and close to the oxidation potential of CO to $CO_2$, allowing for recovery of the electrode [24,25]. This self-recovery was the reason for the maximum coverage of the catalyst with CO in regular PEMFC operation and a flattening of the relative potential drop for lower anodic catalyst loadings with higher CO concentrations, which is partially visible in Figure 7.

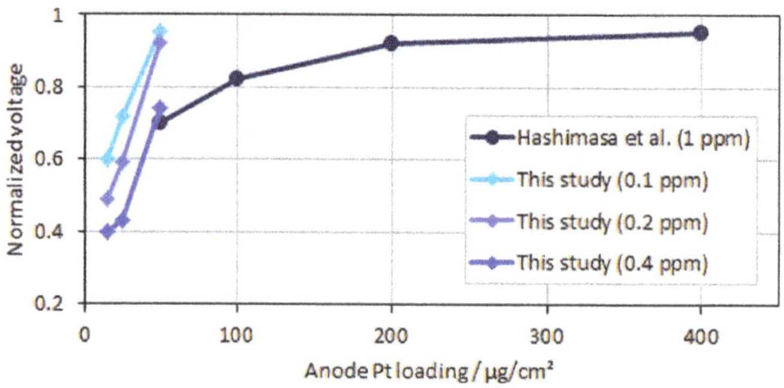


**Figure 7.** Normalized voltages over anodic loading; data adapted from Hashimasa et al. [12].

In the figure, relative voltage drops due to CO poisoning over the anode Pt loading from the study by Hashimasa et al. and the present study are compared. Although the test cells and the operational parameters between the two studies were different (70% fuel usage in the single cell by the Japanese Automobile Research Institute, JARI, versus 8.3% fuel usage in the differential single cell employed in the present study), a general trend for voltage decay with lower anodic loadings or higher CO concentrations can be seen. The onset of the mentioned flattening of the relative voltage drop at maximum CO coverage is visible for the lowest anodic loading and the highest tested CO concentration, where the relative change between MEA types B and C was less significant compared to types A and B.

In general, CO contamination is fairly easy to mitigate by providing $O_2$ to the anode via the air-bleeding technique [3]. This technique not only mitigates CO poisoning, but also partially mitigates poisoning from other contaminants, such as $H_2S$ [16]. However, as discussed above, air bleeding also comes with disadvantages, such as a reduction in fuel efficiency and potential effects on the integrity of the ionomer in the CLs and membrane. Therefore, to minimize potentially amplified side effects from such mitigation strategies, a reduction of the limit for CO in the $H_2$ Quality Standard could be a reasonable option considering the severity of CO poisoning on ultralow anodic loadings, as they likely will be employed in the near future in automotive PEMFCs.

### 3.3. *Effect of $H_2S$ on Ultralow-Loaded Anode CLs*

In contrast to CO, $H_2S$ poisons catalyst surfaces irreversibly through dissociative adsorption on Pt via chemical or electrochemical reaction pathways, as indicated by Equations (4) and (5), respectively. The elemental sulfur on Pt cumulatively occupies active catalyst sites also via linear or bridge bonds, which eventually leads to a complete breakdown of the PEMFC performance [6,14,16]:

$$Pt + H_2S \leftrightarrow Pt - S + H_2, \tag{4}$$

$$Pt + H_2S \leftrightarrow Pt - S + 2H^+ + 2e^-. \tag{5}$$

Higher catalyst loadings provide a higher nominal ECSA and therefore a larger buffer versus such a breakdown. This decrease in tolerance with a reduction in platinum loading is partially visible in Figure 8, which shows the operation times until the breakdown was observed for ultralow-loaded anodes.

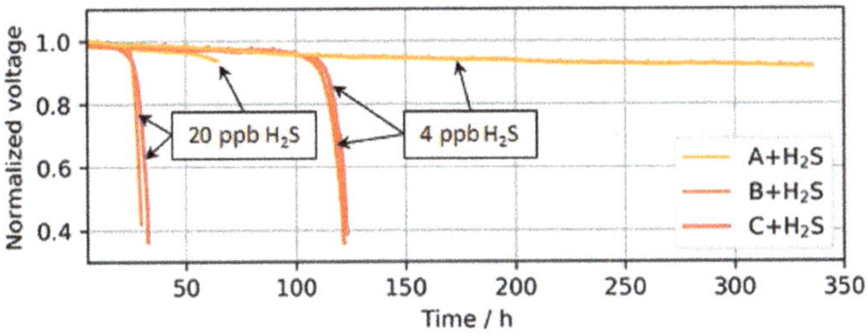


**Figure 8.** Voltage breakdowns induced by 4 and 20 ppb of $H_2S$ during operation at a constant load of 1.0 A/cm$^2$. The operation of MEA type A was purposely stopped after ~340 h and ~70 h, while MEA types B and C stopped automatically after voltage breakdowns were observed.

The voltage breakdowns for the highest tested anodic loading (50 µg/cm$^2$) were not fully observed. In the case of 4 ppb of $H_2S$, the test was purposely interrupted after 340 h of contaminant introduction, as a voltage breakdown was not expected anymore. However, subsequent CVs revealed an almost completely sulfur-blocked ECSA, which is shown in the following sections. In the case of 20 ppb of $H_2S$, the test station automatically stopped at the onset of the breakdown after about ~70 h, but the start of the breakdown was still visible.

Interestingly, for both MEA types with ultralow anodic loadings (MEA types B and C), voltage breakdowns were detected after almost similar poisoning times for both tested $H_2S$ concentrations of 4 and 20 ppb. In Figure 9, which compares the accumulated $H_2S$ supplied until a 30-mV voltage loss was detected in the present study versus the study by Hashimasa et al., these similar poisoning times are visible as a nonproportional decline in the amount of $H_2S$ supplied with the reduction in anodic loading.

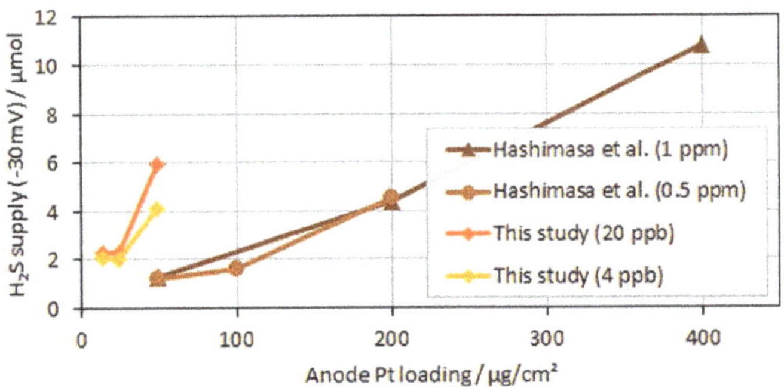


**Figure 9.** $H_2S$ supplied to the cell until voltage dropped by 30 mV over anodic loading; data adapted from Hashimasa et al. [14].

Although Hashimasa et al. described their observed decline as proportional to the reduction in the loading, their data actually rather showed a slight flattening of the curve with the decrease in the anodic loading, comparable to the data from the presented study. Again, although the test cells and the operational parameters were different (70% fuel usage in JARI's single cell versus 8.3% fuel usage in the differential single cell in the present study), the general trend was still visible.

One explanation could be that some of the $H_2S$ adsorbed on the surfaces of the test bench and cell components before actually reaching the CCM and catalyst sites. Depending on the chronology of the tests, this latency could create delays in the voltage breakdown. On the other hand, in the present study, the CVs of lower-loaded anodes also revealed a higher degree of self-recovery from simple shut-down (SD) and start-up (SU) processes.

For these self-recovery tests, the ECSA of the anode CLs exposed to $H_2S$ were determined at the BOL after a simulated SD/SU process, after $H_2S$ poisoning, and again after an SD/SU process. The SD/SU included a short purge with dry nitrogen to avoid open circuit voltage (OCV) in $H_2$/air-atmosphere, a cooldown of the cell to 20 °C, a wait time of 3 h, and finally again heating of the cell to 80 °C and the introduction of neat $H_2$/air to the cell, which was kept at a fixed potential of 0.8 V during the heating. Figure 10 presents these anode CVs for the three different anodic loadings.

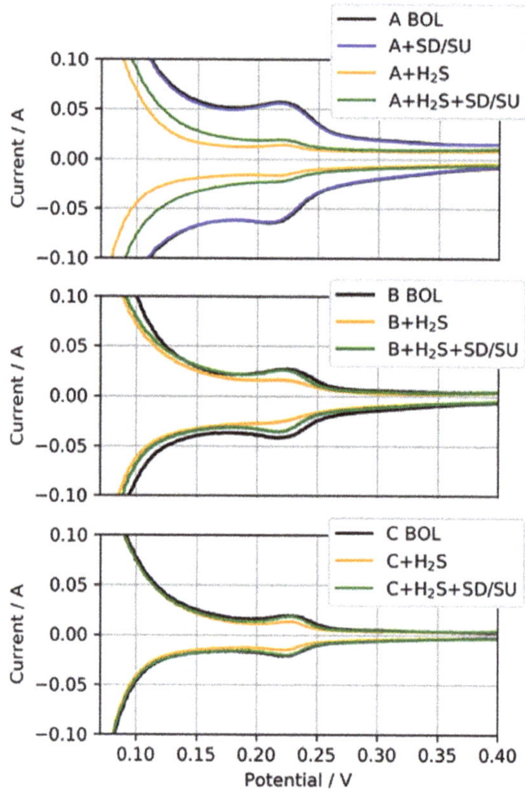


**Figure 10.** Anode CVs for MEA types A, B, and C (50, 25, and 15 μg/cm$^2$) at the beginning of life (BOL), after $H_2S$ contamination (+$H_2S$), and after a subsequent shut-down/start-up (SD/SU) process (+$H_2S$ + SD/SU). For MEA type A, the CV after SD/SU just before the contamination is also shown (A + SD/SU).

Clearly visible is the difference between the CVs at the BOL and after $H_2S$ contamination (black to yellow CV) for all three MEA types, indicating the reduction of ECSA due to sulfur adsorbed on Pt. For MEA type A, the CV after SD/SU and before $H_2S$ contamination (blue CV) is additionally shown to exemplarily demonstrate that the SD/SU process did not significantly affect the CV measurement and ECSA determination, as both CVs overlapped quite well. However, when the SD/SU process was carried out after $H_2S$ contamination, the CV and therefore the ECSA gained in area compared to the poisoned ECSA (yellow to green CV), indicating a partial recovery from previously deactivated ECSA. This self-recovery was increasingly observed with the reduction in the anodic loading. Table 5 presents the nominal ECSAs and percentage changes between the test SD/SU steps.

**Table 5.** ECSA at the BOL and relative change after shut-down/start-up processes (SD/SU) before and after contamination with $H_2S$ based on narrowed boundaries (integration between 150 and 300 mV). Note: the nominal ECSA was lower by about 60–70% than what would be typically expected for the specific catalyst material, while the relative ECSA changes were amplified to some degree due to the narrowed voltage boundaries and therefore the smaller area for integration.

| MEA Type | ECSA ($m^2$/g Pt) | | | |
|---|---|---|---|---|
| | **BOL** | **After SD/SU** | **After $H_2S$** | **After $H_2S$ + SD/SU** |
| A | 20.5 | 20.0 (98%) | 4.4 (22%) | 7.1 (35%) |
| B | 24.4 | 24.2 (99%) | 16.8 (69%) | 20.6 (84%) |
| C | 19.6 | 19.1 (97%) | 14.0 (71%) | 18.4 (94%) |

While only about 35% of the ECSA from MEA type A (50 μg/$cm^2$) could be recovered, 84% and almost a full recovery of 94% could be achieved for MEA types B (25 μg/$cm^2$) and C (15 μg/$cm^2$), respectively, through a simple SD/SU process.

The reason for the different behavior of ultralow-loaded anodes with respect to their tolerance versus $H_2S$ contamination and the improved self-recovery during SD/SU processes might have a dimensional character in combination with the scavenging effect of water versus contaminants [26]. Studies in the literature investigating the recovery of sulfur-poisoned electrodes have often employed CV-like processes to increase the potential and thereby oxidize adsorbed sulfur either on cathode or anode electrodes [27,28]. During this oxidation, sulfur oxides such as sulfur dioxide (and in combination with water-soluble anions such as sulfate ($SO^{2-}{}_4$) or sulfite ($SO^{2-}{}_3$)) develop as shown in Equations (6)–(8) [16]:

$$Pt - S + O_2 \leftrightarrow Pt + SO_2, \tag{6}$$

$$Pt - S + 3H_2O \leftrightarrow Pt + SO_3^{2-} + 6H^+ + 4e^-, \tag{7}$$

$$Pt - S + 4H_2O \leftrightarrow Pt + SO_4^{2-} + 8H^+ + 6e^-. \tag{8}$$

Presumably, during an SD/SU process, the catalyst surfaces and adsorbed species relax, the local potential varies depending on the local equilibrium and the available species on Pt, and chemical reactions occur to the point of the formation of sulfur anions in the presence of water. It should be noted that the potential of the anodic electrode prior to and during the SD can affect the reduction state of the sulfur species, which eventually facilitates their oxidation or desorption [29]. As the different anodic loadings tested in this study were achieved through variations in CL thickness, the anode of MEA type C consequently had the lowest thickness, while the active cell area remained the same for all samples. During an SD, water condensates and eventually is driven out through hydrophobic pores of the microporous and gas diffusion layer (MPL/GDL) or collects in pores and areas, which are energetically favorable. As the interface between the MPL and CLs also contains such pores [30], sulfur in proximity to this interface might dissolve in these water accumulations in the form of soluble sulfur anions [26]. As the active cell area and therefore the CL/MPL interface area should be the same on average for all three MEA types, while the anode CL volumes are different, a higher fraction of anions could get removed for lower-loaded and therefore thinner anode electrodes. These anions dissolved in water eventually are flushed out once the PEMFC is started again. This works better so long as sulfur is weakly bonded to the Pt surface via linear bonds. With time, adsorbed sulfur develops stronger bonds to active sites and is bound more strongly to the catalyst, leading finally to the observed voltage breakdowns of the PEMFCs. Thinner CLs may also be associated with a changed ionomer structure, and the potentials within the layer are generally more homogeneous [31]. However, the differentiation of this effect is beyond the scope of this paper.

Consequently, the reduction of the anodic catalyst material down to ultralow loadings seemed to come with a nonproportional reduction in tolerance versus $H_2S$ poisoning and an improved self-recovery during SD/SU processes. Hence, lowering the ISO limit for sulfur-containing compounds might not be necessary with regard to ultralow-loaded anode electrodes. However, these effects should be further confirmed in large- or full-scale cell tests using realistic automotive fuel utilizations.

## 4. Conclusions

The key findings from this study are that the $H_2$ Quality Standard ISO 14687-2:2012 eventually requires partial adaption to accommodate future automotive PEMFC designs, including ultralow-loaded anodic CLs, and that ultralow-loaded anodes exhibited an improved self-recovery from sulfur poisoning from simple SD/SU processes.

As expected, CO poisoning induced significant performance losses at an increasing rate and severity with decreases in the platinum loading. At an ISO concentration of 0.2 ppm CO in the fuel, the cell voltage was about 40–50% lower compared to operation with neat $H_2$ for ultralow anodic loadings, which raises the question of whether the CO limit in the $H_2$ Quality Standard needs to be reduced with regard to future anodic loadings.

When $H_2S$ was in the fuel, the ultralow-loaded anodic CLs exhibited a nonlinear reduction as opposed to the expected linear reduction in tolerance to the reduction in platinum loading. Simultaneously, these anodic CLs recovered to larger degrees from sulfur poisoning during the SD/SU processes. It is hypothesized that the nonlinear reduction in tolerance and improved self-recovery arose due to the decrease in the ratio between the CL thickness (and coherent ECSA) and the geometric cell area. As the ultralow-loaded anodes were also the thinner CLs, larger fractions of sulfur adsorbed on catalyst surfaces in proximity to pores at the CL–MPL interface could have dissolved in the water present in the form of anions, which were driven out of the cell during operation or during the SU of the PEMFCs.

However, to confirm these findings, the performance of ultralow-loaded anodic CLs in the presence of impurities should be further investigated, ideally in large- or full-scale PEMFCs using automotive fuel consumption rates.

**Author Contributions:** conceptualization, S.P.; data curation, S.P.; writing—original draft, S.P.; writing—review and editing, S.P., K.A.F., and N.Z.

**Funding:** This work was supported by the German Federal Ministry for Economy and Energy within the project HAlMa, contract no. 03ET6098A.

**Acknowledgments:** The authors thank Jean St.-Pierre at the University of Hawaii (Manoa) for technical discussions and suggestions as well as Greenerity GmbH for providing the CCM materials.

**Conflicts of Interest:** The authors declare no conflicts of interest. The funders had no role in the design of the study; in the collection, analysis, or interpretation of the data; in the writing of the manuscript; or in the decision to publish the results.

## References

1. Cheng, X.; Shi, Z.; Glass, N.; Zhang, L.; Zhang, J.; Song, D.; Liu, Z.S.; Wang, H.; Shen, J. A Review of PEM Hydrogen Fuel Cell Contamination: Impacts, Mechanisms, and Mitigation. *J. Power Sources* **2007**, *165*, 739–756. [CrossRef]
2. Zamel, N.; Li, X. Effect of Contaminants on Polymer Electrolyte Membrane Fuel Cells. *Prog. Energy Combust. Sci.* **2011**, *37*, 292–329. [CrossRef]
3. Klages, M.; Tjønnås, J.; Zenith, F.; Halvorsen, I.J.; Scholta, J. Dual Control of Low Concentration CO Poisoning by Anode Air Bleeding of Low Temperature Polymer Electrolyte Membrane Fuel Cells. *J. Power Sources* **2016**, *336*, 212–223. [CrossRef]
4. Iezzi, R.C.; Santos, R.D.M.; da Silva, G.C.; Paganin, V.A.; Ticianelly, E.A. CO Tolerance and Stability of Proton Exchange Membrane Fuel Cells with Nafion® and Aquivion® Membranes and Mo-Based Anode Electrocatalysts. *Braz. Chem. Soc.* **2018**, *29*, 1094–1104. [CrossRef]
5. Lee, S.J.; Mukerjee, S.; Ticianelli, E.A.; McBreen, J. Electrocatalysis of CO Tolerance in Hydrogen Oxidation Reaction in PEM Fuel Cells. *Electrochim. Acta* **1999**, *44*, 3283–3293. [CrossRef]

6. Mohtadi, R.; Lee, W.-K.; Cowan, S.; Van Zee, J.W.; Murthy, M. Effects of Hydrogen Sulfide on the Performance of a PEMFC. *Electrochem. Solid State Lett.* **2003**, *6*, A272–A274. [CrossRef]
7. Gubán, D.; Tompos, A.; Bakos, I.; Vass, A.; Pászti, Z.; Szabó, E.G.; Sajó, I.E.; Borbáth, I. Preparation of CO-Tolerant Anode Electrocatalysts for Polymer Electrolyte Membrane Fuel Cells. *Int. J. Hydrogen Energy* **2017**, *42*, 13741–13753. [CrossRef]
8. Inaba, M.; Sugishita, M.; Wada, J.; Matsuzawa, K.; Yamada, H.; Tasaka, A. Impacts of Air Bleeding on Membrane Degradation in Polymer Electrolyte Fuel Cells. *J. Power Sources* **2008**, *178*, 699–705. [CrossRef]
9. Hengge, K.; Gänsler, T.; Pizzutilo, E.; Heinzl, C.; Beetz, M.; Mayrhofer, K.J.J.; Scheu, C. Accelerated Fuel Cell Tests of Anodic Pt/Ru Catalyst via Identical Location TEM: New Aspects of Degradation Behavior. *Int. J. Hydrogen Energy* **2017**, *42*, 25359–25371. [CrossRef]
10. Li, H.; Gazzarri, J.; Tsay, K.; Wu, S.; Wang, H.; Zhang, J.; Wessel, S.; Abouatallah, R.; Joos, N.; Schrooten, J. PEM Fuel Cell Cathode Contamination in the Presence of Cobalt Ion (Co2+). *Electrochim. Acta* **2010**, *55*, 5823–5830. [CrossRef]
11. Kongkanand, A.; Mathias, M.F. The Priority and Challenge of High-Power Performance of Low- Platinum Proton-Exchange Membrane Fuel Cells. *J. Phys. Chem.* **2016**, *7*, 1127–1137. [CrossRef] [PubMed]
12. Hashimasa, Y.; Matsuda, Y.; Akai, M. Effects of Platinum Loading on PEFC Power Generation Performance Deterioration by Carbon Monoxide in Hydrogen Fuel. *ECS Trans.* **2010**, *26*, 131–142. [CrossRef]
13. Matsuda, Y.; Shimizu, T.; Mitsushima, S. Adsorption Behavior of Low Concentration Carbon Monoxide on Polymer Electrolyte Fuel Cell Anodes for Automotive Applications. *J. Power Sources* **2016**, *318*, 1–8. [CrossRef]
14. Hashimasa, Y.; Matsuda, Y.; Imamura, D.; Motoaki, A. PEFC Power Generation Performance Degradation by Hydrogen Sulfi de and Ammonia—Eff Ects of Lowering Platinum Loading. *Electrochemistry* **2011**, *79*, 343–345. [CrossRef]
15. Kakati, B.K.; Kucernak, A.R.J.; Fahy, K.F. Using Corrosion-like Processes to Remove Poisons from Electrocatalysts: A Viable Strategy to Chemically Regenerate Irreversibly Poisoned Polymer Electrolyte Fuel Cells. *Electrochim. Acta* **2016**, *222*, 888–897. [CrossRef]
16. Lopes, T.; Paganin, V.A.; Gonzalez, E.R. The Effects of Hydrogen Sulfide on the Polymer Electrolyte Membrane Fuel Cell Anode Catalyst: H2S-Pt/C Interaction Products. *J. Power Sources* **2011**, *196*, 6256–6263. [CrossRef]
17. Bonnet, C.; Franck-Lacaze, L.; Ronasi, S.; Besse, S.; Lapicque, F. PEM Fuel Cell Pt Anode Inhibition by Carbon Monoxide: Non-Uniform Behaviour of the Cell Caused by the Finite Hydrogen Excess. *Chem. Eng. Sci.* **2010**, *65*, 3050–3058. [CrossRef]
18. Elgrishi, N.; Rountree, K.J.; McCarthy, B.D.; Rountree, E.S.; Eisenhart, T.T.; Dempsey, J.L. A Practical Beginner's Guide to Cyclic Voltammetry. *J. Chem. Educ.* **2018**, *95*, 197–206. [CrossRef]
19. Camara, G.A.; Ticianelli, E.A.; Mukerjee, S.; Lee, S.J.; McBreen, J. The CO Poisoning Mechanism of the Hydrogen Oxidation Reaction in Proton Exchange Membrane Fuel Cells. *J. Electrochem. Soc.* **2002**, *149*, A748–A753. [CrossRef]
20. Igarashi, H.; Fujino, T.; Watanabe, M. Hydrogen Electro-Oxidation on Platinum Catalysts in the Presence of Trace Carbon Monoxide. *J. Electroanal. Chem.* **1995**, *391*, 119–123. [CrossRef]
21. Akemann, W.; Friedrich, K.A.; Stimming, U. Potential-Dependence of CO Adlayer Structures on Pt(111) Electrodes in Acid Solution: Evidence for a Site Selective Charge Transfer. *J. Chem. Phys.* **2000**, *113*, 6864–6874. [CrossRef]
22. St-Pierre, J. PEMFC Contamination Model: Competitive Adsorption Followed by an Electrochemical Reaction. *J. Electrochem. Soc.* **2009**, *156*, B291–B300. [CrossRef]
23. St-Pierre, J. Proton Exchange Membrane Fuel Cell Contamination Model: Competitive Adsorption Followed by a Surface Segregated Electrochemical Reaction Leading to an Irreversibly Adsorbed Product. *J. Power Sources* **2010**, *195*, 6379–6388. [CrossRef]
24. Schiller, C.A.; Richter, F.; Gülzow, E.; Wagner, N. Relaxation Impedance as a Model for the Deactivation Mechanism of Fuel Cells Due to Carbon Monoxide Poisoning. *Phys. Chem. Chem. Phys.* **2001**, *3*, 2113–2116. [CrossRef]
25. Wagner, N.; Schulze, M. Change of Electrochemical Impedance Spectra during CO Poisoning of the Pt and Pt–Ru Anodes in a Membrane Fuel Cell (PEFC). *Electrochim. Acta* **2003**, *48*, 3899–3907. [CrossRef]
26. St-Pierre, J.; Wetton, B.; Zhai, Y.; Ge, J. Liquid Water Scavenging of PEMFC Contaminants. *J. Electrochem. Soc.* **2014**, *161*, E3357–E3364. [CrossRef]

27. Sethuraman, V.A.; Weidner, J.W. Analysis of Sulfur Poisoning on a PEM Fuel Cell Electrode. *Electrochim. Acta* **2010**, *55*, 5683–5694. [CrossRef]
28. Gould, B.D.; Bender, G.; Bethune, K.; Dorn, S.; Baturina, O.A.; Rocheleau, R.; Swider-Lyons, K.E. Operational Performance Recovery of SO[Sub 2]-Contaminated Proton Exchange Membrane Fuel Cells. *J. Electrochem. Soc.* **2010**, *157*, B1569. [CrossRef]
29. O'Brien, J.A.; Hinkley, J.T.; Donne, S.W.; Lindquist, S.E. The Electrochemical Oxidation of Aqueous Sulfur Dioxide: A Critical Review of Work with Respect to the Hybrid Sulfur Cycle. *Electrochim. Acta* **2010**, *55*, 573–591. [CrossRef]
30. Prass, S.; Hasanpour, S.; Sow, P.K.; Phillion, A.B.; Mérida, W. Microscale X-Ray Tomographic Investigation of the Interfacial Morphology between the Catalyst and Micro Porous Layers in Proton Exchange Membrane Fuel Cells. *J. Power Sources* **2016**, *319*, 82–89. [CrossRef]
31. Vidakovic-Koch, T.; Hanke-Rauschenbach, R.; Gonzales Martinez, I.; Sundmacher, K. Catalyst Layer Modelling. In *Handbook of Electrochemical Energy*; Springer: Berlin/Heidelberg, Germany, 2017; pp. 259–285.

**Sample Availability:** Not available.

molecules

MDPI

*Article*

# Impact of the Cathode Pt Loading on PEMFC Contamination by Several Airborne Contaminants

**Jean St-Pierre * and Yunfeng Zhai**

Hawaii Natural Energy Institute, University of Hawaii—Manoa, Honolulu, HI 96822, USA; yunfeng@hawaii.edu
* Correspondence: jsp7@hawaii.edu; Tel.: +1-808-956-3909

Academic Editor: Shangfeng Du
Received: 1 February 2020; Accepted: 23 February 2020; Published: 27 February 2020

check for updates

**Abstract:** Proton exchange membrane fuel cells (PEMFCs) with 0.1 and 0.4 mg Pt $cm^{-2}$ cathode catalyst loadings were separately contaminated with seven organic species: Acetonitrile, acetylene, bromomethane, iso-propanol, methyl methacrylate, naphthalene, and propene. The lower catalyst loading led to larger cell voltage losses at the steady state. Three closely related electrical equivalent circuits were used to fit impedance spectra obtained before, during, and after contamination, which revealed that the cell voltage loss was due to higher kinetic and mass transfer resistances. A significant correlation was not found between the steady-state cell voltage loss and the sum of the kinetic and mass transfer resistance changes. Major increases in research program costs and efforts would be required to find a predictive correlation, which suggests a focus on contamination prevention and recovery measures rather than contamination mechanisms.

**Keywords:** proton exchange membrane fuel cells; durability; contamination; cathode; catalyst loading

---

## 1. Introduction

Vehicles propelled by proton exchange membrane fuel cells (PEMFCs) are already commercially available. However, opportunities still exist to improve the technology because it is not expected to mature within the foreseeable future [1]. For instance, research activities are still ongoing to reduce cost while maintaining durability with a lower amount of Pt catalyst [2]. Contaminants in air jeopardize PEMFC operation by increasing the cell voltage degradation rate [3] if the intake filter [4] is saturated or damaged. Therefore, risks associated with contamination of low Pt loaded PEMFCs need to be assessed to support commercialization. Furthermore, fuel cell design robustness could be improved by integrating additional mitigation approaches derived from contamination mechanisms.

Only a few publications discuss the impact of the anode catalyst loading during PEMFC exposures to reformate fuel contaminants, such as CO, $CO_2$, $H_2S$, $NH_3$, and halogenated compounds. All of these species are included in the hydrogen fuel standard [5]. For CO and $H_2S$, a lower Pt or PtRu catalyst loading generally leads to an increase in the anode overpotential [6–10]. However, it was reported that for $H_2S$, the catalyst loading effect disappears for values equal to or below 25 μg $cm^{-2}$ [10]. An effect was not observed with the weak contaminant $CO_2$, which is attributed to a concentration that was substantially lower (1%) [7] than in a typical reformate (10–20%) [11]. The same situation was noted for $NH_3$, which is assigned to a rapid conversion to $NH_4^+$ in the presence of protons or water [12,13], followed by ion exchange with ionomer $H^+$ and transport to the cathode away from the anode under the influence of the electric field [14,15]. For halogenated compounds, a decrease in the Pt catalyst loading of both electrodes led to a faster degradation rate in the presence of HCl in both reactant stream humidifiers [16]. The effect of the anode Pt catalyst loading was exploited to develop sensor cells that are more sensitive to contamination by CO, $H_2S$, and $NH_3$ in $H_2$. These sensors were either based on a PEMFC [17] or a $H_2$ pump [18] design. Only two PEMFC contamination documents focusing on the cathode catalyst loading effect were found [19,20]. However, contamination data in [19] are not directly comparable because both the catalyst layer design and catalyst loading were concurrently altered. The authors also refer to 10 ppb $SO_2$ data obtained by another group that showed more severe fuel cell damage with a catalyst loading decrease from 0.4 to 0.3 mg Pt cm$^{-2}$. In contrast, the effect of 2,6-diaminotoluene, a species that leaches out of the fuel cell system balance of plant materials, was more impactful after the Pt catalyst loading was lowered from 0.4 to 0.1 mg Pt cm$^{-2}$ [20]. In comparison to the anode, the higher cathode potential is expected to affect the contamination mechanism with, for example, a different Pt surface charge, altered contaminant adsorbates and reaction intermediates, catalyst coverage, and cell voltage loss. This situation is exacerbated with a catalyst loading change, which affects the overpotential of the irreversible oxygen reduction reaction and the cathode potential. Information about chemical and electrochemical reactions for specific contaminants may be available in the literature. However, the presence of relevant cathode reactants, oxygen and water, may not be considered. For instance, novel intermediates or products were not detected with chlorobenzene in air [21]. However, the presence of acetylene in air led to small amounts of methane [22] that were not expected based on acetylene chemistry and electrochemistry. Therefore, tests completed under these significantly different operating conditions are needed in part because contaminant reactions are not currently predictable in assessing catalyst coverage and cell voltage loss.

This report documents the impact of the cathode Pt catalyst loading effect for PEMFCs contaminated with seven model organic airborne species, which were previously evaluated and selected from a larger pool of 21 contaminants [23]: Acetonitrile (nitrile), acetylene (alkyne), bromomethane (halocarbon), iso-propanol (alcohol), methyl methacrylate (ester), naphthalene (polycyclic aromatic), and propene (alkene). Cell voltage transients obtained under galvanostatic conditions were recorded for this analysis. Additionally, impedance spectroscopy data were acquired to facilitate the development of predictive correlations and contamination mechanisms.

## 2. Results and Discussion

### 2.1. Cell Voltage Transients

Figure 1a depicts voltage transients for cells temporarily exposed to 20 ppm $CH_3CN$. The cell voltage for the first 5 h is constant and higher for the 0.4 mg Pt cm$^{-2}$ catalyst loading. This observation is consistent with previously published data for Gore catalyst coated membranes with the same cathode catalyst loadings and gas diffusion layers (Sigracet 25 BC) [24]. After approximately 5 h of operation, acetonitrile was injected into the cell, which led to a rapid cell voltage decrease that progressively slowed until a steady state was reached. For acetonitrile, the cell voltage loss was larger for the 0.1 mg Pt cm$^{-2}$ catalyst loading. Subsequently, the acetonitrile injection was interrupted, which quickly initiated a voltage recovery that gradually decelerated until a new steady state was reached. For acetonitrile, the cell voltage after recovery coincided with the value before contaminant injection. Acetonitrile contamination and recovery transients are qualitatively and quantitatively consistent with prior results [23,25–28]. At irregular intervals and during all baseline, contamination, and recovery stages, cell voltage transients were minimally disrupted for a short period by impedance spectroscopy measurements and the superimposition of a current signal of a small amplitude and variable frequency.

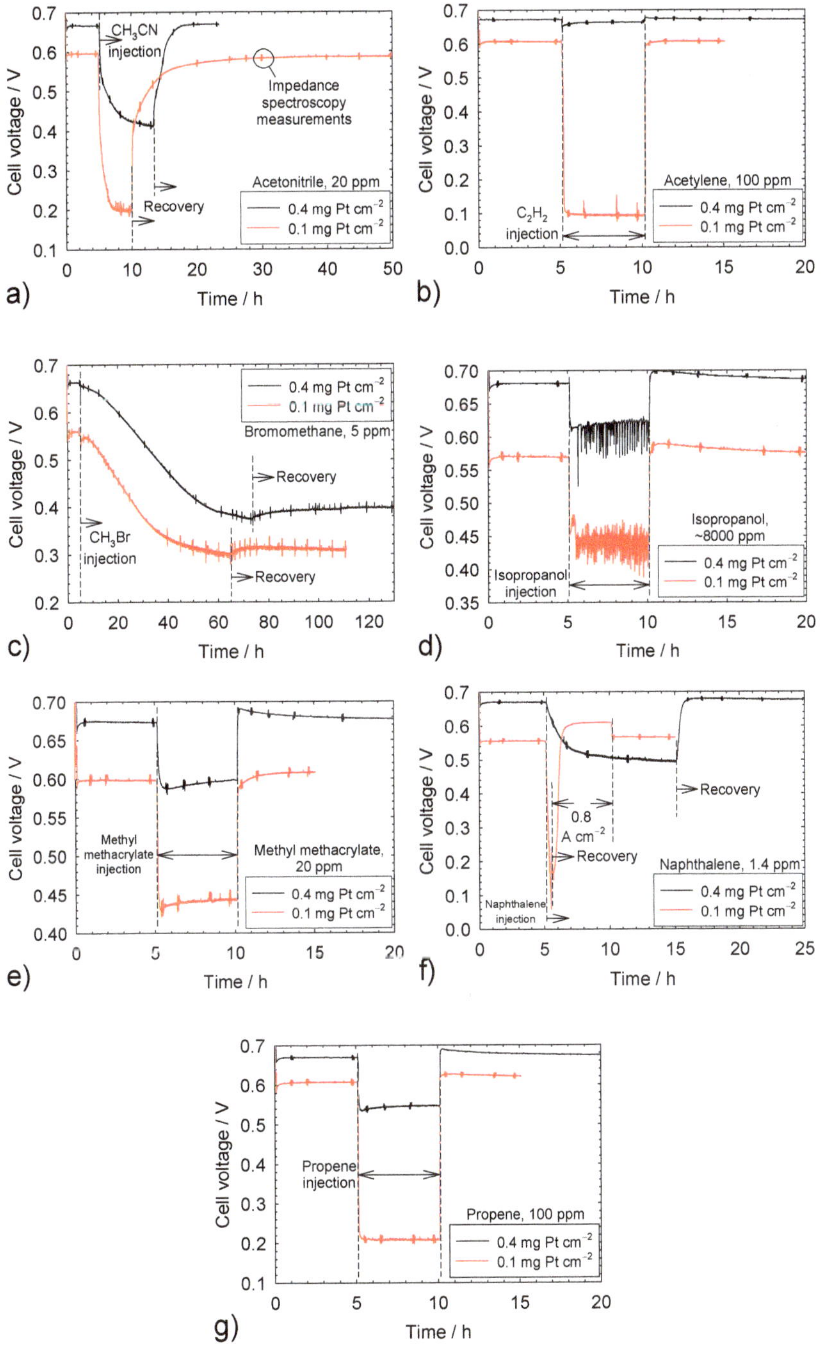


**Figure 1.** Cell voltage transients resulting from a temporary contaminant injection. (**a**) Acetonitrile; (**b**) acetylene; (**c**) bromomethane; (**d**) isopropanol; (**e**) methyl methacrylate; (**f**) naphthalene; (**g**) propene.

Figure 1b–g illustrates voltage transients for the other contaminants. Most of these transients share common features, including a similar initial baseline voltage, a relatively rapid voltage decrease until a steady state is reached, and a complete voltage recovery after contaminant injection was stopped. However, bromomethane transients were significantly slower and only a small fraction of the voltage loss was recovered (Figure 1c). This behavior is the result of a rapid bromomethane hydrolysis within the cell, producing methanol and bromide [28–30]. The effective bromomethane concentration is lower than the nominal value, whereas bromide is prevented from penetrating the ionomer by Donnan exclusion [12]. This situation delays the stronger and inhibiting adsorption of bromomethane and bromide on the catalyst surface. The removal of bromide from the catalyst surface is equally hindered due to an unfavorable cathode potential that is significantly higher than the potential of zero charge, preventing bromide desorption and Donnan exclusion, which explains the incomplete voltage recovery. During isopropanol contamination, the voltage is characterized by rapid fluctuations (Figure 1d), which were not observed for lower isopropanol concentrations [23,31]. These fluctuations are attributed to isopropanol, a surfactant commonly used to disperse Pt/C catalyst particles in solution [32], which adsorbs on carbon materials (gas diffusion layer, catalyst support) [33] and modifies liquid water management (buildup and release of liquid water drops), as previously proposed for acetylene [34]. A higher number of buildup and release events of water drops and a higher voltage fluctuation frequency for the lower catalyst loading (Figure 1d) may be related to the lower cathode potential (cell voltage compensated by a similar ohmic drop), which leaves a higher proportion of isopropanol surfactant unoxidized (oxidation initiated at a potential above 0.32 V vs. the reversible hydrogen electrode (RHE) [35]) and more hydrophilic carbon surfaces. The effect of naphthalene was rapid and severe for the 0.1 mg Pt $cm^{-2}$ catalyst loading (Figure 1f). As a result, the current density was temporarily lowered and the contaminant injection was interrupted before a steady state was obtained to avoid an automatic test station shutdown. Contamination and recovery transients are qualitatively and quantitatively consistent with the prior results for acetylene [23,36–39], bromomethane [23,28–30], isopropanol [23,31], methyl methacrylate [23,31], naphthalene [23,40], and propene [23,28,31,41].

Table 1 summarizes steady-state cell voltages before, during, and after contamination as well as the cell voltage change during and after contamination for both catalyst loadings. The cell voltage decrease during the contamination period is generally higher for the 0.1 mg Pt $cm^{-2}$ catalyst loading (23% to 89% in comparison to 1.2% to 43%). After the recovery period, the cell voltage change is minimal and independent of the catalyst loading, varying from −1.7% to 2%, with the exception of bromomethane (−40% to −45%). The larger cell voltage loss during contamination for the low catalyst loading is an important consideration for the selection of tolerance limits for commercially relevant catalyst loadings. Data obtained with a 0.4 mg Pt $cm^{-2}$ catalyst loading were used to derive tolerance limits [42]. The data of Table 1 suggest that these tolerance limits require a revision for a 0.1 mg Pt $cm^{-2}$ catalyst loading and additional tests carried out over a range of concentrations. In contrast, International Organization for Standardization (ISO) tolerance limits for hydrogen contaminants [5,43], which do not take account of the catalyst loading effect, were deemed too strict for formaldehyde and formic acid, a low anode catalyst loading of 0.05 mg Pt $cm^{-2}$, and automotive operating conditions (high fuel utilization, fuel recirculation) [44]. The formaldehyde tolerance limit was recently modified from 10 [43] to 200 ppb [5].

The magnitude of the cell voltage change during contamination (Table 1) with catalyst loading is species-dependent. For instance, the catalyst loading hardly affected the cell voltage loss for bromomethane (−43% and −47%), whereas for acetylene, the cell voltage loss substantially increased from −1.2% to −85% with a catalyst loading decrease. This observation is attributed to different contamination mechanisms. Impedance spectroscopy data obtained during contamination by all Table 1 species and with a 0.4 mg Pt $cm^{-2}$ catalyst loading revealed that kinetic, ohmic, and mass transport overpotentials were impacted [42]. These and additional impedance spectroscopy data acquired with a 0.1 mg Pt $cm^{-2}$ catalyst loading were analyzed to evaluate the existence of a correlation between these resistances and the cell voltage loss due to contamination at the steady state.

**Table 1.** Steady-state cell voltages at the end of each contamination period, and steady-state cell voltage changes during and after contamination.

| Contaminant | Catalyst Loading/mg Pt cm$^{-2}$ | Cell Voltage/V | | | Cell Voltage Percentage Change/% | |
|---|---|---|---|---|---|---|
| | | Before Contamination [1] | During Contamination | After Contamination | During Contamination | After Contamination |
| Acetonitrile | 0.1 | 0.597 | 0.198 | 0.587 | −67 | −1.7 |
|  | 0.4 | 0.666 | 0.412 | 0.670 | −38 | 0.60 |
| Acetylene | 0.1 | 0.607 | 0.093 | 0.608 | −85 | 0.16 |
|  | 0.4 | 0.672 | 0.664 | 0.672 | −1.2 | 0 |
| Bromomethane | 0.1 | 0.561 | 0.299 | 0.309 | −47 | −45 |
| —Br | 0.4 | 0.663 | 0.376 | 0.398 | −43 | −40 |
| Isopropanol | 0.1 | 0.570 | 0.439 | 0.577 | −23 | 1.2 |
| 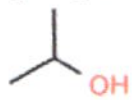 | 0.4 | 0.681 | 0.623 | 0.687 | −8.5 | 0.88 |
| Methyl methacrylate | 0.1 | 0.599 | 0.444 | 0.608 | −26 | 1.5 |
| 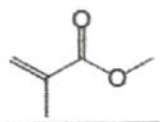 | 0.4 | 0.673 | 0.598 | 0.678 | −11 | 0.74 |
| Naphthalene | 0.1 | 0.557 | 0.060 [2] | 0.566 | −89 | 1.6 |
| 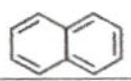 | 0.4 | 0.671 | 0.495 | 0.677 | −26 | 0.89 |
| Propene | 0.1 | 0.609 | 0.209 | 0.621 | −66 | 2.0 |
|  | 0.4 | 0.670 | 0.547 | 0.674 | −18 | 0.60 |

[1] For 0.1 mg Pt cm$^{-2}$, mean = 0.586 V and standard deviation = 0.022 V. For 0.4 mg Pt cm$^{-2}$, mean = 0.671 V and standard deviation = 0.006 V. [2] Not at steady state because the cell voltage was still decreasing.

## 2.2. *Impedance Spectra*

Figure 2a shows impedance spectra (Nyquist representation) for a 0.1 mg Pt cm$^{-2}$ catalyst loading, before, during, and after acetonitrile contamination. All three spectra share the same features and have two main loops that are respectively attributed to oxygen reduction (medium frequencies) and oxygen mass transfer (low frequencies) [45]. A third loop ascribed to hydrogen oxidation is barely visible as a hump at high frequencies [45]. The high-frequency intercept represents the ohmic resistance, which is mostly caused by the poorly conducting membrane [45]. Multiple explanations were proposed for the inductive impedance values at the lowest and highest frequencies, including electrical cables [46,47] for high frequencies, and processes involving side reactions with intermediate species [47], oxide growth [48], or a slow ionomer water uptake/release [49] for low frequencies. Most of these considerations were either ignored because they did not focus on relevant aspects (electrical cables) or were easily dismissed because, in the absence of contaminants, the cathode potential was too low for Pt oxidation and the sub-saturated air stream did not yield an inductive behavior. For the 0.4 mg Pt cm$^{-2}$ catalyst loading, the average cell voltage of 0.671 V (Table 1) compensated with an ohmic loss of 0.1 V for a worst-case scenario (1 A cm$^{-2}$ × 0.1 Ω cm$^2$ from the high-frequency intercepts in Figure 2a) leads to a cathode potential of 0.771 V vs. RHE, which is lower than the smallest Pt oxidation potential of 0.837 V vs. RHE [50]. Acetonitrile contamination causes an increase in the high-frequency intercept and a diameter increase for both main loops (Figure 2a). An increase in ohmic loss was only observed with acetonitrile, owing to the production of ammonium cations by hydrolysis, which displace protons as the main charge carriers in the ionomer [28,51]. In relative terms, this effect is significantly smaller than the kinetic and mass transfer effects, with an approximate doubling of both oxygen reduction and transport loop diameters. However, because the effect is cumulative, a larger change is observed for a longer exposure duration [26]. After contamination, the high-frequency intercept returns to its original value, and both main loops decrease in size to a diameter that is smaller than the original value. These impedance spectra agree with prior results [25–28]. However, smaller kinetic and mass transfer loops are inconsistent with a complete cell voltage recovery (Figure 1a, Table 1). This observation is possibly due to subtle structural or other changes that are not detectable by cell voltage measurements, such as Pt surface reconstruction in the presence of foreign species [52,53]. The oxygen reduction and mass transfer resistances before, during, and after contamination were generally obtained by curve-fitting an equivalent circuit developed for a PEMFC contaminated with $SO_2$ (Figure 3a) [54]. Resistances during contamination for acetonitrile and a 0.1 mg Pt cm$^{-2}$ catalyst loading were derived using a modified equivalent circuit that accounts for the inductive behavior at low frequencies (Figure 3b) [55,56]. Resistances during contamination for acetonitrile (0.4 mg Pt cm$^{-2}$) and propene (0.1 mg Pt cm$^{-2}$) were obtained using a modified version of the Figure 3b equivalent circuit by omitting the cathode resistance Rk (Figure 3b) to limit the number of parameters (Figure 3c). The impedance spectra are accurately represented by the equivalent circuit models (Figure 2a–f). The resistance values are discussed later in this section.

Most of the other impedance spectra for both catalyst loadings and all contaminants are equally well represented by the equivalent circuits shown in Figure 3a,c. For this reason, only a selection is given in Figure 2. A few spectra could not be fitted with any of the equivalent circuits in Figure 3a–c for a few 0.1 mg Pt cm$^{-2}$ catalyst loading cases. For acetylene, the impedance spectrum during contamination was approximately a single loop of a large diameter that could not be fitted to a two-loop equivalent circuit. For isopropanol, cell voltage fluctuations during contamination (Figure 1d) created a low frequency artefact that also prevented the use of the equivalent circuits of Figure 3a or Figure 3c. For naphthalene, the cell voltage transient was interrupted before a steady state was obtained (Figure 1f), which also led to a low-frequency artefact that could not be fitted to the equivalent circuits of Figure 3a–c. The impedance spectra agree with the prior results for acetonitrile [25–28], acetylene [36–38], bromomethane [28–30], isopropanol [31], methyl methacrylate [31], naphthalene [40], and propene [28,31,41].

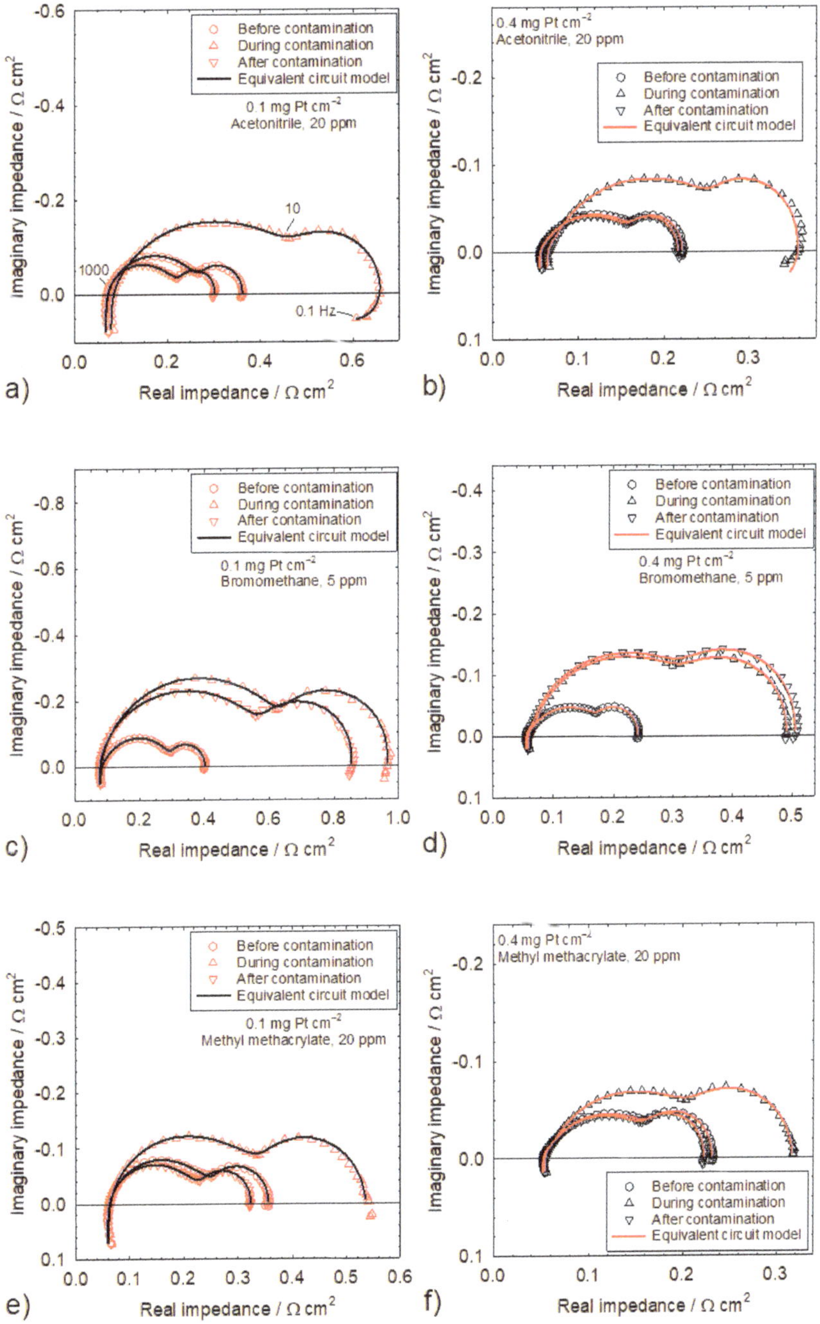


**Figure 2.** Impedance spectra before, during, and after contamination by acetonitrile in (**a**) and (**b**), bromomethane in (**c**) and (**d**), and methyl methacrylate in (**e**) and (**f**) for Pt catalyst loadings of 0.1 mg $cm^{-2}$ in (**a**), (**c**), and (**e**), and 0.4 mg $cm^{-2}$ in (**b**), (**d**), and (**f**).

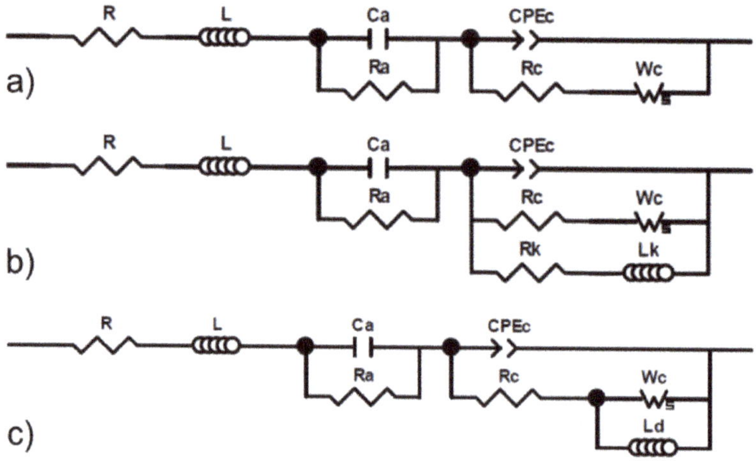


**Figure 3.** Equivalent circuit models for a proton exchange membrane fuel cell (PEMFC). (**a**) The model previously derived for $SO_2$ contamination and used for all 7 organic contaminants investigated in this work; (**b**) the model previously derived to capture low-frequency inductive data in the absence of contaminants and used to fit data during acetonitrile contamination (0.1 mg Pt cm$^{-2}$); (**c**) the modified model (**b**) to capture low-frequency inductive data obtained during acetonitrile (0.4 mg Pt cm$^{-2}$) and propene (0.1 mg Pt cm$^{-2}$) contamination.

Table 2 collects kinetic and mass transfer resistances before, during, and after contamination for both catalyst loadings. Dimensionless kinetic and mass transfer resistances during and after contamination are also given in Table 2. The dimensionless kinetic and mass transfer resistances concurrently increase during contamination and are ≥1.05, with the exception of the 0.93 dimensionless mass transfer resistance for isopropanol and a 0.4 mg Pt cm$^{-2}$ catalyst loading. The isopropanol anomaly may be related to water management, as discussed in the previous section. A hypothesized connection between kinetic and mass transfer resistances during contamination [34] was recently substantiated [57]. Contaminant adsorbates covering the catalyst surface increase the effective current density closer to the limiting value and mass transfer losses in the ionomer layer covering the catalyst. This situation is similar to a decrease in catalyst loading, which has been shown to also increase mass transfer losses [58,59]. The dimensionless kinetic and mass transfer resistances after recovery, with the exception of bromomethane, indicate an incomplete recovery that is less extensive for the lower catalyst loading. For the dimensionless kinetic resistance, values are ≥0.832 and ≥0.95 for, respectively, 0.1 and 0.4 mg Pt cm$^{-2}$ catalyst loadings. For the dimensionless mass transfer resistance, values are ≥0.842 and ≥0.88 for, respectively, 0.1 and 0.4 mg Pt cm$^{-2}$ catalyst loadings. These results are in contrast with the data of Figure 1 and Table 1, showing a complete recovery within experimental error, with the exception of bromomethane. The discrepancy between the recovery extents of cell voltage and kinetic and mass transfer resistances is due to the higher sensitivity of impedance measurements and the movement of the reaction front (current density and catalyst layer effectiveness redistributions over the catalyst layer thickness). The hydrogen peroxide yield is enhanced in the presence of acetonitrile, acetylene, methyl methacrylate, naphthalene, and propene [60–63]. The elevated level of hydrogen peroxide in turn promotes ionomer degradation [64] and structural modifications to the catalyst layer that are relatively more impactful for the lower catalyst loading. Therefore, in view of the lower cell voltage and cathode potential for a lower catalyst loading (Figure 1, Table 1), a higher hydrogen peroxide yield [60–63] and ionomer degradation are expected. Tafel plots obtained before and after contamination with acetylene (Figure 4) support this hypothesis, with a larger cell voltage loss for the 0.1 mg Pt cm$^{-2}$ catalyst loading (7.9 mV in comparison to 2.9 mV).

**Table 2.** Steady-state kinetic and mass transfer resistances at the end of each contamination period, and steady-state dimensionless resistances during and after contamination.

| Contaminant | Catalyst Loading/mg Pt cm$^{-2}$ | Kinetic/Mass Transfer Resistances/Ω cm$^2$ | | | Dimensionless Resistance During/After Contamination [1] | |
|---|---|---|---|---|---|---|
| | | Before Contamination | During Contamination | After Contamination | Kinetic | Mass Transfer |
| Acetonitrile | 0.1 | 0.118/0.095 | 0.396/0.187 | 0.108/0.080 | 3.36/0.915 | 1.97/0.842 |
| | 0.4 | 0.104/0.056 | 0.210/0.079 | 0.099/0.056 | 2.02/0.952 | 1.41/1.00 |
| Acetylene | 0.1 | 0.133/0.107 | - [2] | 0.117/0.106 | - | - |
| | 0.4 | 0.107/0.061 | 0.112/0.064 | 0.109/0.059 | 1.05/1.02 | 1.05/0.967 |
| Bromomethane | 0.1 | 0.103/0.109 | 0.139/0.341 | 0.285/0.288 | 1.35/2.77 | 3.13/2.64 |
| | 0.4 | 0.116/0.062 | 0.265/0.157 | 0.274/0.167 | 2.28/2.36 | 2.53/2.69 |
| Isopropanol | 0.1 | 0.102/0.100 | - [3] | 0.123/0.090 | - | - |
| | 0.4 | 0.100/0.070 | 0.117/0.065 | 0.095/0.068 | 1.17/0.950 | 0.929/0.971 |
| Methyl methacrylate | 0.1 | 0.121/0.104 | 0.164/0.189 | 0.104/0.098 | 1.36/0.860 | 1.82/0.942 |
| | 0.4 | 0.111/0.063 | 0.152/0.104 | 0.107/0.059 | 1.37/0.964 | 1.65/0.937 |
| Naphthalene | 0.1 | 0.115/0.108 | - [3] | 0.119/0.097 | - | - |
| | 0.4 | 0.106/0.075 | 0.288/0.152 | 0.101/0.066 | 2.72/0.953 | 2.03/0.880 |
| Propene | 0.1 | 0.137/0.107 | 0.189/0.118 | 0.114/0.107 | 1.38/0.832 | 1.10/1.00 |
| | 0.4 | 0.117/0.063 | 0.152/0.134 | - [4] | 1.30/- | 2.13/- |

[1] Resistance during/after contamination divided by the resistance before contamination. [2] Equivalent circuit models do not fit due to a side surface reaction involving intermediates. [3] Artefact created by flooding or rapid change in cell voltage. [4] Data was not recorded by error.

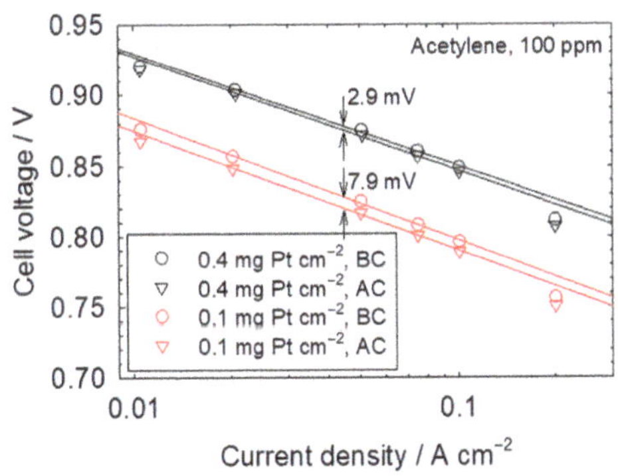


**Figure 4.** Tafel plots before contamination (BC) and after contamination (AC) with 100 ppm acetylene for 0.1 and 0.4 mg Pt cm$^{-2}$ catalyst loadings. The change in cell voltage between plots at a current density of 0.0447 A cm$^{-2}$, a value located in the middle of the range used to correlate data (0.02 to 0.1 A cm$^{-2}$), ignores the slight change in slope.

### *2.3. Contaminant Effect Prediction*

The steady-state cell voltage loss during contamination was correlated with the sum of the kinetic and mass transfer resistance changes during contamination (Figure 5). A significant correlation was not identified, as significant deviations from Ohm's law were noted. Furthermore, it is difficult to argue that there is a catalyst loading effect because the two data sets largely overlap. The absence of a correlation is not surprising, considering the effects of cell design and operating conditions on contamination. Several parameters were mentioned in an earlier attempt to correlate the effect of contaminants on oxygen reduction kinetics [65], including contaminant partial pressure and temperature, exposed

Pt surface features (crystal faces, edges), Pt state (reduced or oxidized), phase in contact with the Pt surface (air, ionomer), adsorption isotherms for $O_2$, contaminants, and related intermediates and products, and elementary chemical and electrochemical reactions and associated rate constants for $O_2$ reduction and contaminant oxidation or reduction. This list is enlarged by factors affecting ohmic and mass transfer losses, including cation and neutral molecules' absorption isotherms influencing ionomer and membrane ionic conductivity and oxygen permeability by swelling and changing the distance between sulfonate groups, and contaminant scavenging by liquid water modifying the effective contaminant concentration [12,13,66–71]. Although cell design and operating conditions were maintained as constant as possible, with the exception of catalyst loading and contaminant concentration, the change in cell resistance is insufficiently precise to capture all contamination nuances and accurately predict the cell voltage loss (Figure 5). An accurate correlation for the cell voltage loss would be useful. However, given the amount of information that will be required and the complexity associated with the derivation of a detailed mathematical model of contamination, a focus on preventive and recovery measures may be more fruitful. This suggestion is reinforced by considering practical aspects, contaminant mixtures [28], and long-term effects [26] that increase the number of contamination parameters and the difficulty in predicting contaminant effects.

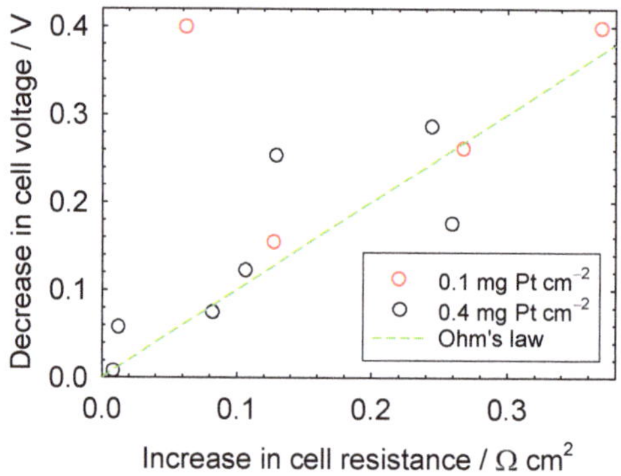


**Figure 5.** Cell voltage loss as a function of the sum of the changes in kinetic and mass transfer resistance.

## 3. Materials and Methods

A single modified Fuel Cell Technologies cell with an active area of 50 $cm^2$ and triple/double serpentine channels for the cathode/anode was used for all experiments. Gore PRIMEA M715 catalyst-coated membranes with a Pt loading of 0.1 or 0.4 mg Pt $cm^{-2}$ (50 % Pt/C) on each side were inserted between SGL Carbon Sigracet 25 BC gas diffusion layers. The cell was operated with a FCATS™ G050 series test station (Green Light Power Technologies). After cell activation, operating conditions were set to air/$H_2$, 2/2 stoichiometry, 48.3/48.3 $kPa_g$ outlet pressure, 50%/100% relative humidity, 80 °C, and 1 A $cm^{-2}$. Contaminant concentrations varied between 1.4 and ~8000 ppm: Acetonitrile (20 ppm), acetylene (100 ppm), bromomethane (5 ppm), isopropanol (~8000 ppm), methyl methacrylate (20 ppm), naphthalene (1.4 ppm), and propene (100 ppm). Contaminant concentrations were individually and empirically adjusted based on prior experience to cause a perceptible to significant cell voltage decrease at the steady state for the 0.4 mg Pt $cm^{-2}$ catalyst loading, and to leave a sufficient cell voltage window for an additional decrease induced by the lower 0.1 mg Pt $cm^{-2}$ catalyst loading. Contaminants were injected after the air humidifier using air-based gas mixtures for most cases. However, isopropanol and naphthalene were respectively evaporated and sublimated by employing a thermally controlled and calibrated liquid/solid holder. Contaminant injection was initiated after 5 h with an exposure that lasted from less than 1 to ~70 h until a steady state was achieved. After the contamination injection was interrupted, the self-induced recovery was recorded until a steady state was obtained, which necessitated between 5 and ~60 h.

During the galvanostatic experiments, impedance spectra were acquired at irregular intervals by superimposing 0.1 Hz to 10 kHz (10 points per decade) current perturbations that caused a voltage change of ~5 mV. The Solartron SI1260 impedance/gain-phase analyzer was operated with ZPlot® software (Version 2.9c, Scribner Associates, Southern Pines, NC, USA). Measurement accuracy was improved by utilizing Stanford Research SR560 low-noise preamplifiers and by winding up both load-bank cables, which have an equal length, to reduce their inductance. The ZView® software (Version 3.5e, Scribner Associates) was employed for fitting impedance spectra to equivalent circuit models. Polarization curves were only recorded before and after acetylene contamination. Polarization curves were measured by decreasing the current density from 2 to 0 (open circuit voltage) A $cm^{-2}$ in a stepwise fashion, allowing a stabilization time of 15 min at each stage, and otherwise using contamination test operating conditions.

## 4. Conclusions

The effect of Pt catalyst loading on the steady-state cell voltage loss was characterized for seven organic airborne contaminants. Impedance spectroscopy was used to gain mechanistic insight. The steady-state cell voltage loss is mostly attributed to a concurrent increase in both kinetic and mass transfer resistances that is reminiscent of the effect of a decrease in catalyst loading in the absence of a contaminant. Low Pt catalyst loadings generally lead to a larger steady-state cell voltage loss. A significant correlation between the steady-state cell voltage loss and the sum of the kinetic and mass transfer resistance changes was not found, and would only be improved with major increases in cost and effort. For this reason, it is proposed to focus activities on contamination prevention and recovery measures.

For a commercially relevant cathode catalyst loading of 0.1 mg Pt $cm^{-2}$, it would be advantageous to expand the current database to other contaminants and contaminant concentrations for the derivation of tolerance limits to support the design of air filters. Although tolerance limits were previously derived for single contaminants rather than for more practically relevant mixtures [42], for very low contaminant concentrations, tolerance limits may still prove useful because the catalyst surface coverage by contaminant adsorbates may be so small that the different species may not interact. In other words, the effects of all contaminants may be additive. It would also be useful to verify this hypothesis with diluted, single, and multiple contaminant mixtures.

**Author Contributions:** Conceptualization, J.S.-P. and Y.Z.; methodology, Y.Z.; formal analysis, J.S.-P. and Y.Z.; investigation, Y.Z.; resources, Y.Z.; writing—original draft preparation, J.S.-P.; writing—review and editing, J.S.-P. and Y.Z.; visualization, J.S.-P.; supervision, J.S.-P.; project administration, J.S.-P. All authors have read and agreed to the published version of the manuscript.

**Funding:** This research was funded by the United States Department of Energy, grant number DE-EE0000467, and the Office of Naval Research, grant number N00014-17-1-2206. The APC was funded by the Office of Naval Research.

**Acknowledgments:** The authors are grateful to the Hawaiian Electric Company for their ongoing support of the operations of the Hawaii Sustainable Energy Research Facility.

**Conflicts of Interest:** The authors declare no conflict of interest. The funders had no role in the design of the study; in the collection, analyses, or interpretation of data; in the writing of the manuscript, or in the decision to publish the results.

## References

1. Nagula, M. Forecasting of Fuel Cell Technology in Hybrid and Electric Vehicles Using Gompertz Growth Curve. *J. Stat. Manag. Syst.* **2016**, *19*, 73–88. [CrossRef]
2. Ogawa, T.; Takeuchi, M.; Kajikawa, Y. Comprehensive Analysis of Trends and Emerging Technologies in All Types of Fuel Cells Based on a Computational Method. *Sustainability* **2018**, *10*, 458. [CrossRef]
3. St-Pierre, J. Air Impurities. In *Polymer Electrolyte Fuel Cell Durability*; Büchi, F.N., Inaba, M., Schmidt, T.J., Eds.; Springer: New York, NY, USA, 2009; pp. 289–321.

4. Kennedy, D.M.; Cahela, D.R.; Zhu, W.H.; Westrom, K.C.; Nelms, R.M.; Tatarchuk, B.J. Fuel Cell Cathode Air Filters: Methodologies for Design and Optimization. *J. Power Sources* **2007**, *168*, 391–399. [CrossRef]
5. *Hydrogen Fuel Quality—Product Specification, ISO 14687:2019(E)*, 1st ed.; International Organization for Standardization: Geneva, Switzerland, 2019.
6. Gasteiger, H.A.; Panels, J.E.; Yan, S.G. Dependence of PEM Fuel Cell Performance on Catalyst Loading. *J. Power Sources* **2004**, *127*, 162–171. [CrossRef]
7. Ahluwalia, R.K.; Wang, X. Effect of CO and $CO_2$ Impurities on Performance of Direct Hydrogen Polymer-Electrolyte Fuel Cells. *J. Power Sources* **2008**, *180*, 122–131. [CrossRef]
8. Hashimasa, Y.; Matsuda, Y.; Imamura, D.; Akai, M. PEFC Power Generation Performance Degradation by Hydrogen Sulfide and Ammonia—Effects of Lowering Platinum Loading. *Electrochemistry* **2011**, *79*, 343–345. [CrossRef]
9. Hafttananian, M.; Ramiar, A.; Shabani, B.; Ranjbar, A.A. Nonlinear Algorithm of PEM Fuel Cell Catalyst Poisoning Progress in the Presence of Carbon Monoxide in Anode Fuel: A Computational Study Using OpenFOAM. *Electrochim. Acta* **2017**, *246*, 348–364. [CrossRef]
10. Prass, S.; Friedrich, K.A.; Zamel, N. Tolerance and Recovery of Ultralow-Loaded Platinum Anode Electrodes upon Carbon Monoxide and Hydrogen Sulfide Exposure. *Molecules* **2019**, *24*, 3514. [CrossRef]
11. de Bruijn, F.A.; Papageorgopoulos, D.C.; Sitters, E.F.; Janssen, G.J.M. The Influence of Carbon Dioxide on PEM Fuel Cell Anodes. *J. Power Sources* **2002**, *110*, 117–124. [CrossRef]
12. St-Pierre, J. PEMFC Contamination Model: Foreign Cation Exchange with Ionomer Protons. *J. Power Sources* **2011**, *196*, 6274–6283. [CrossRef]
13. St-Pierre, J. PEMFC Contaminant Tolerance Limit—Foreign Cations in Ionomers. *Int. J. Hydrogen Energy* **2011**, *36*, 5527–5535. [CrossRef]
14. Weber, A.Z.; Delacourt, C. Mathematical Modelling of Cation Contamination in a Proton-Exchange Membrane. *Fuel Cells* **2008**, *8*, 459–465. [CrossRef]
15. Kienitz, B.L.; Baskaran, H.; Zawodzinski, T.A., Jr. Modeling the Steady-State Effects of Cationic Contamination on Polymer Electrolyte Membranes. *Electrochim. Acta* **2009**, *54*, 1671–1679. [CrossRef]
16. Steinbach, A.J.; Hamilton, C.V., Jr.; Debe, M.K. Impact of Micromolar Concentrations of Externally-Provided Chloride and Sulfide Contaminants on PEMFC Reversible Stability. *Electrochem. Soc. Trans.* **2007**, *11*, 889–902. [CrossRef]
17. Noda, Z.; Hirata, K.; Hayashi, A.; Taniguchi, S.; Nakazato, N.; Seo, A.; Yasuda, I.; Ariura, S.; Shinkai, H.; Sasaki, K. PEFC-Type Impurity Sensors for Hydrogen Fuels. *Int. J. Hydrogen Energy* **2012**, *37*, 16256–16263. [CrossRef]
18. Noda, Z.; Hirata, K.; Hayashi, A.; Takahashi, T.; Nakazato, N.; Saigusa, K.; Seo, A.; Suzuki, K.; Ariura, S.; Shinkai, H.; et al. Hydrogen Pump-Type Impurity Sensors for Hydrogen Fuels. *Int. J. Hydrogen Energy* **2017**, *42*, 3281–3293. [CrossRef]
19. Talke, A.; Misz, U.; Konrad, G.; Heinzel, A.; Klemp, D.; Wegener, R. Influence of Urban Air on Proton Exchange Membrane Fuel Cell Vehicles - Long Term Effects of Air Contaminants in an Authentic Driving Cycle. *J. Power Sources* **2018**, *400*, 556–565. [CrossRef]
20. Mehrabadi, B.A.T.; Dinh, H.N.; Bender, G.; Weidner, J.W. Effect of System Contaminants on the Performance of a Proton Exchange Membrane Fuel Cell. *J. Electrochem. Soc.* **2016**, *163*, F1527–F1534. [CrossRef]
21. Zhai, Y.; Baturina, O.; Ramaker, D.; Farquhar, E.; St-Pierre, J.; Swider-Lyons, K. Chlorobenzene Poisoning and Recovery of Platinum-Based Cathodes in Proton Exchange Membrane Fuel Cells. *J. Phys. Chem. C* **2015**, *119*, 20328–20338. [CrossRef]
22. Zhai, Y.; St-Pierre, J. Acetylene Contamination Mechanisms in the Cathode of Proton Exchange Membrane Fuel Cells. *ChemElectroChem* **2017**, *4*, 655–670. [CrossRef]
23. St-Pierre, J.; Zhai, Y.; Angelo, M.S. Effect of Selected Airborne Contaminants on PEMFC Performance. *J. Electrochem. Soc.* **2015**, *161*, F280–F290. [CrossRef]
24. Yu, H.; Bonville, L.; Maric, R. Analysis of $H_2$/Air Polarization Curves: The Influence of Low Pt Loading and Fabrication Process. *J. Electrochem. Soc.* **2018**, *165*, F272–F284. [CrossRef]
25. Zhai, Y.; St-Pierre, J. Acetonitrile Contamination in the Cathode of Proton Exchange Membrane Fuel Cells and Cell Performance Recovery. *Appl. Energy* **2019**, *242*, 239–247. [CrossRef]
26. Zhai, Y.; Ge, J.; Qi, J.; St-Pierre, J. Effect of Acetonitrile Contamination on Long-Term Degradation of Proton Exchange Membrane Fuel Cells. *J. Electrochem. Soc.* **2018**, *165*, F3191–F3199. [CrossRef]

27. Reshetenko, T.V.; St-Pierre, J. Study of the Acetonitrile Poisoning of Platinum Cathodes on Proton Exchange Membrane Fuel Cell Spatial Performance Using a Segmented Cell System. *J. Power Sources* **2015**, *293*, 929–940. [CrossRef]
28. Qi, J.; Zhai, Y.; St-Pierre, J. Effect of Contaminant Mixtures in Air on Proton Exchange Membrane Fuel Cell Performance. *J. Power Sources* **2019**, *413*, 86–97. [CrossRef]
29. Reshetenko, T.V.; Artyushkova, K.; St-Pierre, J. Spatial Proton Exchange Membrane Fuel Cell Performance Under Bromomethane Poisoning. *J. Power Sources* **2017**, *342*, 135–147. [CrossRef]
30. Zhai, Y.; Baturina, O.; Ramaker, D.; Farquhar, E.; St-Pierre, J.; Swider-Lyons, K. Bromomethane Contamination in the Cathode of Proton Exchange Membrane Fuel Cells. *Electrochim. Acta* **2016**, *213*, 482–489. [CrossRef]
31. Reshetenko, T.V.; St-Pierre, J. Effects of Propylene, Methyl Methacrylate and Isopropanol Poisoning on Spatial Performance of a Proton Exchange Membrane Fuel Cell. *J. Power Sources* **2018**, *378*, 216–224. [CrossRef]
32. Wang, M.; Park, J.H.; Kabir, S.; Neyerlin, K.C.; Kariuki, N.N.; Lv, H.; Stamenkovic, V.R.; Myers, D.J.; Ulsh, M.; Mauger, S.A. Impact of Catalyst Ink Dispersing Methodology on Fuel Cell Performance Using in-Situ X-ray Scattering. *ACS Appl. Energy Mater.* **2019**, *2*, 6417–6427. [CrossRef]
33. Reddy, P.M.K.; Krushnamurty, K.; Subrahmanyam, C. Surface Modification of Carbon Fabric for Isopropanol Removal from Gas Stream. *Microelectron. Eng.* **2014**, *126*, 60–64. [CrossRef]
34. St-Pierre, J.; Ge, J.; Zhai, Y.; Reshetenko, T.V.; Angelo, M. PEMFC Cathode Contamination Mechanisms for Several VOCs—Acetonitrile, Acetylene, Bromomethane, Iso-propanol, Methyl Methacrylate, Naphthalene and Propene. *Electrochem. Soc. Trans.* **2013**, *58*, 519–528. [CrossRef]
35. Lee, C.-G.; Umeda, M.; Uchida, I. Cyclic Voltammetric Analysis of $C_1$-$C_4$ Alcohol Electrooxidations with Pt/C and Pt-Ru/C Microporous Electrodes. *J. Power Sources* **2006**, *160*, 78–89. [CrossRef]
36. Zhai, Y.; St-Pierre, J. Proton Exchange Membrane Fuel Cell Cathode Contamination—Acetylene. *J. Power Sources* **2015**, *279*, 165–171. [CrossRef]
37. Reshetenko, T.V.; St-Pierre, J. Study of Acetylene Poisoning of Pt Cathode on Proton Exchange Membrane Fuel Cell Spatial Performance Using a Segmented Cell System. *J. Power Sources* **2015**, *287*, 401–415. [CrossRef]
38. Zhai, Y.; St-Pierre, J. Impact of Operating Conditions on the Acetylene Contamination in the Cathode of Proton Exchange Membrane Fuel cells. *J. Power Sources* **2017**, *372*, 134–144. [CrossRef]
39. Zhai, Y.; St-Pierre, J. Tolerance and Mitigation Strategies of Proton Exchange Membrane Fuel Cells Subject to Acetylene Contamination. *Int. J. Hydrogen Energy* **2018**, *43*, 17475–17479. [CrossRef]
40. Reshetenko, T.V.; St-Pierre, J. Study of the Aromatic Hydrocarbons Poisoning of Platinum Cathodes on Proton Exchange Membrane Fuel Cell Spatial Performance Using a Segmented Cell System. *J. Power Sources* **2016**, *333*, 237–246. [CrossRef]
41. St-Pierre, J.; Virji, M.B.V. Cell Performance Distribution in a Low-Temperature Proton Exchange Membrane Fuel Cell Stack during Propene Contamination. *J. Appl. Electrochem.* **2016**, *46*, 169–181. [CrossRef]
42. St-Pierre, J.; Zhai, Y.; Ge, J.; Angelo, M.; Reshetenko, T.; Molter, T.; Bonville, L.; Pasaogullari, U.; Collins, W.; Wessel, S. The Effect of Airborne Contaminants on Fuel Cell Performance and Durability. In *DOE Hydrogen and Fuel Cells Program—2013 Annual Progress Report*; DOE/GO-102013-4260; United States Department of Energy: Washington DC, USA, 2013; pp. V-1–V-6.
43. *Hydrogen Fuel—Product Specification—Part 2: Proton Exchange Membrane (PEM) Fuel Cell Applications for Road Vehicles, ISO 14687-2:2012(E)*, 1st ed.; International Organization for Standardization: Geneva, Switzerland, 2012.
44. Viitakangas, J.; Ihonen, J.; Koski, P.; Reinikainen, M.; Aarhaug, T.A. Study of Formaldehyde and Formic Acid Contamination Effect on PEMFC. *J. Electrochem. Soc.* **2018**, *165*, F718–F727. [CrossRef]
45. Malevich, D.; Halliop, E.; Peppley, B.A.; Pharoah, J.G.; Karan, K. Investigation of Charge-Transfer and Mass-Transport Resistances in PEMFCs with Microporous Layer Using Electrochemical Impedance Spectroscopy. *J. Electrochem. Soc.* **2009**, *156*, B216–B224. [CrossRef]
46. Cruz-Manzo, S.; Chen, R.; Rama, P. Inductive Effect on the Fuel Cell Cathode Impedance Spectrum at High Frequencies. *J. Fuel Cell Sci. Technol.* **2012**, *9*, 051002. [CrossRef]
47. Pivac, I.; Barbir, F. Inductive Phenomena at Low Frequencies in Impedance Spectra of Proton Exchange Membrane Fuel Cells—A Review. *J. Power Sources* **2016**, *326*, 112–119. [CrossRef]
48. Setzler, B.P.; Fuller, T.F. A Physics-Based Impedance Model of Proton Exchange Membrane Fuel Cells Exhibiting Low-Frequency Inductive Loops. *J. Electrochem. Soc.* **2015**, *162*, F519–F530. [CrossRef]

49. Schneider, I.A.; Bayer, M.H.; Wokaun, A.; Scherer, G.G. Impedance Response of the Proton Exchange Membrane in Polymer Electrolyte Fuel Cells. *J. Electrochem. Soc.* **2008**, *155*, B783–B792. [CrossRef]
50. Gu, Y.; St-Pierre, J.; Joly, A.; Goeke, R.; Datye, A.; Atanassov, P. Aging Studies of Pt/Glassy Carbon Model Electrocatalysts. *J. Electrochem. Soc.* **2009**, *156*, B485–B492. [CrossRef]
51. Zhai, Y.; Ge, J.; St-Pierre, J. The Ionic Conductivity and Catalyst Activity Effects of Acetonitrile on Proton Exchange Membrane Fuel Cells. *Electrochem. Commun.* **2016**, *66*, 49–52. [CrossRef]
52. Butcher, D.R.; Zhu, Z.; Mao, B.; Wang, H.; Liu, Z.; Salmeron, M.; Somorjai, G.A. Mobility on the Reconstructed Pt(100)-hex Surface in Ethylene and in its Mixture with Hydrogen and Carbon Monoxide. *Chem. Commun.* **2013**, *49*, 6903–6905. [CrossRef]
53. Tao, F.; Dag, S.; Wang, L.-W.; Liu, Z.; Butcher, D.R.; Bluhm, H.; Salmeron, M.; Somorjai, G.A. Break-Up of Stepped Platinum Catalyst Surfaces by High CO Coverage. *Science* **2010**, *327*, 850–853. [CrossRef]
54. Zhai, Y.; Bethune, K.; Bender, G.; Rocheleau, R. Analysis of the $SO_2$ Contamination Effect on the Oxygen Reduction Reaction in PEMFCs by Electrochemical Impedance Spectroscopy. *J. Electrochem. Soc.* **2012**, *159*, B524–B530. [CrossRef]
55. Wang, X.; Hsing, I.-M.; Leng, Y.-J.; Yue, P.-L. Model Interpretation of Electrochemical Impedance Spectroscopy and Polarization Behavior of $H_2$/CO Mixture Oxidation in Polymer Electrolyte Fuel Cells. *Electrochim. Acta* **2001**, *46*, 4397–4405. [CrossRef]
56. Wagner, N.; Gülzow, E. Change of Electrochemical Impedance Spectra (EIS) with Time during CO-Poisoning of the Pt-Anode in a Membrane Fuel Cell. *J. Power Sources* **2004**, *127*, 341–347. [CrossRef]
57. Bethune, K.; St-Pierre, J.; LaManna, J.M.; Hussey, D.S.; Jacobson, D.L. Contamination Mechanisms of Proton Exchange Membrane Fuel Fells—Mass Transfer Overpotential Origin. *ACS Appl. Energy Mater.*. In preparation.
58. Weber, A.Z.; Kusoglu, A. Unexplained Transport Resistances for Low-Loaded Fuel-Cell Catalyst Layers. *J. Mater. Chem. A* **2014**, *2*, 17207–17211. [CrossRef]
59. Kongkanand, A.; Mathias, M.F. The Priority and Challenge of High-Power Performance of Low-Platinum Proton-Exchange Membrane Fuel Cells. *J. Phys. Chem. Lett.* **2016**, *7*, 1127–1137. [CrossRef] [PubMed]
60. Ge, J.; St-Pierre, J.; Zhai, Y. PEMFC Cathode Catalyst Contamination Evaluation with a RRDE—Acetylene. *Electrochim. Acta* **2014**, *133*, 65–72. [CrossRef]
61. Ge, J.; St-Pierre, J.; Zhai, Y. PEMFC Cathode Catalyst Contamination Evaluation with a RRDE—Acetonitrile. *Electrochim. Acta* **2014**, *134*, 272–280. [CrossRef]
62. Ge, J.; St-Pierre, J.; Zhai, Y. PEMFC Cathode Catalyst Contamination Evaluation with a RRDE—Methyl Methacrylate. *Int. J. Hydrogen Energy* **2014**, *39*, 18351–18361. [CrossRef]
63. Ge, J.; St-Pierre, J.; Zhai, Y. PEMFC Cathode Catalyst Contamination Evaluation with a RRDE—Propene and Naphthalene. *Electrochim. Acta* **2014**, *138*, 437–446. [CrossRef]
64. Zatoń, M.; Rozière, J.; Jones, D.J. Current Understanding of Chemical Degradation Mechanisms of Perfluorosulfonic Acid Membranes and their Mitigation Strategies: A Review. *Sustain. Energy Fuels* **2017**, *1*, 409–438. [CrossRef]
65. St-Pierre, J.; Zhai, Y.; Ge, J. Relationships between PEMFC Cathode Kinetic Losses and Contaminants' Dipole Moment and Adsorption Energy on Pt. *J. Electrochem. Soc.* **2016**, *163*, F247–F254. [CrossRef]
66. St-Pierre, J. PEMFC Contamination Model: Neutral Species Sorption by Ionomer. *Electrochem. Soc. Trans.* **2011**, *41*, 307–315. [CrossRef]
67. Okada, T.; Dale, J.; Ayato, Y.; Asbjørnsen, O.A.; Yuasa, M.; Sekine, I. Unprecedented Effect of Impurity Cations on the Oxygen Reduction Kinetics at Platinum Electrodes Covered with Perfluorinated Ionomer. *Langmuir* **1999**, *15*, 8490–8496. [CrossRef]
68. Okada, T.; Ayato, Y.; Dale, J.; Yuasa, M.; Sekine, I.; Asbjornsen, O.A. Oxygen Reduction Kinetics at Platinum Electrodes Covered with Perfluorinated Ionomer in the Presence of Impurity Cations $Fe^{3+}$, $Ni^{2+}$ and $Cu^{2+}$. *Phys. Chem. Chem. Phys.* **2000**, *2*, 3255–3261. [CrossRef]
69. Okada, T.; Ayato, Y.; Satou, H.; Yuasa, M.; Sekine, I. The Effect of Impurity Cations on the Oxygen Reduction Kinetics at Platinum Electrodes Covered with Perfluorinated Ionomer. *J. Phys. Chem. B* **2001**, *105*, 6980–6986. [CrossRef]

70. Okada, T.; Satou, H.; Yuasa, M. Effects of Additives on Oxygen Reduction Kinetics at the Interface between Platinum and Perfluorinated Ionomer. *Langmuir* **2003**, *19*, 2325–2332. [CrossRef]
71. St-Pierre, J.; Wetton, B.; Zhai, Y.; Ge, J. Liquid Water Scavenging of PEMFC Contaminants. *J. Electrochem. Soc.* **2014**, *161*, E3357–E3364. [CrossRef]

**Sample Availability:** Samples of the compounds are not available from the authors.

*molecules*

MDPI

*Article*

# Performance Recovery after Contamination with Nitrogen Dioxide in a PEM Fuel Cell

**Yasna Acevedo Gomez, Göran Lindbergh and Carina Lagergren ***

Applied Electrochemistry, Department of Chemical Engineering, KTH Royal Institute of Technology, 10044 Stockholm, Sweden; yasna@kth.se (Y.A.G.); gnli@kth.se (G.L.)

* Correspondence: carinal@kth.se; Tel.: +46-8-790-6507

Academic Editors: Jean St-Pierre and Shangfeng Du

Received: 24 December 2019; Accepted: 1 March 2020; Published: 2 March 2020

check for updates

**Abstract:** While the market for fuel cell vehicles is increasing, these vehicles will still coexist with combustion engine vehicles on the roads and will be exposed to an environment with significant amounts of contaminants that will decrease the durability of the fuel cell. To investigate different recovery methods, in this study, a PEM fuel cell was contaminated with 100 ppm of $NO_2$ at the cathode side. The possibility to recover the cell performance was studied by using different airflow rates, different current densities, and by subjecting the cell to successive polarization curves. The results show that the successive polarization curves are the best choice for recovery; it took 35 min to reach full recovery of cell performance, compared to 4.5 h of recovery with pure air at 0.5 A $cm^{-2}$ and 110 mL $min^{-1}$. However, the performance recovery at a current density of 0.2 A $cm^{-2}$ and air flow 275 mL $min^{-1}$ was done in 66 min, which is also a possible alternative. Additionally, two operation techniques were suggested and compared during 7 h of operation: air recovery and air depletion. The air recovery technique was shown to be a better choice than the air depletion technique.

**Keywords:** PEM fuel cell; performance; recovery; nitrogen dioxide; contamination

## 1. Introduction

As the world is heading towards clean energy sources, the proton exchange membrane (PEM) fuel cell plays an important role, being a good alternative for the transportation sector and stationary power systems. Automobile manufacturers have been releasing electric vehicles as a viable solution to decrease greenhouse gas emissions [1]. The fuel cell vehicle is becoming popular and may be the right solution to replace internal combustion engine (ICE) vehicles in the near future [2,3]. However, the durability of the fuel cell is still an issue, where one aspect is pollutants in the air that seriously affect the performance. It is well known that the air contains unwanted contaminants that come from ICE vehicles, agriculture, and industries. As the fuel cell market grows, fuel cell vehicles must coexist with ICE vehicles on the roads. The coexistence of these two types of vehicles may lead to a dramatic decrease in the fuel cell performance, thus a recovery strategy must be considered in a real traffic situation.

Among the contaminants in air, nitrogen dioxide is one that seriously affects the performance of the PEM fuel cell but has not been completely studied in the literature. In our previous study [4], severe degradation of the cell performance was shown at different concentrations of $NO_2$. For all the tests, the same total dosage of $NO_2$ was added, but the possibility for the cell performance to recover after contamination differed. At higher concentrations of 50, 100, and 200 ppm $NO_2$, the performance could only be partially recovered. In the study, a mechanism for $NO_2$ contamination was proposed based on cyclic voltammetry (CV) observation in which $NO_2$ is oxidized to $NO_3^-$ at 1.05 V, then in the negative sweep reduced to $NO_2^-$ at 0.68 V, followed by a subsequent reduction of $NO_2^-$ to $N_2O$ and/or

$NH_2OH$ at potentials lower than 0.5 V. The proposed mechanism was confirmed by the detection of NO as intermediate species and $N_2O$ by simultaneous mass spectrometry.

Other authors have shown that the contamination can be fully recovered in some cases [5,6], almost recovered in other cases [6–8], or not recovered [5,9], depending on the $NO_2$ concentration, exposure time, and operating conditions. Misz et al. [6] and Jing et al. [8] tested the contamination with 1 ppm $NO_2$ over 1 and 100 h, respectively; the shorter exposure time resulted in fully recovered performance while performance following the longer exposure time was almost recovered after cyclic voltammetry scan as a recovery process. It is seen that long-term exposure produces an unrecoverable effect. Mohtadi et al. [9] and Uribe et al. [5] contaminated the fuel cell with 5 ppm $NO_2$ over 12 and 15 h, respectively. In these cases, the result from Mohtadi et al. [9] was partially recovered cell performance, while the result from Uribe et al. [5] was fully recovered performance. The difference of these two recovery processes was that Mohtadi et al. [9] operated the cell in the range of 0.68–0.7 V and Uribe et al. [5] at 0.5 V. Our previous results [4], in agreement with the results of Chen et al. [10] and Lin et al. [11], showed that, at lower potentials in the negative sweep, reduction of nitrite occurs, and thus it is removed from the Pt-catalyst. Higher concentrations were tried by Yang et al. [7] (10, 140, and 1480 ppm) and Misz et al. [6] (10 and 15 ppm), in which performance recovery was almost reached in all of the cases after approximately 1 h with $NO_2$. When it comes to long-term operation, Uribe et al. [5] showed that performance following contamination of 0.4 ppm $NO_2$ for around 520 h was not recovered, probably due to the low amount of catalyst they used (17 μg Pt $cm^{-2}$) that was quickly damaged.

St-Pierre et al. [12] simulated performance recovery after 500 h of exposure to 0.1 ppm $NO_2$. Even if they used dry air conditions in their simulation, where the performance was dramatically affected, the performance was recovered and reached its initial value. This result is contradictory to the one obtained by Uribe et al. [5] and may be due to different conditions, but unfortunately the operating conditions used were not specified in Uribe's report.

The aim of this study was to contribute to the improvement of the durability of the fuel cell by trying different operating conditions that influence the recovery process after $NO_2$ contamination. These processes included successive polarization curves and recovery at different flow rates and current densities. In real traffic situations, exposure to high amounts of $NO_2$ is unavoidable, and recovery methods that can be applied online in real fuel cell vehicles are desired. Therefore, two such realistic operation techniques were suggested and compared: consecutive recovery with air and air depletion.

## 2. Results and Discussion

### 2.1. Performance of the Contaminated MEA

The degradation of fuel cell performance upon contamination with 100 ppm of $NO_2$ in air and its subsequent recovery of performance is shown in Figure 1a. The sequence of experiments was to run the cell in a galvanostatic mode at 0.5 A $cm^{-2}$ with clean air for 30 min without contaminant, followed by the introduction of 100 ppm $NO_2$ in the cathode air flow for 3 h, and then recovery of performance with pure air. Polarization curves (Figure 1b) and electrochemical impedance spectroscopy (EIS) (Figure 1c) were done at the beginning of life (BOL), after contamination with $NO_2$, and after recovery with air. Figure 1a shows the dramatic performance degradation of 197 mV after 3 h of contamination. However, after switching off the $NO_2$ contaminant and running the cell with clean air, the fuel cell performance was completely recovered in 4.5 h.

The polarization curves in Figure 1b show a clear contamination of $NO_2$, mainly at lower current densities, where the Pt-catalyst active sites are affected by $NO_2$. In this part of the curve, the contamination is related to the electrode kinetics, most likely at the cathode, which is the main contributor to the performance loss and where the contaminant is introduced. In the graph, it is also shown that the performance was completely recovered when pure air was added at 110 mL $min^{-1}$ with a current density of 0.5 A $cm^{-2}$. Furthermore, the recovered performance was better than at beginning of life (BOL) at high current densities, which may indicate better conductivity in the membrane due to water being produced by the ORR, while at the same time intermediate species are being reduced in the actual potential range, as mentioned in our previous publication [4].

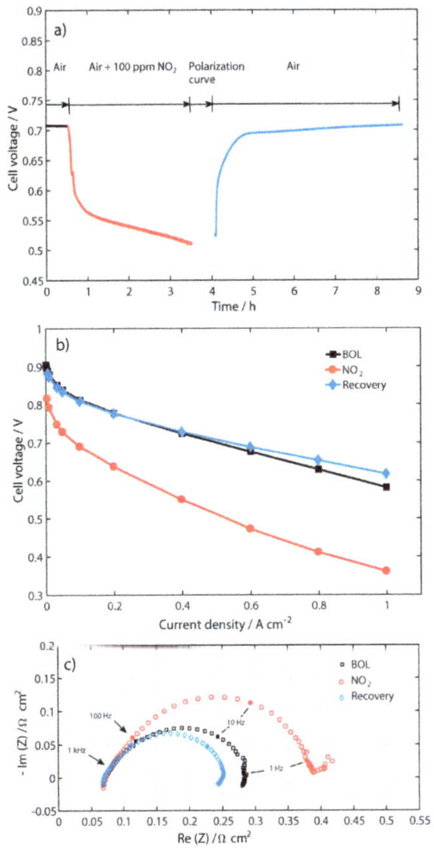


**Figure 1.** (**a**) Performance during contamination with 100 ppm $NO_2$ and recovery at 0.5 A cm$^{-2}$ and air flow 110 mL min$^{-1}$; (**b**) polarization curves; and (**c**) galvanostatic EIS measurements at 0.5 A cm$^{-2}$ for BOL, with $NO_2$, and after recovery in cathode air flow.

To better diagnose the performance limitation after contamination with $NO_2$ and the respective recovery in the fuel cell, EIS spectra were recorded at 0.5 A cm$^{-2}$ and shown in Figure 1c. After contamination with $NO_2$, a second semicircle is beginning to be formed at lower frequencies. However, this second semicircle disappears after the recovery process with pure air. Additionally, the polarization resistance decreases, which is in accordance with the polarization curve in Figure 1b and may be related to a better access to platinum sites after the recovery process. There is no change in the high frequency resistance (HFR), showing that the membrane resistance was not affected by contamination with $NO_2$ and the recovery process.

Based on the degradation and time for performance recovery shown in Figure 1a, different air flow rates, different constant current densities, and successive polarization curves were tested during the recovery of the contaminated MEA to investigate and understand the recovery process of this contaminant.

## *2.2. Recovery at Different Air Flow Rates*

To find a shorter performance recovery time for the MEA contaminated with $NO_2$, different air flow rates (110, 165, 220, and 275 mL $min^{-1}$) were tested for the recovery process at a constant current density of 0.5 A $cm^{-2}$, as shown in Figure 2a. The time required to reach the same cell voltage as before contamination is defined as the recovery time. The same contamination sequence as described above was used and the air flow rate was changed to the desired value for the recovery of the performance. Figure 2a shows that all the curves reached their initial values after the recovery process, but after different periods of time. The faster recovery time was found to be at the highest flow rate, 275 mL $min^{-1}$. This is a clear sign that the $NO_2$ contaminant is not as well attached to the Pt-catalyst surface as sulfur compounds are [9]. As soon as clean air is introduced into the recovery process, most of the $NO_2$ is removed from the Pt-catalyst. This is shown in Figure 2a by the sharp increase in cell voltage (~180 mV) within about 30 min, after which a slower relaxation period occurs until steady state is reached.

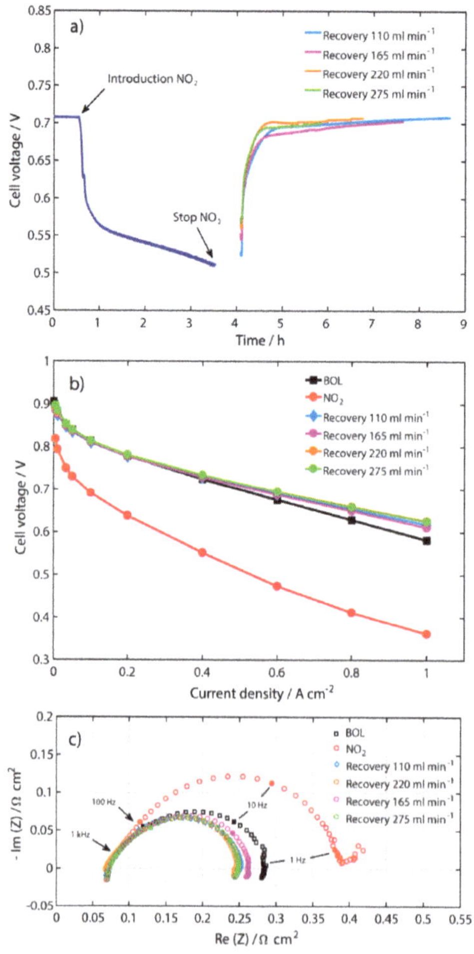


**Figure 2.** (**a**) Performance with 100 ppm $NO_2$ and after recovery at different airflows 110, 165, 220, and 275 mL $min^{-1}$ at 0.5 A $cm^{-2}$; (**b**) polarization curves; and (**c**) galvanostatic EIS measurements at 0.5 A $cm^{-2}$ for the same conditions as in (**a**).

Figure 2b shows that all polarization curves overlap up to the current density of 0.4 A $cm^{-2}$. As the current density increases further, small differences in potential can be seen between the curves at different flow rates, where the two highest air flow rates show the best performance. The performance after the recovery process is higher than at BOL for all the different air flow rates, in the same way as in Figure 1b.

The EIS spectra after the recovery process at different air flow rates are depicted in Figure 2c. It can be seen that there is no significant difference in the HFR where the spectra intercept the real axis. After the recovery process, all of the spectra show a lower polarization resistance when compared with BOL, which is in accordance with Figure 2b. It might be possible that some Pt-sites were activated after the recovery process with pure air.

### 2.3. *Recovery at Different Current Densities*

Another strategy investigated was to recover the contaminated MEA at different current densities, as shown in Figure 3a. The same contamination procedure was done as in Figure 1a and the different controlled current densities for recovery process were 0.2, 0.5, 0.75, and 1 A $cm^{-2}$. It can be seen that all of the performance recovery measurements reached a steady state at their respective current densities. As also seen in the experiments with different air flow rates, the voltage increases abruptly after the $NO_2$ is switched off and replaced with clean air, which here again may be related to the rapid removal of $NO_2$ from the Pt-catalyst. The necessary time to reach steady state after the recovery process was different for the different current densities, and decreased as the current density increased above 0.5 A $cm^{-2}$. Surprisingly, the time to reach steady state at the recovery current density of 0.2 A $cm^{-2}$ was the shortest. This is a sign of a different mechanism that occurs at this specific current density. From our previous study using cyclic voltammetry in inert media with no water production [4], it was seen that around the range of potential that this current density corresponds to (0.65–0.76 V), reduction of $NO_3^-$ to $NO_2^-$ may occur. However, in the present experiments, water is produced at the cathode side and may react with $NO_2$ producing $HNO_3$ and NO, as shown in Equation (1). It can be pointed out that nitric acid in water is normally present as $NO_3^-$ [13].

$$3\,NO_2 + H_2O \rightarrow 2H^+ + 2\,NO_3^- + NO \quad (1)$$

The range of potentials in which the performance is recovered at the current density of 0.2 A $cm^{-2}$, i.e. 0.65–0.76 V (Figure 3a), is almost the same as the one in the inert media [4]; therefore, $NO_3^-$ may be reduced to $NO_2^-$ in the present experiments as well. Additionally, NO contamination is similar to CO contamination in that both contaminants affect the catalyst layer and, at low current densities in presence of $O_2$, NO is removed from the catalyst. This may explain the faster recovery at lower current densities (0.2 A $cm^{-2}$), while at higher current densities the NO contamination is more severe and oxygen is predominantly producing water through ORR. This suggests that chemical reactions may be present and followed by electrochemical reaction, in the same way as discussed by Chen et al. [10].

Figure 3b shows the respective polarization curves after the recovery process at different current densities. The performance after the recovery process done at the current densities of 0.5, 0.75, and 1 A $cm^{-2}$ overlapped with that at BOL until 0.4 A $cm^{-2}$ in the polarization curve. At higher current densities, they still overlapped each other but they differed from the BOL, in a similar way as in Figure 2b. However, the behavior of the performance after the recovery process done at 0.2 A $cm^{-2}$ was different. It is seen that this performance was not fully recovered; even though it reached a steady state during the recovery process, as shown in Figure 3a, it still had 15 mV left to full recovery. For this recovery current density, a better performance than at BOL was seen at current densities higher than 0.6 A $cm^{-2}$. A possible reason it did not fully recover at the current density of 0.2 A $cm^{-2}$ may be the formation of intermediate species around 0.7 V that may have affected the performance.

Figure 3c shows the EIS spectra after the recovery process at different current densities. As in Figure 2c, there is no significant difference in the HFR and the polarization resistance decreases after the recovery process. The lowest polarization resistance was observed for the recovery at 0.5 A cm$^{-2}$.

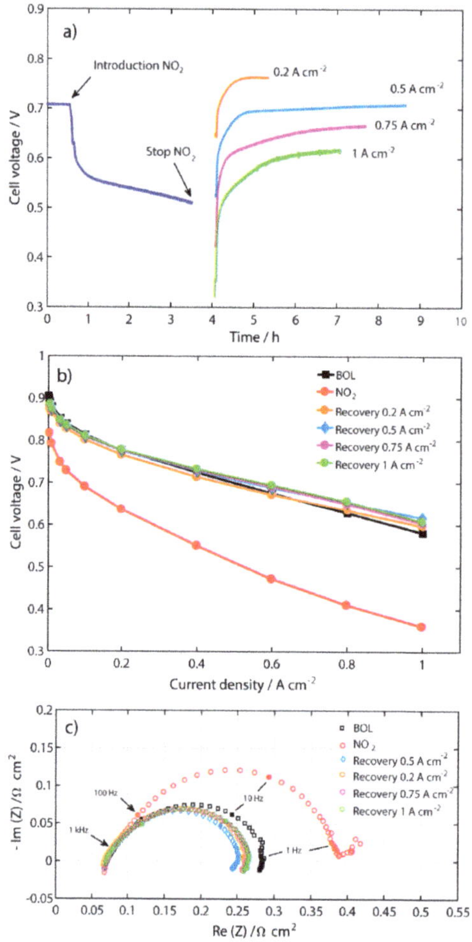


**Figure 3.** (**a**) Performance with 100 ppm $NO_2$ and recovery at current densities 0.2, 0.5, 0.75, and 1 A cm$^{-2}$ and constant 110 mL min$^{-1}$ air flow; (**b**) polarization curves; and (**c**) galvanostatic EIS measurements at 0.5 A cm$^{-2}$ at BOL, after contamination, and after recovery at each current density.

## 2.4. Other Types of Recovery

Thus far, it has been shown that the recovery process time after contamination with $NO_2$ can be shortened. A summary of times for recovery, from the used recovery methods, is shown in Table 1. The two shortest times were found to be at current density of 0.2 A cm$^{-2}$ and air flow of 275 mL min$^{-1}$. Therefore, these two operating conditions were combined to potentially obtain an even shorter recovery time (Figure 4a). Additionally, successive polarization curves after contamination with $NO_2$ were tried as a recovery method. For this experiment, the polarization curves were conducted in galvanodynamic mode at a step rate of 5 mA s$^{-1}$. Figure 4b shows the polarization curves for the latter two recovery methods compared with the polarization curves at 0.5 A cm$^{-2}$ and 110 mL min$^{-1}$, at BOL, and directly after contamination with 100 ppm $NO_2$. The recovery time for the 0.2 A cm$^{-2}$ and 275 mL min$^{-1}$

air flow was 66 min, i.e. the time was reduced by 10 min compared with the recovery process at 0.2 A cm$^{-2}$ and 110 mL min$^{-1}$. This indicates that the airflow rate is an important parameter; it seems that $NO_2$ can be removed from the Pt-catalyst by the air, and/or that $O_2$ is participating in chemical and electrochemical reactions in the removal of $NO_2$ species, as mentioned in the Section 2.3. The recovery time when performing successive polarization curves was 35 min, which was found to be the fastest way to recover the performance of the fuel cell contaminated with $NO_2$.

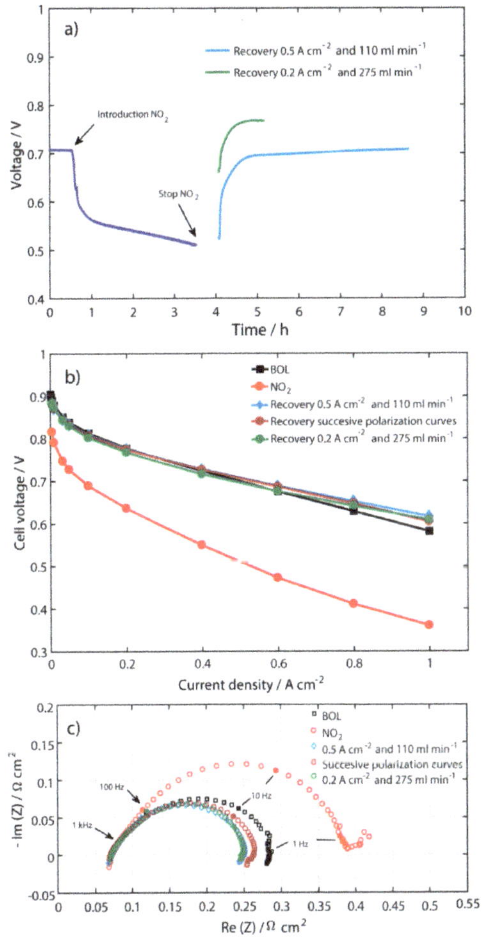


**Figure 4.** (**a**) Performance with 100 ppm $NO_2$, recovery at 0.5 A cm$^{-2}$ and 110 ml min$^{-1}$, and recovery at 0.2 A cm$^{-2}$ and 275 ml min$^{-1}$; (**b**) Polarization curves at BOL, after contamination with $NO_2$, after recovery at 0.5 A cm$^{-2}$ and 110 mL min$^{-1}$, after recovery by successive polarization curves with 110 mL min$^{-1}$ as a constant air flow, and after recovery at 0.2 A cm$^{-2}$ and 275 mL min$^{-1}$; and (**c**) EIS spectra for the measurements done in (**b**).

It is worth mentioning that none of the polarization curves after contamination with 100 ppm $NO_2$ reached values around 0.2 V. The lowest potential (0.35 V) was reached at a current density of 1 A cm$^{-2}$. Therefore, the reduction of $NO_2^-$ to $N_2O$ and/or $NH_2OH$ [4] may not be present in these set of experiments.

**Table 1.** Summary of the performance recovery time after introduction of 100 ppm $NO_2$ to the cathode air flow. Pure $H_2$ was used at the anode.

| Recovery at Different Air Flows Rates at 0.5 A cm$^{-2}$ | | Recovery at Different Current Densities at 110 mL min$^{-1}$ | |
|---|---|---|---|
| mL min$^{-1}$ | min | A cm$^{-2}$ | min |
| 110 | 274 | 0.2 | 76 |
| 165 | 213 | 0.5 | 274 |
| 220 | 161 | 0.75 | 217 |
| 275 | 106 | 1 | 181 |

A theoretical prediction for the recovery of $NO_2$ was made by St-Pierre et al. [12]; however, in their prediction, they did not include all processes in the fuel cell that may be affected by degradation, such as ohmic losses and mass transport, which explains the results obtained. It would be interesting to investigate performance recovery in a wider current density range.

The EIS measurements in Figure 4c show that, even though the shortest recovery time was reached by the successive polarization curves, the spectra of the experiment at 0.5 A cm$^{-2}$ and 110 mL min$^{-1}$ together with the spectra at 0.2 A cm$^{-2}$ and 275 mL min$^{-1}$ were those that had the lowest polarization resistance.

### *2.5. Comparison of Two Operation Techniques*

Finally, two operation techniques for the cathode were applied and compared by introducing 50 ppm of $NO_2$ in different ways, as shown in Figure 5a, with the goal to suggest online application in a fuel cell car. This concentration was chosen because it is more probable to find 50 ppm $NO_2$ in air than 100 ppm or higher concentrations. In both experiments, the cell was first stable for 30 min, keeping the same potential. The experiment done with air recovery (blue line) consisted of introducing $NO_2$ with balance of air to the cathode for 20 min, and then recovering the performance with clean air for 2 h. The same sequence was repeated three times. In the experiment with air depletion (orange line), the $NO_2$ contaminated air was fed to the cathode during 20 min, after which the air gas flow was switched off until the potential reached 0.01 V. At that point, the gas was switched on again. This experiment was made with the purpose to sweep the cell voltage within a wide range in order to let the fuel cell to recover quickly. The procedure was repeated 21 times to be comparable in time with the air recovery technique. Figure 5a shows a complete reversibility during the air recovery technique, in which all cycles reached the initial value (0.7 V). In both techniques, a lower cell voltage is seen after the 20 min with $NO_2$ compared to the first contamination cycle, but no significant difference is shown between the cycles. The outcome of the air recovery is in accordance with the results of Mohtadi et al. [9], who obtained a complete recovery after three cycles with 5 ppm of $NO_2$. On the other hand, the cell performance obtained by Yang et al. [7] did not reach the initial value after recovery. However, they used a different pressure (0.5 bar), and it is known that the pressure is an important parameter concerning recovery of a fuel cell contaminated by $NO_2$ [6].

At the end of each experiment, a polarization curve was recorded (see Figure 5b). The figure shows that the strategy with air depletion resulted in a lower performance after 7 h of operation, which might be caused by deterioration of the electrode due to peroxide formation at low electrode potentials [14–16]. On the other, the polarization curve after air recovery revealed a complete recovery of the Pt-catalyst, and even better performance at current densities higher than 0.4 A cm$^{-2}$. The air recovery technique suggests that $NO_2$ is only attached to the Pt-catalyst of the electrode and that it can be easily removed by air, apparently, without affecting other components.

EIS was also conducted at the end of each experiment (Figure 5c). The figure shows no significant difference between the two strategies, although the HFR of the air depletion spectrum increased only corresponding to about 3 mV when compared to beginning of life, but this is in the range of error.

These results show that it is possible to operate a specific technique online in a fuel cell vehicle in order to deal with $NO_2$ air pollution. However, the technique must be adapted to a more realistic drive cycle. Operating parameters such as air flow rate and current density can also possibly be incorporated in a recovery method to keep good performance after $NO_2$ contamination.

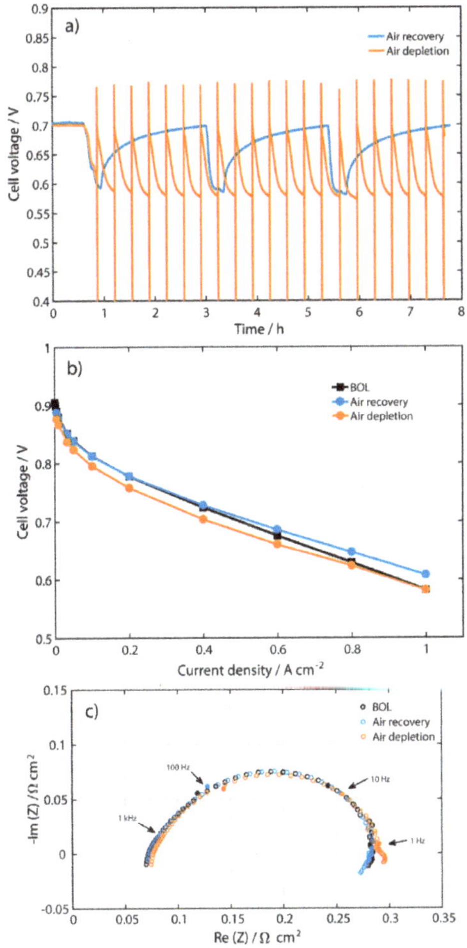


**Figure 5.** (**a**) Transient cell voltage when introducing 50 ppm $NO_2$ to the cathode air flow at 0.5 A $cm^{-2}$ for two different strategies, namely air recovery (blue) and air depletion (orange); (**b**) polarization curves; and (**c**) galvanostatic EIS measurement at 0.5 A $cm^{-2}$ at BOL and after testing the two strategies.

## 3. Materials and Methods

The experimental set up used in this investigation was the same as used in our previous study with $NO_2$ [4]. A commercial fuel cell hardware from Fuel Cell Technologies, Inc., and a commercial membrane electrode assembly (MEA) (Gore™Primea® 5641), with catalyst loadings of 0.45 mg $cm^{-2}$ Pt-alloy on the anode and 0.4 mg $cm^{-2}$ Pt on the cathode, were used in all of the experiments. The same type of gas diffusion layer (GDL) (Carbel™) was used at both anode and cathode. The geometric electrode area used was 1.5 $cm^2$. The cell temperature was kept at 80 °C and 1 atm, and the humidification of the gases was 90% RH. The gas cylinder used was the same as in [4], and the contamination flow was controlled by an Alicat Scientific mass flowmeter.

The electrochemical characterization procedure was the same as in our previous study [4]. For the contamination step, a galvanostatic measurement was done, followed by polarization curve measurement and electrochemical impedance spectroscopy (EIS) by use of a Solartron Interface SI1287 potentiostat together with a 1255 frequency response analyzer, controlled by CorrWare software. For the EIS, an AC amplitude of 60 mA (roughly corresponding to 3–15 mV depending on frequency and operating conditions) was used in the frequency range between 10 kHz and 30 mHz. It was assumed, in all experiments, that the electrical bulk and contact resistances were not affected by the introduction of $NO_2$, and that the high frequency resistance is related to the resistance of the membrane.

## 4. Conclusions

The results show that it is possible to find adequate performance recovery methods that can be applied in a fuel cell car in a real traffic situation where large amounts of $NO_2$ are present. In the experiments done in galvanostatic mode at 0.5 A cm$^{-2}$ with air flow of 110 mL min$^{-1}$, a significant potential drop was observed due to the presence of $NO_2$ in the cathode air. This performance loss was however totally recovered after 4.5 h with clean air. The study shows that it is possible to significantly decrease the time for performance recovery by running successive polarization curves or by applying 0.2 A cm$^{-2}$ and an air flow of 275 mL min$^{-1}$. Two operation techniques that can be used online in a fuel cell vehicle were also tested: air recovery and air depletion. The air recovery technique was found to be the best option for recovery of performance. Therefore, we assume that air can pull out the $NO_2$ molecules that surround the Pt-catalyst to free up the active site at higher current densities; however, at the current density of 0.2 A cm$^{-2}$, possibly a different contamination mechanism occurs.

**Author Contributions:** Conceptualization, Y.A.G.; methodology, Y.A.G.; validation, Y.A.G. and C.L.; formal analysis, Y.A.G.; investigation, Y.A.G.; resources, G.L. and C.L.; writing—original draft preparation, Y.A.G.; writing—review and editing, Y.A.G., C.L., and G.L.; visualization, Y.A.G. and C.L.; supervision, C.L. and G.L.; project administration, Y.A.G. and C.L.; and funding acquisition, C.L. and G.L. All authors have read and agreed to the published version of the manuscript.

**Funding:** This research was funded by the European project BIOGAS2PEM-FC (FP7), grant number 314940 and the Swedish governmental initiative StandUp for Energy.

**Acknowledgments:** The materials for this work were provided by Powercell AB. The experimental set up was built with the help of Mr. Hongkuan Wang.

**Conflicts of Interest:** The authors declare no conflict of interest. The funders had no role in the design of the study; in the collection, analyses, or interpretation of data; in the writing of the manuscript, or in the decision to publish the results.

## References

1. Das, H.S.; Tan, C.W.; Yatim, A.H.M. Fuel Cell Hybrid Electric Vehicles: A Review on Power Conditioning Units and Topologies. *Renew. Sustain. Energy Rev.* **2017**, *76*, 268–291. [CrossRef]
2. Das, V.; Padmanaban, S.; Venkitusamy, K.; Selvamuthukumaran, R.; Blaabjerg, F.; Siano, P. Recent Advances and Challenges of Fuel Cell Based Power System Architectures and Control – A Review. *Renew. Sustain. Energy Rev.* **2017**, *73*, 10–18. [CrossRef]
3. Daud, W.R.W.; Rosli, R.E.; Majlan, E.H.; Hamid, S.A.A.; Mohamed, R.; Husaini, T. PEM Fuel Cell System Control: A review. *Renew. Energy* **2017**, *113*, 620–638. [CrossRef]
4. Acevedo Gomez, Y.; Lagergren, C.; Lindbergh, G. Effect of Nitrogen Dioxide Impurities on PEM Fuel Cell Performance. *Submitt. Int. J. Hydrog. Energy* **2019**.
5. Uribe, F.; Smith, W.; Wilson, M.; Valerio, J.; Rockward, T. Electrodes for Polymer Electrolyte Membrane Operation on Hydrogen/Air and Reformate/Air. Available online: https://pdfs.semanticscholar.org/0de2/236e59c04ef161c5928e751893784784be87.pdf (accessed on 18 September 2018).
6. Misz, U.; Talke, A.; Heinzel, A.; Konrad, G. Sensitivity Analyses on the Impact of Air Contaminants on Automotive Fuel Cells. *Fuel Cells* **2016**, *16*, 444–462. [CrossRef]
7. Yang, D.; Ma, J.; Xu, L.; Wu, M.; Wang, H. The Effect of Nitrogen Oxides in Air on the Performance of Proton Exchange Membrane Fuel Cell. *Electrochim. Acta* **2006**, *51*, 4039–4044. [CrossRef]

8. Jing, F.; Hou, M.; Shi, W.; Fu, J.; Yu, H.; Ming, P.; Yi, B. The Effect of Ambient Contamination on PEMFC Performance. *J. Power Sources* **2007**, *166*, 172–176. [CrossRef]
9. Mohtadi, R.; Lee, W.K.; Van Zee, J.W. Assessing Durability of Cathodes Exposed to Common Air Impurities. *J. Power Sources* **2004**, *138*, 216–225. [CrossRef]
10. Chen, M.; Du, C.; Zhang, J.; Wang, P.; Zhu, T. Effect, Mechanism and Recovery of Nitrogen Oxides Poisoning on Oxygen Reduction Reaction at Pt/C Catalysts. *J. Power Sources* **2011**, *196*, 620–626. [CrossRef]
11. Lin, C.-Y.; Hung, W.-T.; Wu, C.-T.; Ho, K.-C. Electrochemical Reduction of NO2 at a Pt/Membrane Electrode—Application to Amperometric $NO_2$ Sensing. *Sens. Actuators B: Chem.* **2009**, *136*, 32–38. [CrossRef]
12. St-Pierre, J.; Jia, N.; Rahmani, R. PEMFC Contamination Model: Competitive Adsorption Demonstrated with NO2. *J. Electrochem. Soc.* **2008**, *155*, B315–B320. [CrossRef]
13. Villamena, F.A. Chapter 2 - Chemistry of Reactive Species. In *Reactive Species Detection in Biology*; Villamena, F.A., Ed.; Elsevier: Boston, MA, USA, 2017; pp. 13–64.
14. Lopes, T.; Chlistunoff, J.; Sansiñena, J.-M.; Garzon, F.H. Oxygen Reduction Reaction on a Pt/Carbon Fuel Cell Catalyst in the Presence of Trace Quantities of Ammonium Ions: An RRDE Study. *Int. J. Hydrog. Energy* **2012**, *37*, 5202–5207. [CrossRef]
15. Ge, J.; St-Pierre, J.; Zhai, Y. PEMFC Cathode Catalyst Contamination Evaluation with a RRDE- Propene and Naphthalene. *Electrochim. Acta* **2014**, *138*, 437–446. [CrossRef]
16. Garsany, Y.; Baturina, O.A.; Swider-Lyons, K.E. Impact of Sulfur Dioxide on the Oxygen Reduction Reaction at Pt/Vulcan Carbon Electrocatalysts. *J. Electrochem. Soc.* **2007**, *154*, B670–B675. [CrossRef]


**Sample Availability:** Samples of the tested MEAS are available from the authors.

molecules

MDPI

*Article*

# Automotive Subzero Cold-Start Quasi-Adiabatic Proton Exchange Membrane Fuel Cell Fixture: Design and Validation

**Antonio O. Pistono and Cynthia A. Rice ***

Department of Chemical Engineering, Tennessee Technological University, Cookeville, TN 38505, USA; aopistono21@gmail.com

* Correspondence: crice@tntech.edu

Academic Editors: Jean St-Pierre and Shangfeng Du

check for updates

Received: 30 January 2020; Accepted: 17 March 2020; Published: 19 March 2020

**Abstract:** Subzero automotive cold-starts of proton exchange membrane fuel cell (PEMFC) stacks require accelerated thermal rises to achieve nominal operating conditions and close-to-instantaneous usable output power. Advances in the material, structure and operational dependence on the balance between the maximum power output and the electrochemical conversion of hydrogen and oxygen into water requires validation with subzero cold-starts. Herein are presented the design and validation of a quasi-adiabatic PEMFC to enable single-cell evaluation, which would provide a more cost-effective option than stack-level testing. At −20 °C, the operational dependence of the preconditioned water content (3.2 verse 6.2) for a galvanic cold-start (<600 mA cm$^{-2}$) was counter to that of a laboratory-scale isothermal water fill test (10 mA cm$^{-2}$). The higher water content resulted in a faster startup to appreciable power output within 0.39 min versus 0.65 min. The water storage capacity, as determined from the isothermal water fill test, was greater, for the lower initial water content of 3.2, than 6.2, 17.4 ± 0.3 mg versus 12.8 ± 0.4 mg, respectively. Potentiostatic cold-starts produced usable power in 0.09 min. The versatility and reproducibility of the single cell quasi-adiabatic fixture avail it to future universal cold-start stack relevant analyzes involving operational parameters and advanced materials, including: applied load, preconditioning, interchanging flow field structures, diffusion media, and catalyst coated membranes.

**Keywords:** proton exchange membrane fuel cells; subzero cold-starts; automotive; isothermal water fill tests

## 1. Introduction

Automotive proton exchange membrane fuel cell (PEMFC) stacks are required to withstand the same environmental extremes, including subzero temperatures, as the internal combustion engine. The U.S. Department of Energy's 2020 automotive PEMFC requirements are survivability from −40 °C and cold-starts from −30 °C. A PEMFC subzero cold-start is defined as the initiation of PEMFC operation to meet the required nominal operating temperature and power. As of 2015, the −20 °C cold-start target of 0.5 min to 50% rated power has only been met for a PEMFC stack when using one-and-a-half times the targeted parasitic shutdown/start-up energy [1]. A PEMFC stack is comprised of as many as 400 non-reactive repeat units (flow fields, coolant channels, and current collectors), each encasing the membrane electrode assembly (MEA) component. These non-reactive components behave as thermal sinks, scavenging generated heat during cold-starts. Presently, a common energy-intensive strategy is to use resistive heating to cold-start a PEMFC stack [2]. At nominal operating temperatures, a fine balance is maintained between the rate of water production and evaporative removal from the PEMFC stack. The process of subzero cold-starting of a PEMFC stack is challenging in that a balance must be attained between the rate of heat generation and product-water redistribution. At subzero temperatures, product-water accumulation and ice formation result in mass transport losses that can lead to failure if the PEMFC stack does not self-heat to above 0 °C before oxygen is completely blocked from the accessible reaction sites. As cathode catalysts and catalyst layers advance and become more efficient to meet cost targets, more parasitic power is required to self-heat non-reactive components and in addition to increased material mechanical performance/durability issues due to subzero operation. Therefore, optimization and material validation require a single-cell rapid testing platform.

Subscale single PEMFC cold-starts in standard laboratory fixtures are limited by the thermal mass of the endplates. The testing of short-stacks, of 20–30 repeat units, is cost prohibitive and strongly influenced by performance losses since the endplates behave as thermal sinks. According to the literature, the dominant subzero PEMFC studies investigating the influence of material and operational parameters have been restricted to subscale single-cell freeze-thaw testing [3–14], isothermal water fill tests [15–37], and non-isothermal water fill tests [3,21,33,38–47]. United Technology Corporation (UTC) Power and its corporate research facility co-developed two quasi-adiabatic single PEMFC fixtures in the mid-2000s with geometric active areas of 25 cm$^2$ and 320 cm$^2$ [48–50]. The results demonstrated that it was possible under a galvanic load cold-start to replicate the center cells in a stack's voltage and thermal profile in the quasi-adiabatic fixture. Balliet and Newman validated their two-dimensional liquid water transport cold-start model to the UTC-Power's quasi-adiabatic PEMFC fixture performance profiles [49]. However, the inadequate structural integrity of the quasi-adiabatic PEMFC fixture limited its reusability. The present co-author, cited in references [48,50], changed her surname after these studies were published. Published stack results are minimal and limited mostly to modeling [51–56].

Subzero PEMFC cold-starts are possible due to non-frozen water found in the membrane and catalyst layers of the MEA that support proton conduction ($\sigma_{H^+}$) and rapid exothermic heat generation [57,58]. Ohmic heat generation contributes to cold-start performance and is dependent on the initial water concentration. A portion of water in the ionic domains of the membrane and catalyst layers remains non-frozen at subzero temperatures due to colligative and supercooling effects [59]. Liquid water is retained within the catalyst//ionomer aggregate interfaces due to attractive forces of the charged ionomer end-chain sites ($-SO_3^-$), allowing interconnected transport through the agglomerates between the membrane and catalyst layers. The vapor-saturated water content (λ) at the aggregate interfaces is typically less than 14 and reaches a maximum of 22 for liquid saturated [54,60]. The hydrogen fuel supplied to the anode catalyst layer is electrooxidized to protons and electrons, $H_2 \rightarrow 2H^+ + 2e^-$. The protons are transported through the hydrated ionic domains from the anode layer through the membrane to the cathode layer. In the presence of supplied oxygen, protons and electrons recombining within the cathode catalyst layer to form product water and heat, $O_2 + 4H^+ + 4e^- \rightarrow 2H_2O$.

Herein, a single cell quasi-adiabatic PEMFC fixture was designed to be structurally engineered for reproducible subscale cold-starts. Subzero isothermal water fill tests are shown to inadequately advance the understanding of the operational impact of the initial water content ($\lambda_{initial}$) on automotive stack relevant cold-starts. At −20 °C, isothermal water fill tests under 10 mA cm$^{-2}$ applied loads were compared to cold-starts with loads set to 600 mA cm$^{-2}$.

## 2. Results and Discussion

### 2.1. Quasi-Adiabatic PEMFC Fixture Design

The design constraints required for a single-cell subzero cold-start PEMFC fixture are (i) thermal isolation of the flow fields and MEA from the endplates, (ii) impervious humidified gas manifolds, (iii) structural uniformity of the active area under axial load, and (iv) high electrical conductivity between the anode and cathode sides of the PEMFC through an external circuit. Material compatibility issues of the multi-layered testing PEMFC fixture are exacerbated due to frequent thermal cycling between subzero temperatures and up to 80 °C. Figure 1 shows one side of the symmetric hardware (flow fields, heater, gas and coolant manifold, insulation, and end plates) used in the quasi-adiabatic fixture. The geometric active surface area of the quasi-adiabatic fixture was scaled down from the prototypic cell size found in automotive stacks (16 versus 320 $cm^2$ active area) to simplify development and conservatively minimize any expected issues with flow field deflection; however, it is expected the active area and other geometries are scalable to develop an optimized fixture.

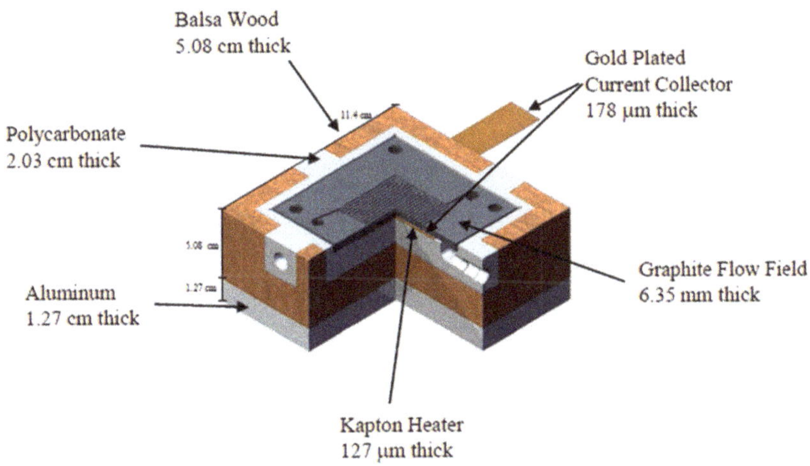


**Figure 1.** One half of symmetric quasi-adiabatic single-cell hardware. The parallel coolant channels are machined into the backside of the graphite flow field plate.

2.1.1. Humidified Gas Manifolds

The nature of material properties makes porous structures more thermally insulated, thus satisfying constraint (i) above, but failing constraint (ii) that requires humidified gas containment. Therefore, two distinct layers were required to insulate and distribute humidified gasses to and from both sides of the MEA. Several ridged, high-density plastics were considered for the humidified gas manifold, see Table 1. All the plastics presented in Table 1 have acceptable high densities (>1200 kg $m^{-3}$) for excluding $H_2$ leakage and sufficiently high temperature limits (>121 °C) for thermal stability within the operating range of a PEMFC (−40 °C ↔ 90 °C). The manifold material must be machinable and not brittle for threaded fittings to connect humidified inlet and outlets. The ideal plastic would have a high compressive strength and compressive modulus to maintain uniformed axial load across the PEMFC fixture during thermal cycling and subsequent rebuilds (constraint (iii)), while having low thermal conductivity (constraint (i)) to retain heat generated by the MEA during PEMFC operation. An additional requirement of constraint (ii) is that the manifold material would not allow water adsorption, as it would freeze, fracture the material, and cause structural failure below 0 °C. Table 1 highlights the maximum and minimum material property values for the high-density plastics considered, underlined and underlined in shaded gray box, respectively. Only materials with low water adsorption (<1%) were considered to ensure the structure integrity of the fixture. UTC-Power adiabatic fixtures used high-density polyamide-imide (Pyropel-HD) as the internal gas manifold material with the maximum compressive modulus of all the plastics from Table 1. However, due to cost and embrittlement issues that made the inlet and outlet connector junctions prone to stress breakage, it was not selected. The manifold material was selected from the remaining plastics in Table 1 by optimizing the lowest range for both thermal conductivity and water uptake. Polyvinylidene fluoride was not selected although it had the lowest water uptake (0% of lower limit in range), as the thermal conductivity was on the higher end (47.8% of lower limit in range). Polycarbonate (PC 1000) was selected as the internal gas manifold material for the fixture as it had the lowest combination of thermal conductivity (11.0% of lower limit in range) and water uptake (8.9% of lower limit in range).

**Table 1.** High-density polymers considered for the gas manifold of the quasi-adiabatic PEMFC fixture. The highest and lowest material properties values of each type are underlined and the highest is further identified by a shaded gray box. Not available (N/A).

| Polymer Type | Trade Name | Supplier | Density (kg m$^{-3}$) | Maximum Operating (°C) | Compressive Strength (MPa) | Compressive Modulus (GPa) | Thermal Conductivity (W m$^{-1}$ K$^{-1}$) | Water Uptake (%) |
|---|---|---|---|---|---|---|---|---|
| Polyvinylidene fluoride | Symalit unfilled | Quadrant | *1780* | 149 | *68.9 (10% def)* | *1.1* | 0.216 | *0.05* |
| Polyimide | D7000 PI | Quadrant | 1380 | 240 | *145 (5% nom strain)* | N/A | 0.22 | *4* |
| Polyethersulfone | PES | Westlake Plastic | 1370 | N/A | 100 | 2.68 | 0.239 | 1.85 |
| Polyamide-imide | Pyropel-HD | Albany International | 1360 | *288* | N/A | *3.72* | 0.23 | Est 0.58 |
| Polyaryletherketone | AV-848 | Solvay AvaSpire | 1320 | N/A | 118 | N/A | 0.22 | 0.5 |
| Polyetherimide | Duratron U1000 | Quadrant | 1280 | 171 | 152 (10% def) | 3.31 | *0.177* | 1.25 |
| Polysulfone | PSU 1000 | Quadrant | 1240 | 149 | 89.6 (10% def) | 2.59 | *0.259* | 0.6 |
| Polycarbonate | PC 1000 | Quadrant | *1200* | *121* | 79.3 (10% def) | 2.07 | 0.186 | 0.4 |

### 2.1.2. Insulation

Several porous semi-ridged materials were considered to promote retention of the heat generated by the MEA, as listed in Table 2. The density and thermal conductivity of the materials were an order of magnitude lower than that of the humidified gas manifold, thus improving the insulating properties of the material. All the materials presented in Table 2 have acceptably low densities (< 480 kg m$^{-3}$) that allow them to be insulative and sufficiently high temperature limits (> 149 °C) that enhance thermal stability within the range of PEMFC operation (−40 °C ↔ 90 °C). However, the decreased density made the material more porous and susceptible to entrainment of condensed water from the interior of the environmental chamber. The first three materials listed (polyisocyanurate, cellular glass, and calcium silicate) are common insulators with low compressive strength, ranging from 0.2 ↔ 0.7 MPa, and low thermal conductivity, averaging around 0.043 ± 0.022 W m$^{-1}$ K$^{-1}$. However, their open structure leads to high water uptake that would result in structural deformation nonuniformly altering the axial load across the active cell area, thus negatively impacting sealing of the manifold and electric continuity. The subsequent materials were of two distinctly different types: synthetic nonwoven fibrous polyimide and naturally grown balsa wood. Balsa wood is a common low-cost, low-weight construction material used for applications such as aircraft construction. In each of these families of materials, the compressive strength increases with density, and the thermal conductivity adversely increases as well. The natural strength of balsa woods is due to the multilayering of primary and secondary walls, forming a randomly distributed fiber-reinforced composite that resists out-of-plane deformation [61].

**Table 2.** Semi-ridged materials considered as insulation of the quasi-adiabatic PEMFC fixture. The highest and lowest material properties values of each type are underlined and the highest is further identified by a shaded gray box. Not available (N/A).

| Material | Trade Name | Supplier | Density (kg m$^{-3}$) | Maximum Operating (°C) | Compressive Strength (MPa) | Compressive Modulus (GPa) | Thermal Conductivity (W m$^{-1}$ K$^{-1}$) |
|---|---|---|---|---|---|---|---|
| Polyisocyanurate | TRYMER 2000 XP | ITW Insulation Systems | *32.8* | 149 | 0.16«0.21 | *0.003 << 0.005* | *0.027* |
| Cellular Glass | FOAMGLAS ONE | Pittsburg Corning | 117 | 482 | 0.62 | 0.9 | 0.032«0.054 |
| Calcium Silicate | Thermo-12 Gold | Johns Manville | 230 | *1200* | 0.690 (5% def) | N/A | 0.053«0.058 |
| Nonwoven Polyimide | Pyropel MD-12 | Albany International | 190 | 288 | *0.07* | 0.006 | 0.036 |
| Nonwoven Polyimide | Pyropel MD-18 | Albany International | 290 | 288 | 0.1 | 0.015 | 0.041 |
| Nonwoven Polyimide | Pyropel MD-30 | Albany International | *480* | 288 | 0.41 | 0.1 | 0.049 |
| Balsa | SB.50 | AIREX AG BALTEK | 109 | 163 | 5.5 | 1.6 | 0.048 |
| Balsa | SB.100 | AIREX AG BALTEK | 148 | 163 | 9.2 | 2.5 | 0.066 |
| Balsa | SB.150 | AIREX AG BALTEK | 285 | 163 | 22 | 4.4 | *0.084* |

The percentage of water uptake was evaluated for both Pyropel and balsa wood by immersion in water. Immersion was a severe scenario as the only source of water that could contact the insulating material would be condensed water from the environmental chamber. Water uptake was found to be 245% over the dry weight for Pyropel MD-18, Figure 2. The as-received balsa wood water uptake was 50%, compared to the initial mass. Sealing the wood surface with a thin coating of polyurethane reduced the water uptake to only 14%. Coated balsa wood was selected as the insulating material due to the combination of relatively high compressive strength, low water uptake, and low thermal conductivity.

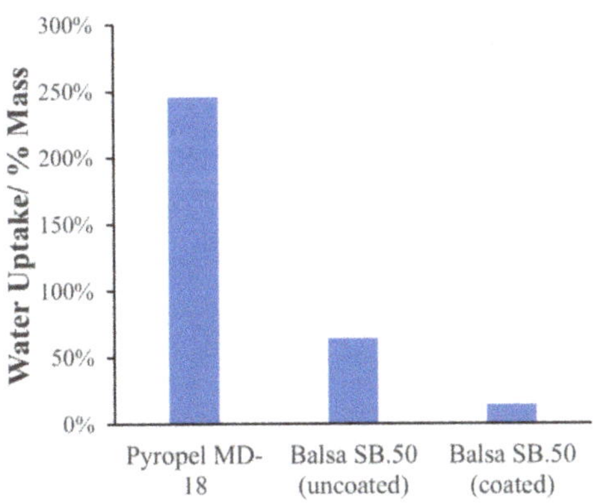


**Figure 2.** Water uptake of down-selected insulation materials for 15-min water immersion.

2.1.3. Uniformity of Applied Axial Load

The peripheral manifolds and insulation parts of the assembly surrounding the MEA must apply uniform axial load across the faces of the flow fields to retain gasses and ensure electrical continuity between the catalyst layers and the current collectors. To compare the quasi-adiabatic fixture's compression uniformity across the MEA with the standard fixture used in previous studies (Figure 3), pressure paper was used instead of the membrane and catalyst layers. Note that the vertical line in Figure 3a is an artifact from the pressure paper and should be disregarded. The standard

PEMFC fixture had a flow field geometry identical to that of the quasi-adiabatic PEMFC fixture. The compression of the quasi-adiabatic fixture (Figure 3a) was mostly uniform across the face of the flow fields, but intensity was less than that of the standard fixture (Figure 3b). The torque was not increased above 70 in·lbs due to concerns with deflection of the endplates compromising the contact in the center of the MEA.

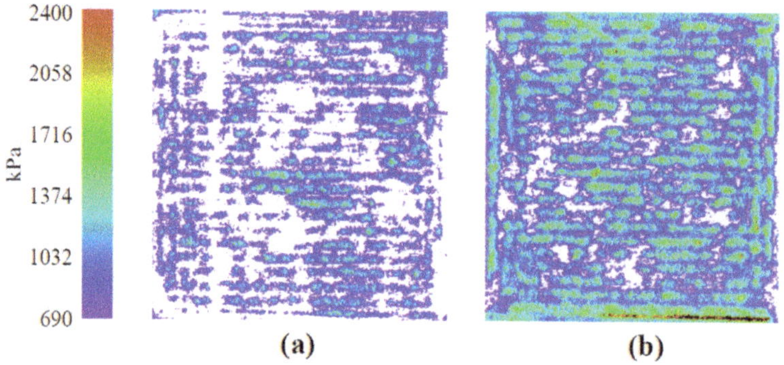


**Figure 3.** Compression paper for contact intensity and uniformity between the 16 $cm^2$ flow fields (**a**) quasi-adiabatic fixture and (**b**) standard fixture. Each fixture was torqued to 70 in·lbs.

## 2.2. *PEMFC Testing*

Use of both the quasi-adiabatic fixture and a standard fixture allowed for characterization of the MEA at normal operating conditions (both fixtures), cold-starts from −20 °C, and isothermal water fill tests at −20 °C.

### 2.2.1. Operating Performance

In Figure 4, the conditioned PEMFC $H_2$/Air polarization curves of the quasi-adiabatic and a standard PEMFC fixture for the same MEA, black and gray, respectively, are compared. The 76 m$\Omega$ $cm^2$ additional resistance of the MEA in the quasi-adiabatic fixture (Figure 4a) is related to the compression issues shown in Figure 3. The cell voltages are corrected using current interrupt resistance (iR-corrected) to compensate for electronic resistance losses due to the reduced axial loading of the quasi-adiabatic PEMFC fixture. The iR-corrected polarization profiles are nearly identical in the ohmic and mass transfer regions. The resistance of the quasi-adiabatic PEMFC fixture is 45% greater than that of the standard fixture.

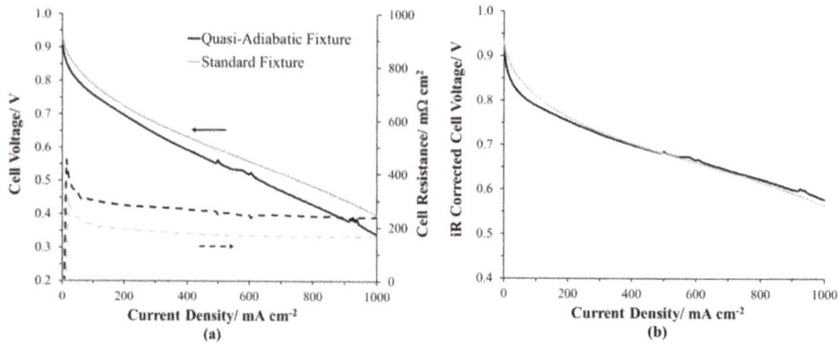


**Figure 4.** $H_2$/21% $O_2$ polarization curves at 80 °C cell temperature and 44.3% relative humidity under ambient pressure for the quasi-adiabatic fixture and a standard fixture (**a**) cell voltage and (**b**) compensated iR free cell voltage.

### 2.2.2. Water Fill Tests

There are five stages to a low-applied load water fill test: (i) initial supply of reactant gasses elevating the cell voltage, (ii) hydration of the interconnected ionomer domains of the MEA once the initial load is applied, (iii) maintenance of quasi-steady-state cell voltage during the attainment of maximum ionomer hydration, (iv) filling of the large non-hydrophilic pores of the cathode catalyst layer reducing $O_2$ mass transport, and (v) freeze-out due to ice blockage in the cathode catalyst layer [18]. During these water fill tests, the product water was restricted from entering the diffusion media under the applied test conditions of low water-production rate, low-heat generation rate, and minimal flow rates (no convective transport of water). Under an applied load of 10 mA cm$^{-2}$, the water storage capacity was evaluated for two different pre-conditioned $\lambda_{initial}$(3.2 and 6.2) at −20 °C. The $\lambda_{initial}$ was selected to match multiple published water fill tests—the lower setpoint ($\lambda_{initial}$ = 3.2) represents a dry PEMFC scenario, while the higher setpoint ($\lambda_{initial}$ = 6.2) is closer to an operating PEMFC. Both $\lambda_{initial}$ condition profiles had an initial jump in voltage although the high $\lambda_{initial}$ resulted in a higher initial cell voltage (Figure 5a) due to a lower initial resistance (Figure 5b). The maximum cell voltage was reached at similar times (>1 min) and values (0.81 V). The run with the lower $\lambda_{initial}$ stayed in the quasi-steady-state cell voltage stage >5 min longer due to a high ionomer fill capacity caused by starting at a lower ionomer hydration level and water movement from the ionomer caused by resistive heating [15]. The freeze-out stage was identical for both pre-conditioning hydration levels. The water storage capacity, calculated using Faraday's Law, of the higher $\lambda_{initial}$ of 6.2 preconditioned water fill tests was only 12.8 ± 0.4 mg while that of the $\lambda_{initial}$ runs of 3.2 was 17.4 ± 0.3 mg.

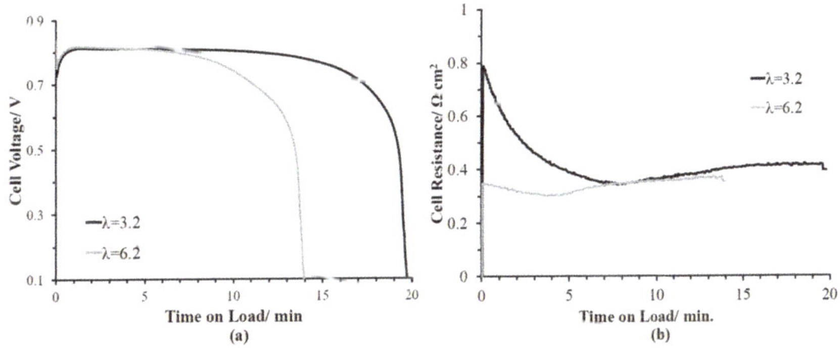


**Figure 5.** Preconditioned initial water content runs (3.2 and 6.2) isothermal water fill test at 10 mA cm$^{-2}$, −20 °C and $H_2$/21% $O_2$ (0.05/0.1 lpm, respectively). (**a**) Cell voltage and (**b**) cell resistance versus time on load.

### 2.2.3. Cold-Starts

The impact of adjacent cell heating, $\lambda_{initial}$, and galvanic versus potentiostat applied loads were investigated on −20 °C cold-starts. The galvanically applied load of 600 mA cm$^{-2}$ was selected to match the work published by Balliet and Newman on UTRC's quasi-adiabatic PEMFC fixture [49]. The applied galvanic load establishes the PEMFCs maximum attainable current density with the cell voltage approaching 0 V. In a PEMFC stack, it is common for the end cells to reach negative voltages during the first few seconds of a subzero cold-start because the overall voltage of the stack is positive. Table 3 summarizes the cold-start conditions investigated within this study: $\lambda_{initial}$, adjacent cell heat-adjustment factor, and applied load. The heat-adjustment factor was included to supplement heat that would be provided from neighboring cells in a stack [49,51], as well as thermal losses in the quasi-adiabatic fixture. The heat comes from resistive heating pads located adjacent to the coolant loops of the flow fields enshrouded by the balsa wood. Each type of cold-start was preformed twice to ensure reproducibility. Representative cold-start profiles are shown in Figure 6.

Figure 6a,b compare the power density and cathode flow channel temperature versus time, respectively, for the four types of cold-start presented herein. The power densities were initially low for all of the galvanic cold-starts while the potentiostatic start instantaneously had power because the cell voltage was maintained above 0 V. For all cold-start conditions reported herein, the current density increased with time on load at subzero temperatures until the set point of 600 mA $cm^{-2}$ could be supported by the cell voltage. Once the current density exceeded the set point, it was adjusted back down by the Scribner fuel cell software, allowing the cell voltage to rise to higher values. The current density improves as the temperature of the MEA increases, due to increased reaction kinetics and proton conduction through the ionomer. To correlate the subzero dependent current density and cell voltage response during a cold-start, the time scale origin was positioned such that it corresponded with the time the cell temperature reached 0 °C, as shown in Figure 6c. The cell-resistance profiles, proportional to proton conduction, were similar for all the cold-starts (Figure 6d) due to similar ionomer water contents, with the exception of the lower adjacent cell heating adjustment factor of 1×.

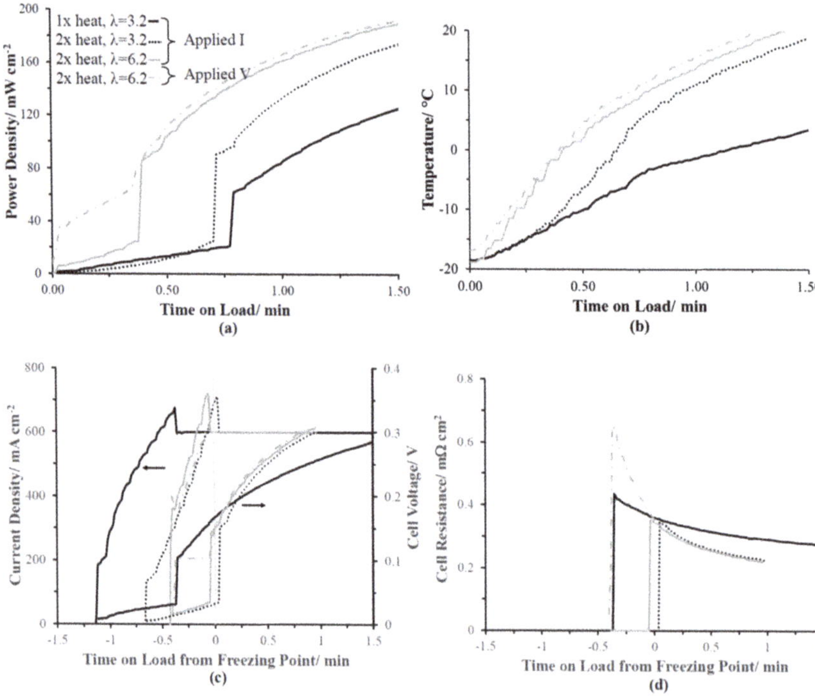


**Figure 6.** Preconditioned initial water content runs (3.2 and 6.2), cold-start tests under set galvanic load of 600 mA $cm^{-2}$, or applied potentiostatic hold of 0.1 V at −20 °C and $H_2$/21% $O_2$ (0.5/0.75 lpm, respectively). (**a**) Power density and (**b**) cathode channel temperature versus time on load. (**c**) Current density and (**d**) cell resistance vs. time on load from 0 °C.

### 2.2.4. Heat Adjustment Factor

In Figure 6, doubling the predicted adjacent cell heating adjustment factor (1× → 2×) significantly impacted the cold-start response profile for an $\lambda_{initial}$ of 3.2 is shown. The output of the heating pads (Q in Watts) was scaled by a multiplier of either 1 or 2 to equation 1 to compensate for the load-dependent fraction of heat that would be lost under applied load (current (i) in Amps) times the overpotential calculated from the difference between the thermoneutral voltage (1.48 V) and the cell voltage (V in volts) [49].

$$Q = Ai(1.48 - V) \quad (1)$$

where A is the geometric surface area of the PEMFC.

The resulting key cold-start performance metrics are summarized in Table 3 and include the average of both runs with standard deviation. The initial applied current density increased by a factor of 3.75 for the higher heating adjustment factor. The non-zero rise in voltage for the 1× heating adjustment factor occurred at a cathode flow field channel temperature around −3.1 °C, while for 2×, the transition was near 4 °C. The thermocouple point of contact is unknown within the cathode flow field channel and most likely a combination of the solid flow field temperature and the exterior of the cathode diffusion media. The mass of the flow field channels acts as a heat sink, reducing the internal temperature of the cathode flow channel, and hence, yielding the negative non-zero temperature transition for the lower heating adjustment factor. The increased heat adjustment factor suggests that a lower non-reactive thermal mass and lower thermal mass would yield a more successful cold-start. The sluggish heating profile of the lower heating adjustment factor increased the required cold-start time until usable power was available. After 1 min into the cold-start, the lower heating adjustment factor (1×) power density output was only 35% compared to the 2× cold-start (Figure 6a).

**Table 3.** Cold-start parameters used in tests from Figure 6 and corresponding select performance metrics.

| | | | | |
|---|---|---|---|---|
| | **Cold-Start Parameters** | | | |
| Heat Adjustment Factor | 1× | 2× | 2× | 2× |
| Initial Water Content | 3.2 | 3.2 | 6.2 | 6.2 |
| Applied Load | 600 mA cm$^{-2}$ | 600 mA cm$^{-2}$ | 600 mA cm$^{-2}$ | 0.1 V |
| | **Cold-Start Performance Metrics** | | | |
| Initial Current Density (mA cm$^{-2}$) | 47 ± 15 | 129 ± 8 | 361 ± 3 | 402 ± 3 |
| Temperature of Cell Voltage Rise (°C) | −3.1 ± 0.1 | 4.0 ± 0.6 | −0.9 ± 1.0 | −1.8 ± 1.0 |
| Time to Cell Voltage Rise (min) | 0.78 ± 0.01 | 0.65 ± 0.09 | 0.39 ± 0.01 | 0.38 ± 0.01 |
| Time to Power > 40 mA cm$^{-2}$ (min) | 0.78 ± 0.01 | 0.65 ± 0.09 | 0.39 ± 0.01 | 0.09 ± 0.01 |
| Time to 20 °C (min) | 5.6 ± 0.1 | 1.6 ± 0.1 | 1.4 ± 0.1 | 1.4 ± 0.1 |

Other heating adjustment factors could be used to match other cell designs and materials. The heating pads output is an independent variable enabling the simulation of inner stack cells (symmetric heating case) or end cells (asymmetric heating case). Anomalous cells, arising from partially blocked coolant/flow field channels or degraded materials, can also give rise to other asymmetric heating cases for contiguous cells. Validation of heating adjustment factors with stack data requires significant resources and is not trivial because in-situ heating fluxes for all cells in a stack due to heterogeneous components and locally variable heat fluxes. Even if heating adjustment factors are empirically matched using single cell and stack data, a significant amount of work and resources are still required. For these reasons, validation of the quasi-adiabatic fixture with stack data was deemed outside the scope of this report.

### 2.2.5. Initial Water Content

Increasing the $\lambda_{initial}$ from 3.2 to 6.2, using a heating adjustment factor of 2×, improved the cold-start performance; however, this result was counter to the two-thirds higher isothermal water storage capacity results (Figure 5) for the lower $\lambda_{initial}$ 3.2. For the higher $\lambda_{initial}$ found upon initially applying the load, the cell voltage could sustain nearly double current density because of the more optimal distribution of interconnected non-frozen water domains within the ionomer previously quantified with subzero electrochemical impedance spectroscopy in Dr. Rice's lab [15]. The measured cold-start time until the voltage increased on average was reduced from 0.65 min to 0.39 min for the higher $\lambda_{initial}$, translating to appreciable power densities sooner.

#### 2.2.6. Applied Load

The type of applied load controlled the onset of appreciable power densities during the initial phase of the cold-start. Jiang and Wang demonstrated that potentiostatic cold-starts maximized the heat output [47]. The potentiostatic hold of 0.1 V multiplied the current density to get instantaneous power densities. The initial current density was 1.8× greater than that of the galvanic applied load under identical conditions.

## 3. Materials and Methods

### *3.1. PEMFC Assembly*

Tests were performed in two different symmetric subscale PEMFC fixtures with an active area of 16 $cm^2$. The proton-conducting membrane used was Nafion HP (Ion Power, New Castle, DE, USA). The anode and cathode catalyst layers were directly sprayed (Badger Airbrush 150) onto the membrane with a final loading of ~ 0.4 $mg_{Pt}$ $cm^{-2}$ (46.6% Pt on high surface area carbon, Tanaka, Chiyoda-ku, Tokyo) and 30 wt% Nafion (1100EW, Ion Power). The microporous side of the hydrophobic gas diffusion layers (SGL25BC, Ion Power) were positioned adjacent to the catalyst layers. The symmetrically sandwiched gas diffusion layers and catalyst layers around the membrane comprise the membrane electrode assembly (MEA). Polytetrafluoroethylene films (Interplast) sealed the perimeter of the compressed gas diffusion layers against the flow fields.

Figure 1 shows one side of the symmetric hardware (flow field, heater, gas and coolant manifold, insulation, and end plates) used in the quasi-adiabatic fixture. The build layup and dimensions are summarized in Table 4. Two dual-sided flow fields/coolant channels were machined out of graphite (BMC-940, MetroMold, Rogers, MN, USA), the flow fields were comprised of opposing triple serpentine channels (width 0.75 mm, depth 1 mm, and land/channel ratio 1.5) to provide reactant transport, and parallel coolant channels (width 2.54 mm, depth 1.52 mm, and land/channel ratio 1) assisted thermal management through heat generated at nominal operating temperatures of the applied load (80 °C circulating 60% ethylene glycol/40% water, Isotemp 9500, Fisher Scientific, Hampton, NH, USA).

**Table 4.** Symmetric lay-up of quasi-adiabatic PEMFC fixture centered around the proton-conducting membrane.

| PEMFC Fixture Component | | | Count | Thickness (cm) | Total Area ($cm^2$) | Seal (cm) |
|---|---|---|---|---|---|---|
| Standard Components | Membrane Electrode Assembly | Membrane (center of build) | 1 | 0.002 | 20.25 | |
| | | Catalyst Layer | 2 | > 0.001 | 16 | |
| | | Gas Diffusion Layer (uncompressed) | 2 | 0.0235 | 16 | 0.0152 |
| | | Flow Field | 2 | 0.635 | 62.4 | |
| | | Electrical Contactor (uncompressed) | 2 | 0.019 | 16 | 0.0102 |
| | | Current Collector | 2 | 0.0178 | 25.8 | |
| Adiabatic Portion | | Heating Pad | 2 | 0.0127 | 25.8 | |
| | | Manifold | 2 | 2.03 | 84.5 | |
| | | Insulator | 2 | 5.08 | 130.6 | |
| | | Endplate | 2 | 1.27 | 130.6 | |

The standard portions of the fuel-cell fixture used in both the standard and quasi-adiabatic fixtures had current collectors (gold-plated copper, electroplated in-house) compressing the flow fields. To maintain electrical continuity between the flow fields and the current collector at the non-reactive interfaces, a compressed non-hydrophobic SGL25AA was placed in the window of the polytetrafluoroethylene seal. Kapton-encased resistive heating arrays (Omega Engineering, KH-608/5-P, Norwalk, CT, USA) were positioned near the coolant side of the flow fields. Aluminum end plates (6061-T6) external to the quasi-adiabatic portion of the fixture and stainless-steel bolts torqued to 40 in-lbs. were used to maintain uniform electrical contact and force across the MEA. The quasi-adiabatic portion of the fixture had internal gas manifolds made of polycarbonate (Quadrant EPP PC 1000,

Reading, PA, USA) and were insulated from the aluminum endplates with spray-polyurethane sealed Balsa wood (Specialized Balsa Wood, LLC, Loveland, CO, USA).

### 3.2. Instrumentation

A Scribner 850e fuel cell test system was the central control unit for the PEMFC testing presented herein. The system monitored cell voltage, temperature and high frequency cell resistance, while establishing reactant gas flow with specific relative humidities (RH), applied load, and isothermal temperature. A Labview program and supporting hardware were used to monitor test station/software communication and perform the necessary actions for the freezing and cold-start sequence. The thermocouple used to monitor the PEMFC temperature was a flexible ultra-fine (insulated 0.24 mm diameter) designed for in-vivo applications (T-type, Physitemp IT-24P) with an accuracy of ± 0.1 °C and located in the cathode flow field channel. Sub-zero temperatures were established using the Isotemp 9500 lab chiller and a ScienTemp 43–1.7 chest freezer equipped with a bulkhead fitting to allow electrical and feed/exit line connections. The membrane resistance was monitored under non-applied load conditions using a Milliohm meter (Agilent Technologies, 4338B, Santa Clara, CA, USA). During cold-starts, the Labview program monitored the current and voltage measured by the test station to emulate adjacent cell heating. The heating pad output was set to be a multiple of the heat that would be generated from adjacent cells in a stack (1× and 2×) and was controlled by two independent-phase angle fired controllers (Eurotherm Corp., Model-984, Worthing, United Kingdom).

### 3.3. Materials Characterization

Water uptake tests were performed on both the manifold and insulation materials by immersing approximately 10 g cubic samples in water at room temperature for 15 min and evaluating mass increases. Contact uniformity under axial load was evaluated using compression paper (super low, Fujifilm, 0.5–2.5 MPa) instead of the membrane and catalyst layers between the flow fields.

### 3.4. PEMFC BOL Conditioning

At the beginning of life (BOL), to hydrate and activate the PEMFC, 10 cathode potential cycles were run at 80 °C (gas feed dew points 75% RH) by maintaining the anode potential at 0 V vs. DHE (100% $H_2$, 0.75 slpm) and varying the cathode potential by switching between 100% $N_2$ (~0.12 V, 1.5 slpm) and air (>0.9 V, 21% $O_2$ in a $N_2$ balance, 1.5 slpm). The RH of the PEMFC feed steams was calculated from the due points ($T_{DP}$) of the saturators for the specific cell temperature (T) according to the August-Roche-Magnus approximation [62] (Equation (2)):

$$\%RH = 100\%\left(\frac{exp^{\left(\frac{17.625T_{DP}}{T_{DP}+243.04}\right)}}{exp^{\left(\frac{17.625T}{T+243.04}\right)}}\right) \tag{2}$$

Then, $H_2$/Air polarization curves were performed from open circuit to 0.3 V until the voltage response profile stabilized. Between all polarization curves, the accumulated surface oxides on the cathode surface were reduced in the presence of $N_2$ to remove surface oxides.

### 3.5. Freeze Pre-Conditioning

The initial water content ($\lambda_{initial}$) was reestablished prior to each subzero test by (i) repeating five $H_2$/Air and $H_2/N_2$ potential cycles at 80 °C (45% RH), (ii) two $H_2$/Air polarization curves at 80 °C (45% RH), and (iii) establishing the equilibrium $\lambda_{initial}$ by purging the cell with symmetric $N_2$ (0.75 slpm) at either 45 °C (45% RH) or 35 °C (75% RH) for >18 h. After the equilibrium purge, the gas feed/exit lines were closed, an electrical shorting strap was placed across the anode and cathode to protect the cathode from high carbon corrosion potentials (>0.6 V), and the cell was frozen to −20 °C. The $\lambda_{initial}$'s were calculated from the feed %RH's using the equation developed by Hinatsu et al. [63] (Equation (3)):

$$\lambda_{initial} = 14.1\left(\frac{\%RH}{100}\right)^3 - 16\left(\frac{\%RH}{100}\right)^2 + 10.8\left(\frac{\%RH}{100}\right) + 0.3 \quad (3)$$

The $\lambda_{initial}$ values used within this study were 3.2 and 6.2 (45% RH and 75% RH, respectively). Prior to all subzero testing, the coolant was purged from the PEMFC coolant channels.

### 3.6. Water Fill Tests

After the completion of all cold-start variations, the MEA was removed from the quasi-adiabatic fixture and installed in a reference PEMFC fixture, and then subzero isothermal water fill tests were performed at −20 °C [15]. Initially, the open circuit voltage was established in the presence of $H_2$/Air (0.05/0.10, 0% RH), on the anode and cathode, respectively. A small constant load of 10 mA $cm^{-2}$ was applied until the cell voltage dropped below 0.1 V. Runs were repeated 2–3 times to ensure accuracy.

### 3.7. Cold-Starts

Subzero cold-starts were performed at −20 °C in the quasi-adiabatic PEMFC fixture. Initially, the open-circuit voltage was established in the presence of $H_2$/Air (0.5 slpm/0.75 slpm, 0% RH), on the anode and cathode, respectively. Under applied load using the upper set point value of 600 mA $cm^{-2}$, the stoichiometry was never less than 2. The applied load was controlled either galvanically or potentiostatically. The galvanic loads were ramped up to the set point in less than 1 min as the non-negative PEMFC voltage could sustain higher currents. The potentiostatic hold was initially set to 0.1 V. The output of the heating pads (Q in Watts) was scaled by a multiplier of either 1× or 2× to Equation (1). Runs were repeated 2–3 times to ensure accuracy.

## 4. Conclusions

Single-cell, −20 °C cold-starts were attained in a quasi-adiabatic fixture, consisting of polycarbonate gas manifolds and balsa wood insulation. This fixture used heating pads placed on the exterior of the internal flow fields to simulate the anticipated heat from adjacent cells in a stack. A 2× heating factor was used due to adjacent cell heating and thermal losses from the flow-field mass. The quasi-adiabatic single-cell fixture can emulate the thermal temperature rise and product water redistribution during cold-starts. Only a limited number of published, stack-level cold-start results, restricted mostly to simulations, are presented in the literature. The majority of the published subzero PEMFC testing is done on single cells and quantifies the water fill capacity before freezeout using a water fill test. The results presented herein succinctly demonstrate the inadequacies of the commonly used lab scale isothermal water fill tests in validating operational and material subzero cold-start capabilities. The higher rate of water production during the galvanic cold-starts (600 mA $cm^{-2}$) showed maximum hydration of the membrane within less than 2 min in contrast to the 4–8 min required in the isothermal water fill test (10 mA $cm^{-2}$). As the internal cell temperature rose above 0 °C during a cold-start, nearly 20 mg of water were produced. However, for the isothermal water fill test, the highest water fill capacity ($\lambda_{initial}$ = 3.2) was only 17.4 mg. The higher $\lambda_{initial}$ of 6.2 had a lower isothermal water storage capacity than that of 3.2, but conversely, a galvanic cold-start resulted in a shortened time to usable power. The potentiostatic cold-start (0.1 V) provided useful power immediately, resulting in superior cold-start performance.

**Author Contributions:** Conceptualization was originally provided by C.A.R. stemming from her work at United Technologies Corporation and material selection was led by A.O.P. Both A.O.P. and C.A.R. developed the methodology, while A.O.P. further refined the testing. All experimental work was performed by A.O.P. under the direct supervision of C.A.R. Data analysis was initially performed by A.O.P. and further processed by C.A.R. All authors have read and agreed to the published version of the manuscript.

**Funding:** This research received no external funding

**Acknowledgments:** The authors would like to thank Chris Wilson from the Mechanical Engineering Department at Tennessee Tech University for discussion on material compliance, and the Center for Manufacturing Research for facilities and financial assistance.

**Conflicts of Interest:** The authors declare no conflict of interest.

## References

1. Papageorgopoulos, D. Fuel Cells R&D Overview—2018 Annual Merit Review and Peer Evaluation Meeting. Available online: https://www.hydrogen.energy.gov/pdfs/review18/fc01_papageorgopoulos_2018_o.pdf (accessed on 29 January 2020).
2. Gu, J.S.; Shin, H.K.; Oh, K.H.; Jeon, Y.H.; Kim, J.H. End cell heater for fuel cell, and fuel cell including same. U.S. Patent 10,468,696, 5 November 2019.
3. Oszcipok, M.; Riemann, D.; Kronenwett, U.; Kreideweis, M.; Zedda, M. Statistic analysis of operational influences on the cold start behaviour of PEM fuel cells. *J. Power Sources* **2005**, *145*, 407–415. [CrossRef]
4. Wilson, M.S.; Valerio, J.A.; Gottesfeld, S. Low platinum loading electrodes for polymer electrolyte fuel cells fabricated using thermoplastic ionomers. *Electrochim. Acta* **1995**, *40*, 355–363. [CrossRef]
5. Cho, E.; Ko, J.-J.; Ha, H.Y.; Hong, S.-A.; Lee, K.-Y.; Lim, T.-W.; Oh, I.-H. Effects of Water Removal on the Performance Degradation of PEMFCs Repetitively Brought to <0 °C. *J. Electrochem. Soc.* **2004**, *151*, A661–A665.
6. McDonald, R.C.; Mittelsteadt, C.; Thompson, E. Effects of Deep Temperature Cycling on Nafion® 112 Membranes and Membrane Electrode Assemblies. *Fuel Cells* **2004**, *4*, 208–213. [CrossRef]
7. St-Pierre, J.; Roberts, J.; Colbow, K.; Campbell, S.; Nelson, A. PEMFC operational and design strategies for sub zero environments. *J. New Mater. Electrochem. Sys.* **2005**, *8*, 163–176.
8. Guo, Q.; Qi, Z. Effect of freeze-thaw cycles on the properties and performance of membrane-electrode assemblies. *J. Power Sources* **2006**, *160*, 1269–1274. [CrossRef]
9. Hou, J.; Yu, H.; Zhang, S.; Sun, S.; Wang, H.; Yi, B.; Ming, P. Analysis of PEMFC freeze degradation at −20°C after gas purging. *J. Power Sources* **2006**, *162*, 513–520. [CrossRef]
10. Kagami, F.; Hishinuma, Y.; Chikahisa, T. Performance and self-starting of a pefc at temperatures below freezing. *J. Therm. Sci. Eng.* **2002**, *10*, 25–33.
11. Kim, S.; Ahn, B.K.; Mench, M.M. Physical degradation of membrane electrode assemblies undergoing freeze/thaw cycling: Diffusion media effects. *J. Power Sources* **2008**, *179*, 140–146. [CrossRef]
12. Mukundan, R.; Kim, Y.S.; Garzon, F.H.; Pivovar, B. Freeze/Thaw Effects in PEM Fuel Cells. *ECS Trans.* **2006**, *1*, 403–413.
13. Song, W.; Hou, J.; Yu, H.; Hao, L.; Shao, Z.-G.; Yi, B. Sub-freezing endurance of PEM fuel cells with different catalyst-coated membranes. *J. Appl. Electrochem.* **2008**, *39*, 609–615. [CrossRef]
14. Hwang, G.S.; Kim, H.; Lujan, R.; Mukundan, R.; Spernjak, D.; Borup, R.L.; Kaviany, M.; Kim, M.H.; Weber, A.Z. Phase-change-related degradation of catalyst layers in proton-exchange-membrane fuel cells. *Electrochim. Acta* **2013**, *95*, 29–37. [CrossRef]
15. Pistono, A.; Rice, C.A. Subzero water storage capacity in proton exchange membrane fuel cells: Effects of preconditioning method. *Int. J. Hydrogen Energy* **2019**, *44*, 22098–22109. [CrossRef]
16. Pistono, A.; Rice, C.A. The Effect of Material Properties on the Subzero Water Storage Capacity of Cathode Catalyst Layers for Proton Exchange Membrane Fuel Cells. *J. Electrochem. Soc.* **2017**, *164*, F582–F590. [CrossRef]
17. Pistono, A.; Rice, C.A.; Lewis, J.; Ramani, V. Subzero Degradation Analysis of Membrane Electrode Assemblies Modified with Additives. *ECS Trans.* **2013**, *50*, 531–537. [CrossRef]
18. Pistono, A.; Rice-York, C.A.; Boovaragavan, V. Electrochemical Impedance Spectroscopy Detection of Saturation Level in a Frozen Polymer Electrolyte Membrane Fuel Cell. *J. Electrochem. Soc.* **2011**, *158*, B233. [CrossRef]
19. Wang, Y.; Mukherjee, P.P.; Mishler, J.; Mukundan, R.; Borup, R.L. Cold start of polymer electrolyte fuel cells: Three-stage startup characterization. *Electrochim. Acta* **2010**, *55*, 2636–2644. [CrossRef]
20. Macauley, N.; Lujan, R.W.; Spernjak, D.; Hussey, D.S.; Jacobson, D.L.; More, K.L.; Borup, R.L.; Mukundan, R. Durability of Polymer Electrolyte Membrane Fuel Cells Operated at Subfreezing Temperatures. *J. Electrochem. Soc.* **2016**, *163*, F1317–F1329. [CrossRef]

21. Stahl, P.; Biesdorf, J.; Boillat, P.; Friedrich, K.A. An Investigation of PEFC Sub-Zero Startup: Evidence of Local Freezing Effects. *J. Electrochem. Soc.* **2016**, *163*, F1535–F1542. [CrossRef]
22. Stahl, P.; Biesdorf, J.; Boillat, P.; Friedrich, K.A. An Investigation of PEFC Sub-Zero Startup: Influence of Initial Conditions and Residual Water. *Fuel Cells* **2017**, *17*, 778–785. [CrossRef]
23. Biesdorf, J.; Forner-Cuenca, A.; Siegwart, M.; Schmidt, T.J.; Boillat, P. Statistical Analysis of Isothermal Cold Starts of PEFCs: Impact of Gas Diffusion Layer Properties. *J. Electrochem. Soc.* **2016**, *163*, F1258–F1266. [CrossRef]
24. Roth, J.; Zurbrügg, M.R.; Irvine, S.; Marone, F.; Stampanoni, M.; Büchi, F.N. Investigation of PEFC freeze start by X-ray tomographic microscopy. *ECS Trans.* **2013**, *58*, 453–462. [CrossRef]
25. Oberholzer, P.; Boillat, P.; Siegrist, R.; Perego, R.; Kästner, A.; Lehmann, E.; Scherer, G.G.; Wokaun, A. Cold-Start of a PEFC Visualized with High Resolution Dynamic In-Plane Neutron Imaging. *J. Electrochem. Soc.* **2011**, *159*, B235–B245. [CrossRef]
26. Dursch, T.J.; Ciontea, M.A.; Radke, C.J.; Weber, A.Z. Isothermal Ice Crystallization Kinetics in the Gas-Diffusion Layer of a Proton-Exchange-Membrane Fuel Cell. *Langmuir* **2012**, *28*, 1222–1234. [CrossRef] [PubMed]
27. Swamy, T.; Kumbur, E.C.; Mench, M.M. Characterization of Interfacial Structure in PEFCs: Water Storage and Contact Resistance Model. *J. Electrochem. Soc.* **2010**, *157*, B77. [CrossRef]
28. Nandy, A.; Jiang, F.; Ge, S.; Wang, C.-Y.; Chen, K.S.; Gaikwad, A.; Desai, D.; Gallaway, J.; Steingart, D. Effect of Cathode Pore Volume on PEM Fuel Cell Cold Start. *J. Electrochem. Soc.* **2010**, *157*, B726. [CrossRef]
29. Miao, Z.; Yu, H.; Song, W.; Hao, L.; Shao, Z.-G.; Shen, Q.; Hou, J.; Yi, B. Characteristics of proton exchange membrane fuel cells cold start with silica in cathode catalyst layers. *Int. J. Hydrogen Energy* **2010**, *35*, 5552–5557. [CrossRef]
30. Mukundan, R.; Lujan, R.; Davey, J.R.; Spendelow, J.; Hussey, D.S.; Jacobson, D.L.; Arif, M.; Borup, R. Ice Formation in PEM Fuel Cells Operated Isothermally at Sub-Freezing Temperatures. *ECS Trans.* **2009**, *25*, 345–355.
31. Thompson, E.L.; Jorne, J.; Gu, W.; Gasteiger, H.A. PEM Fuel Cell Operation at −20 °C. II. Ice Formation Dynamics, Current Distribution, and Voltage Losses within Electrodes. *J. Electrochem. Soc.* **2008**, *155*, B887. [CrossRef]
32. Thompson, E.L.; Jorne, J.; Gu, W.; Gasteiger, H.A. PEM fuel cell operation at −20 °C. I. Electrode and membrane water (charge) storage. *J. Electrochem. Soc.* **2008**, *155*, B625–B634. [CrossRef]
33. Tajiri, K.; Tabuchi, Y.; Wang, C.-Y. Isothermal cold start of polymer electrolyte fuel cells. *J. Electrochem. Soc.* **2007**, *154*, B147–B152. [CrossRef]
34. Ge, S.; Wang, C.-Y. Characteristics of subzero startup and water/ice formation on the catalyst layer in a polymer electrolyte fuel cell. *Electrochim. Acta* **2007**, *52*, 4825–4835. [CrossRef]
35. Oszcipok, M.; Zedda, M.; Riemann, D.; Geckeler, D. Low temperature operation and influence parameters on the cold start ability of portable PEMFCs. *J. Power Sources* **2006**, *154*, 404–411. [CrossRef]
36. Hiramitsu, Y.; Mitsuzawa, N.; Okada, K.; Hori, M. Effects of ionomer content and oxygen permeation of the catalyst layer on proton exchange membrane fuel cell cold start-up. *J. Power Sources* **2010**, *195*, 1038–1045. [CrossRef]
37. Biesdorf, J.; Stahl, P.; Siegwart, M.; Schmidt, T.J.; Boillat, P. When Size Matters: Active Area Dependence of PEFC Cold Start Capability. *J. Electrochem. Soc.* **2015**, *162*, F1231–F1235. [CrossRef]
38. Mayrhuber, I.; Marone, F.; Stampanoni, M.; Schmidt, T.J.; Büchi, F.N. Fast X-ray Tomographic Microscopy: Investigating Mechanisms of Performance Drop during Freeze Starts of Polymer Electrolyte Fuel Cells. *ChemElectroChem* **2015**, *2*, 1551–1559. [CrossRef]
39. Dursch, T.; Trigub, G.; Liu, J.; Radke, C.J.; Weber, A.Z. Non-isothermal melting of ice in the gas-diffusion layer of a proton-exchange-membrane fuel cell. *Int. J. Heat Mass Transf.* **2013**, *67*, 896–901. [CrossRef]
40. Li, J.; Lee, S.; Roberts, J. Ice formation and distribution in the catalyst layer during freeze-start process—CRYO-SEM investigation. *Electrochim. Acta* **2008**, *53*, 5391–5396. [CrossRef]
41. Oszcipok, M.; Hakenjos, A.; Riemann, D.; Hebling, C. Start Up and Freezing Processes in PEM Fuel Cells. *Fuel Cells* **2007**, *7*, 135–141. [CrossRef]
42. Jiang, F.; Fang, W.; Wang, C.-Y. Non-isothermal cold start of polymer electrolyte fuel cells. *Electrochim. Acta* **2007**, *53*, 610–621. [CrossRef]
43. Ishikawa, Y.; Shiozawa, M.; Kondo, M.; Ito, K. Theoretical analysis of supercooled states of water generated below the freezing point in a PEFC. *Int. J. Heat Mass Transf.* **2014**, *74*, 215–227. [CrossRef]

44. Jia, L.; Tan, Z.; Kang, M.; Zhang, Z. Experimental investigation on dynamic characteristics of proton exchange membrane fuel cells at subzero temperatures. *Int. J. Hydrogen Energy* **2014**, *39*, 11120–11127. [CrossRef]
45. Jiao, K.; Li, X. Effects of various operating and initial conditions on cold start performance of polymer electrolyte membrane fuel cells. *Int. J. Hydrogen Energy* **2009**, *34*, 8171–8184. [CrossRef]
46. Ishikawa, Y.; Hamada, H.; Uehara, M.; Shiozawa, M. Super-cooled water behavior inside polymer electrolyte fuel cell cross-section below freezing temperature. *J. Power Sources* **2008**, *179*, 547–552. [CrossRef]
47. Jiang, F.; Wang, C.-Y. Potentiostatic Start-Up of PEMFCs from Subzero Temperatures. *J. Electrochem. Soc.* **2008**, *155*, B743–B751. [CrossRef]
48. Gupta, N.; York, C.; Needham, J.; Hagans, P. Platform for Rapid Prototyping of PEM Fuel Cell Designs with Enhanced Cold-Start Performance and Durability. *ECS Trans.* **2006**, *1*, 383–388.
49. Balliet, R.J.; Newman, J. Cold Start of a Polymer-Electrolyte Fuel Cell II. Model Verification Using Parametric Studies. *J. Electrochem. Soc.* **2011**, *158*, B939. [CrossRef]
50. Patterson, T.; O'Neill, J. *PEM Fuel Cell Freeze Durability and Cold Start Project*; No. DOE/GO/86042-1; Office of Scientific and Technical Information (OSTI): Golden, CO, USA, 2008.
51. Promislow, K.; Wetton, B. A simple, mathematical model of thermal coupling in fuel cell stacks. *J. Power Sources* **2005**, *150*, 129–135. [CrossRef]
52. Zhu, Y.; Lin, R.; Jiang, Z.; Zhong, D.; Wang, B.; Shangguan, W.; Han, L. Investigation on cold start of polymer electrolyte membrane fuel cells with different cathode serpentine flow fields. *Int. J. Hydrogen Energy* **2019**, *44*, 7505–7517. [CrossRef]
53. Yang, Z.; Du, Q.; Huo, S.; Jiao, K. Effect of membrane electrode assembly design on the cold start process of proton exchange membrane fuel cells. *Int. J. Hydrogen Energy* **2017**, *42*, 25372–25387. [CrossRef]
54. Wei, L.; Liao, Z.; Suo, Z.; Chen, X.; Jiang, F. Numerical study of cold start performance of proton exchange membrane fuel cell with coolant circulation. *Int. J. Hydrogen Energy* **2019**, *44*, 22160–22172. [CrossRef]
55. Luo, Y.; Guo, Q.; Du, Q.; Yin, Y.; Jiao, K. Analysis of cold start processes in proton exchange membrane fuel cell stacks. *J. Power Sources* **2013**, *224*, 99–114. [CrossRef]
56. Zhou, Y.; Luo, Y.; Yu, S.; Jiao, K. Modeling of cold start processes and performance optimization for proton exchange membrane fuel cell stacks. *J. Power Sources* **2014**, *247*, 738–748. [CrossRef]
57. Yao, L.; Peng, J.; Zhang, J.-B.; Zhang, Y.-J. Numerical investigation of cold-start behavior of polymer electrolyte fuel cells in the presence of super-cooled water. *Int. J. Hydrogen Energy* **2018**, *43*, 15505–15520. [CrossRef]
58. Zhan, Z.; Zhao, H.; Sui, P.; Jiang, P.; Pan, M.; Djilali, N. Numerical analysis of ice-induced stresses in the membrane electrode assembly of a PEM fuel cell under sub-freezing operating conditions. *Int. J. Hydrogen Energy* **2018**, *43*, 4563–4582. [CrossRef]
59. Johnson, E.F. Thermodynamics and its applications, Ed. Modell, M.; Reid, R.C. Prentice-Hall, englewood cliffs, N.J. (1974). 553 pages. *AIChE J.* **1975**, *21*, 203–204. [CrossRef]
60. Springer, T.E. Polymer Electrolyte Fuel Cell Model. *J. Electrochem. Soc.* **1991**, *138*, 2334. [CrossRef]
61. Newaz, G.; Mayeed, M.S.; Rasul, A. Characterization of balsa wood mechanical properties required for continuum damage mechanics analysis. *Proc. Inst. Mech. Eng. Part L: J. Mater. Des. Appl.* **2014**, *230*, 206–218. [CrossRef]
62. Alduchov, O.A.; Eskridge, R.E. Improved magnus form approximation of saturation vapor pressure. *J. Appl. Meteorol.* **1996**, *35*, 601–609. [CrossRef]
63. Hinatsu, J.T.; Mizuhata, M.; Takenaka, H. Water Uptake of Perfluorosulfonic Acid Membranes from Liquid Water and Water Vapor. *J. Electrochem. Soc.* **1994**, *141*, 1493. [CrossRef]

MDPI
St. Alban-Anlage 66
4052 Basel
Switzerland
Tel. +41 61 683 77 34
Fax +41 61 302 89 18
www.mdpi.com

*Molecules* Editorial Office
E-mail: molecules@mdpi.com
www.mdpi.com/journal/molecules

www.ingramcontent.com/pod-product-compliance
Lightning Source LLC
Chambersburg PA
CBHW061452210926
51850CB00008B/1688
*9783036515441*